ADVANCES IN RESEARCH ON TEACHING

Volume 7 • 1998

EXPECTATIONS IN
THE CLASSROOM

ADVANCES IN RESEARCH ON TEACHING

EXPECTATIONS IN THE CLASSROOM

Editor: **JERE BROPHY**
College of Education
Michigan State University

VOLUME 7 • 1998

JAI PRESS INC.

Greenwich, Connecticut	London, England

Copyright © 1998 JAI PRESS INC.
55 Old Post Road No. 2
Greenwich, Connecticut 06836

JAI PRESS LTD.
38 Tavistock Street
Covent Garden
London WC2E 7PB
England

All rights reserved. No part of this publication may be reproduced, stored on a retrieval system, or transmitted in any way, or by any means, electronic, mechanical, photocopying, recording, filming or otherwise without prior permission in writing from the publisher.

ISBN: 0-7623-0261-5

Manufactured in the United States of America

CONTENTS

LIST OF CONTRIBUTORS	vii
INTRODUCTION Jere Brophy	ix

TEACHER EXPECTATIONS
    Lee Jussim, Alison Smith, Stephanie Madon,
    and Polly Palumbo
	1

THE ANTECEDENTS AND CONSEQUENCES OF TEACHER EFFICACY John A. Ross	49
TEACHER EFFICACY AND THE VULNERABILITY OF THE DIFFICULT-TO-TEACH STUDENT Leslie C. Soodak and David M. Podell	75
EXPECTATIONS AND BEYOND: THE DEVELOPMENT OF MOTIVATION AND LEARNING IN A CLASSROOM CONTEXT Pekka Salonen, Erno Lehtinen, and Erkki Olkinuora	111
SHARED EXPECTATIONS: CREATING A JOINT VISION FOR URBAN SCHOOLS Catherine D. Ennis	151
PREFERENTIAL AFFECT: THE CRUX OF THE TEACHER EXPECTANCY ISSUE Elisha Babad	183

EXPECTANCY EFFECTS IN "CONTEXT": LISTENING
TO THE VOICES OF STUDENTS AND TEACHERS
   Rhona S. Weinstein and Clark McKown   215

WHEN STIGMA BECOMES SELF-FULFILLING
PROPHECY: EXPECTANCY EFFECTS AND THE
CAUSES, CONSEQUENCES, AND TREATMENT OF
PEER REJECTION
   Monica J. Harris, Richard Milich, and
   Cecile B. McAninch   243

RESEARCH ON THE COMMUNICATION OF
PERFORMANCE EXPECTATIONS: A REVIEW
OF RECENT PERSPECTIVES
   Thomas L. Good and Elisa K. Thompson   273

LIST OF CONTRIBUTORS

Elisha Babad	School of Education
	Hebrew University of Jerusalem

Jere Brophy	College of Education
	Michigan State University

Catherine D. Ennis	Department of Kinesiology
	University of Maryland

Thomas L. Good	College of Education
	University of Arizona

Monica J. Harris	Department of Psychology
	University of Kentucky

Lee Jussim	Department of Pyschology
	Rutgers University

Erno Lehtinen	Department of Education
	University of Turku

Stephanie Madon	Department of Psychology
	Rutgers University

Cecile B. McAninch	Department of Behavioral Sciences
	University of Kentucky

Clark McKown	Department of Psychology
	University of California, Berkeley

Richard Milich	Department of Psychology University of Kentucky
Erkki Olkinuora	Department of Education University of Turku
Polly Palumbo	Department of Psychology Rutgers University
David M. Podell	College of Staten Island City University of New York
John A. Ross	Ontario Institute for Studies in Education University of Toronto
Pekka Salonen	Department of Teacher Training in Rauma University of Turku
Alison Smith	Department of Psychology Rutgers University
Leslie Soodak	Department of Educational Psychology Rutgers University
Elisa K. Thompson	College of Education University of Arizona
Rhona S. Weinstein	Department of Psychology University of California, Berkeley

INTRODUCTION

Jere Brophy

This volume is devoted to research on the formation, functioning, and effects of teachers' and students' expectations in classrooms. Expectations are inferences, based on prior experiences, about what is likely to occur in the future. For example, teacher expectations are inferences that teachers make about present and future academic achievement and general classroom behavior of students. Teachers form general expectations for the class as a whole, as well as differentiated expectations for individual students, concerning such issues as what goals, texts, and learning activities are appropriate for the students; the probability that the students will accomplish the curricular outcomes expected at the grade level; the likelihood that poor current performance can be improved through more intensive teaching or better effort on the part of the students; and the probability that the students will respond positively to a given motivational or disciplinary strategy.

Like any expectations, teachers' expectations concerning their students tend to be self-sustaining. They affect both perception, by causing teachers to be alert for what they expect and less likely to notice what they do not expect, and interpretation, by causing teachers to interpret (and perhaps distort) what they see so that it is consistent with their expectations. These mechanisms make it possible for some expectations to persist even though they are at least partially inaccurate.

EARLY RESEARCH

Scholarly interest in teachers' expectations initially concentrated on the hypothesis that these expectations can function as self-fulfilling prophecies—that is, that they can set in motion cycles of teacher behavior and student response that condition students to become what the teacher expects them to be, or at least more so than they would have become if the teacher had not held and acted upon these particular expectations. Robert Merton defined and illustrated the concept of the self-fulfilling prophecy in 1948, and Kenneth Clark (1965) identified low teacher expectations as one cause of the low achievement of students in inner-city schools. However, it was not until the 1968 publication of Robert Rosenthal and Lenore Jacobson's *Pygmalion in the Classroom* that the topic of teacher expectations "arrived" on the educational scene. That study received great publicity and stimulated a flurry of research on teacher expectations and related topics that continued into the 1980s. When I reviewed this literature for an article that appeared 15 years after *Pygmalion in the Classroom* was published, I was able to cite well over 100 studies (Brophy, 1983).

To provide context for the present volume, I will briefly summarize some highlights from that review. First, experimental studies have demonstrated beyond question that expectations, once in place, can have self-fulfilling prophecy effects. In the case of teachers' expectations about students, if a teacher expects a certain pattern of behavior or level of performance from a student, this expectation can cause the teacher to treat the student in ways that reflect this expectation and increase the probability that it will come true. However, demonstrating that it is possible to produce self-fulfilling prophecy effects in experimental situations is not the same as demonstrating that teachers' expectations normally have such effects in classrooms. Classroom studies of relationships between teachers' expectations and students' achievement (adjusted for their prior achievement levels) have confirmed the existence of self-fulfilling prophecy effects but also suggested that the magnitude of these effects is relatively small on the average (making perhaps a 5-10% difference in student achievement outcomes). This is both statistically and practically significant, but it is not nearly as dramatic an effect as the publicity surrounding *Pygmalion in the Classroom* led many educators to expect.

Follow-up studies helped to explain and qualify this finding. First, studies of how teachers form and change expectations indicated that teachers ordinarily use the best evidence available and develop rather accurate expectations. Furthermore, most of the inaccuracies in these expectations are corrected as more information becomes available. Thus, in most classrooms the potential for dramatic self-fulfilling prophecy effects of teacher expectations on student achievement is limited because such effects imply the consistent communication of relatively inaccurate expectations. Dramatic self-fulfilling prophecy effects can be observed only when teachers not only form significantly inaccurate expectations in the first place but rigidly maintain these expectations in the face of contradictory feedback.

Some investigators examined individual differences in teachers' self-fulfilling prophecy effects on student achievement. They found that the majority of teachers develop accurate and reality-based expectations and adjust these expectations in response to changes in students. Furthermore, most of the differential teacher-student interaction patterns observed in their classrooms represent either appropriate proactive attempts to meet differential student needs or at least understandable reactive responses to contrasting student behaviors. As a result, in these classrooms the magnitudes of self-fulfilling prophecy effects of teacher expectations on student achievement approach zero.

However, a minority of teachers develop and maintain inaccurate expectations, and treat students accordingly. Such teachers have been described as prone to bias and as overreactive to labels placed on students (e.g., learning disabled or low achiever). They are likely to create warmer social-emotional relationships with their high expectation students, to teach them more (and more difficult) material, to give them more opportunities to respond and ask questions, and to provide them with more and better feedback about their performance. Meanwhile, they are less likely than other teachers to foster the achievement of their low expectation students through such supportive instruction, and more likely to impede their progress through behaviors such as minimizing and depersonalizing their interactions with these students, accepting or even praising their low quality or incorrect responses, failing to recognize and encourage their progress, and giving up on them quickly rather than patiently guiding them toward understandings when they are confused. As a result, in these classrooms the self-fulfilling prophecy effects of teachers' expectations are likely to account for more than a 5-10 percent difference in student achievement outcomes.

Most of the early work on teacher expectations was focused on the expectations held by individual teachers for different students in their classes. Gradually, however, awareness grew that teacher expectation phenomena occur with respect not only to individual students within classrooms, but also to different ability groups within classrooms and different tracks within schools. Thus, some investigators began looking at expectations and related beliefs connected to grouping and tracking practices. Other investigators looked at relationships between teachers' expectations and other teacher beliefs (e.g., tendencies to attribute student achievement to fixed ability levels versus controllable factors such as quality of instruction) or teachers' coping and problem-solving styles (e.g., giving up versus redoubling effort when initial instruction has not been successful).

Still other investigators focused attention on students and their contributions to expectation phenomena within the teacher-student relationship. Their studies revealed that students differ in the degree to which they pay attention to cues communicating teachers' expectations, read these cues accurately, and accept versus resist the "messages" being communicated and the tendency of these "messages" to shape the students' behavior toward expected outcomes. Some investigators, while acknowledging that teacher behaviors can condition students, emphasized that student behaviors can condition teachers as well. Their studies suggested that some of the differential patterns of teacher-student interaction observed in classrooms are better understood as student effects on teachers than as teacher effects on students (e.g., students who often raise their hands and eagerly seek response opportunities are more likely to be called on frequently than students who "hide" from the teacher, and students whose responses are typically not only correct but thoughtful and creative are more likely to be praised frequently than students whose responses are typically confused or incorrect).

In general, between 1968 and 1983 research on classroom expectation phenomena was notable both for the sheer quantity of studies produced and for progressions from relatively artificial experimental studies to studies focusing on the expectations of in-service teachers concerning the students in their classrooms, from controversy about whether or not expectancy effects exist to the development and testing of models of the processes involved in the formation and communication of expectations, from an exclusive focus on teacher effects on students to consideration of student effects on teachers and the broader dynamics of the teacher-student relationship, from the individual classroom level to the school level and beyond, and from an exclusive focus on student achievement to consideration of a broader range of cognitive, affective, and behavioral student outcomes.

MORE RECENT RESEARCH

In the 15 years since 1983 we have seen an apparent reduction of scholarly interest in expectation phenomena in classrooms, and especially in the self-fulfilling prophecy effects of teacher expectations on student achievement. A relative handful of scholars (represented in this volume by Elisha Babad, Lee Jussim, and Rhona Weinstein) have continued to do programmatic research focused around the "expectations" construct, but this construct has been featured much less frequently in the past 15 years than it had been in the previous 15 years.

Does this mean that most expectation-related issues have been resolved or that scholars have simply lost interest in them? In a word, no. There has been at least as much scholarly interest in expectation-related phenomena recently as there was previously, but this is less obvious than it was before, for two reasons.

First, there has been a remarkable development of theory and research in social cognition in recent years, and one effect of this has been to shift attention from the expectation construct to somewhat more specific constructs such as self-efficacy and attributional inferences. As scientific knowledge domains develop, they become more differentiated, so that phenomena originally conceptualized loosely using relatively general constructs become differentiated and elaborated using more specific constructs. Classroom researchers continue to address many of the same issues relating to the formation, functioning, and effects of teachers' expectations that have been studied by researchers who emphasize the expectations construct, but many contemporary researchers are couching their work within attribution theory, self-efficacy theory, or other recent developments in social cognition. This is especially the case with scholars whose approaches to classroom expectation phenomena are rooted primarily in the discipline of psychology.

The second reason for reduced emphasis on the expectation construct applies primarily to scholars whose approaches are rooted in the discipline of education. These scholars often are interested in issues that can be recognized (and profitably treated) as expectation phenomena, but many of them have not yet begun to frame the issues in this manner. For example, currently there is a great deal of interest in upgrading the general quality of U.S. schools and the performance of U.S. students on international assessments of educational achievement. Among other things, this has led to adoption of national goals and calls for national and state curriculum standards. Furthermore, within subject matter areas, recent concerns about teaching for understanding and promoting authentic learning have led to the publication of standards for curriculum, instruction, and assessment by leading organizations concerned with subject matter teaching. In addition, other reformers have called for programs designed to address affective and citizenship goals by teaching students to monitor and control their own motivation and problem-solving strategies, to initiate and sustain productive social interactions, to work and learn collaboratively, to resolve conflicts through negotiation, and so on.

These various standards and reform proposals embody expectations about what is feasible and cost effective to accomplish with students at a given grade level. However, these expectations are seldom recognized explicitly and tested for feasibility and cost effectiveness. Instead, their appropriateness is usually taken for granted. Proposed standards are treated as unproblematic statements of outcomes that reasonably can be expected, rather than as hypotheses that may involve contradictions or unrealistic expectations. Frequently the proposals are overly ambitious, prescribing both broad coverage of content and an emphasis on teaching for understanding, appreciation, and application of what is taught (which implies an emphasis on depth of development of key ideas over breadth of content coverage). If interest in these curricular issues continues beyond its current stage of promulgation of untested standards and we begin seriously to address issues of what students at a given grade level can be expected to learn and what evidence will be needed to inform these decisions, we will see a mushrooming of attention to expectation phenomena, even if they are not always conceptualized as such.

OVERVIEW OF THE VOLUME

Regardless of what the future may hold for the expectation construct, there is plenty to be said right now about expectation phenomena in the classroom. This volume reflects my belief that the time is right for a reemphasis on expectation phenomena in thinking about schooling, and in particular, teacher-student interaction in classrooms. It is time to take stock, viewing with hindsight the work of the first 15 years and presenting and integrating some of the work done in the last 15 years. Toward that end, the volume offers eight chapters presenting research on various aspects of teacher and student expectations, followed by a discussion chapter.

In Chapter 1, Lee Jussim, Alison Smith, Stephanie Madon, and Polly Palumbo review a broad range of research on teachers' expectations, and in the process, present findings from several of their own studies. The chapter begins with a look back at the controversy surrounding *Pygmalion in the Classroom*, noting that, in addition to clarification of the issues involved, this controversy led to the development of meta-analysis as a statistical tool for summarizing results integrated across multiple studies. The chapter includes attention to the contrasts in purposes and outcomes between experimental and naturalistic classroom studies, alternative explanations (self-fulfilling prophecies, perceptual biases, or accurate assessment and prediction) for the high correlations between teachers' expectations and students' achievement outcomes, and the ways that the magnitudes of classroom expectancy effects tend to vary with student demographics and prior achievement levels, tracking and other school context factors, teacher characteristics, and other variables. The authors also consider whether expectancy effects accumulate across school years, teachers' general beliefs about the appropriateness of standards to which students might be held, and teachers' expectations for whole classes of students. Following this focus on expectation outcomes, the authors shift to expectation processes, reviewing research on how teachers form expectations for different students, how these expectations lead to differential teacher-student interactions, and how students perceive and react to differential treatment. Finally, the authors conclude the chapter by identifying implications from all of this research for school teachers and administrators.

Chapters 2 and 3 represent work on teacher expectation phenomena that has been informed by self-efficacy theory and operationalized around the construct of teacher efficacy. In Chapter 2 John Ross identifies antecedent conditions associated with the waxing and waning of teacher efficacy, catalogs it's influence on teacher actions and on student achievement, examines the results of interventions designed to enhance it, and considers implications for research and practice. Ross reports that teacher efficacy varies with certain personal characteristics of teachers and demographic and organizational characteristics of classrooms and schools, that higher teacher efficacy is associated with the use of more effective instructional and classroom management practices and with more positive cognitive and affective student outcomes, and that some forms of intervention have had positive effects on the self-efficacy perceptions of the teachers who participated in them.

In Chapter 3 Leslie Soodak and David Podell highlight findings from their ongoing program of research on teacher efficacy as it affects teachers' effectiveness in working with students who are difficult to teach because they demonstrate significant learning or behavior problems. They begin by reviewing some of the distinctions introduced by Albert Bandura in developing his self-efficacy theory, then review research on teacher efficacy that has incorporated these distinctions. Next they present findings from their own studies, which indicate that meaningful distinctions can be made between three aspects of teachers' self-efficacy perceptions: (1) teachers' beliefs in their own personal efficacy as teachers, (2) teachers' beliefs that student outcomes are dependent on teacher behavior, not just genetic or cultural background factors that teachers cannot control, and (3) teachers' beliefs that positive learning outcomes achieved by their own students are at least in part attributable to their own teaching skills and efforts. Next, the authors present research indicating that initially high teaching efficacy perceptions tend to decrease during the early in-service years (especially for elementary teachers), suggesting the need for teacher education programs and school districts to provide for both more realistic expectations and more coping support in the preservice and early in-service years. In the remainder of the chapter the authors present their findings indicating that, compared to teachers with higher perceived teaching efficacy, teachers low in perceived teaching efficacy are more likely to refer difficult-to-teach students for removal to special education settings, less likely to suggest solutions to these students' problems that the teachers themselves would have to implement, and less supportive of inclusion policies. In discussing their findings the authors emphasize the need for teacher education methods that develop novice teachers' effective teaching skills and related perceptions of teaching efficacy, especially concerning ability to persevere and succeed with difficult-to-teach students.

In Chapter 4 Pekka Salonen, Erno Lehtinen, and Erkki Olkinuora present findings from their research in Finnish classrooms on the mutual adaptations that occur between teachers and students. They first depict students' characteristics as learners, describing research on student motivation, coping styles, metacognitive regulation of learning strategies, and other variables. Then they describe, and illustrate with interesting case material, how students who consistently display certain learner characteristics tend to condition teachers to display reciprocal characteristics in their interactions with them. Sometimes these student characteristics are carryovers from patterns they have developed at home through interactions with parents or older siblings. The case examples illustrate how certain students are able to manipulate teachers into lowering their expectations or demands, providing the students with too much help, giving answers instead of continuing to work to get the students to come up with the answers themselves, or in other ways defeating their own (the teachers') instructional goals and intentions.

In Chapter 5 Catherine Ennis touches on many of these same themes, this time in the context of urban high schools in the United States. U.S. high school students are less compliant with adult authority figures than younger students. Many come from minority backgrounds and expect teachers who display certain characteristics and a curriculum that they can recognize as relevant to their needs and interests. If urban high school classrooms are to "work," the teacher and students must negotiate shared expectations concerning curriculum, instruction, assessment, and general classroom climate and management. Drawing from interview and observational data, Ennis illustrates some of the teacher attitudes and strategies involved in displaying caring and supportiveness to students and building a community of learners that share a stake in the class's success. She concludes the chapter with guidelines for managing classrooms and motivating students in ways that support academic learning and success expectations in urban high schools.

As the volume progresses, relatively more attention is focused on the perceptions and thinking of students (not just teachers). This becomes more clearly evident beginning in Chapter 6, in which Elisha Babad presents findings from his work on the affective aspects of teacher-student interaction in schools in Israel. He begins by reviewing key findings on the contrasting patterns of interaction that teachers are likely to have with students toward whom they hold contrasting expectations or attitudes. The review includes attention to individual differences in teachers' proneness to interact differently with different kinds of students, as well as information about students' perceptions of this differential treatment. In the rest of the chapter Babad summarizes findings from his line of studies on the teacher's pet phenomenon. He describes, among other findings, how high-bias teachers who are the most likely to show strong expectation effects are also the most likely to have pets, how students tend to be quite aware of the existence of a teacher's pet in their classroom and of how this student is treated differently from the others, and how the presence of a teacher's pet in the class may have negative effects on the other students' attitudes, especially if the pet is an unpopular student.

The emphasis on students' awareness of and responses to differential teacher treatment continues in Chapter 7. Here, Rhona Weinstein and Clark McKown review findings from a series of studies indicating that even younger elementary students are very aware of their teachers' differential treatment of individuals within their classrooms. The students draw inferences about teachers' liking or disliking of students and about teachers' estimations of the students' intelligence from their observations of whom teachers call on to respond to what kinds of questions, what sorts of accomplishments are praised and how this praise is phrased, differences in the kinds of tasks assigned to different students and the degrees of structuring or help given to them as they work on the tasks, and nonverbal cues such as facial expressions and voice tones. Related studies indicate that more equitable cognitive and affective outcomes are observed in classrooms in which teachers minimize this kind of differentiation between students. In the rest of the chapter Weinstein and McKown describe intervention studies designed to maximize the equity as well as the quality of students' educational experiences. Many of the themes emphasized in these treatments are similar to those emphasized by Ennis (showing caring and support to all students and building a community of learners within each classroom). The interventions described by Weinstein and McKown differ from most expectation-related interventions developed to date, however, in that they involve the principal, the parents, the teachers as a collaborating staff, and the students as participants in the whole school community, not just the teacher and students in a particular classroom.

Attention is focused most heavily on students in Chapter 8, in which Monica Harris, Richard Milich, and Cecile McAninch present findings from a series of studies on the labeling effects associated with special education diagnoses. These studies indicate clearly that stigmatizing labels often cue negative expectations and related self-fulfilling prophecy behaviors not only in teachers but in peers. Children who carry such labels are often ignored, rejected, or even mistreated by their peers, and these negative peer reactions are due at least in part (and sometimes entirely) to the stigmatizing label rather than to any cognitive or behavioral deficiency that may be present. One implication of this, the authors point out, is that social skills improvement programs for students may need to include components designed to change the way that peers perceive and interact with the target students, because otherwise the peers' negative perceptions and treatment of the target students may continue even though any problematic aspects of the target students' behavior may have disappeared. The chapter concludes with many additional suggestions for ways in which teachers can minimize the salience of potentially stigmatizing labels, and thus minimize the peer rejection problems that often accompany such stigmatizing.

The volume concludes with Chapter 9, in which Tom Good and Elisa Thompson discuss the material in Chapters 1-8. These authors first synthesize and critique the contents of individual chapters, noting relationships among them. Next, they analyze the strengths and weaknesses of the volume as a whole. Finally, they suggest ways of responding to existing limitations in this area of research by using new conceptualizations and approaches to inquiry.

REFERENCES

Brophy, J. (1983). Research on the self-fulfilling prophecy and teacher expectations. *Journal of Educational Psychology, 75,* 631-661.
Clark, K. (1965). *Dark ghetto.* New York: Harper & Row.
Merton, R. (1948). The self-fulfilling prophecy. *Antioch Review, 8,* 193-210.
Rosenthal, R., & Jacobson, L. (1968). *Pygmalion in the classroom: Teacher expectation and pupils' intellectual development.* New York: Holt, Rinehart & Winston.

TEACHER EXPECTATIONS

Lee Jussim, Alison Smith, Stephanie Madon,
and Polly Palumbo

Teacher expectations. The term has been known to throw some social scientists into paroxysms of righteous indignation at teachers for their supposed role in creating racial and social inequalities. Others (equally righteously) condemn teacher expectation research as hopelessly flawed and are appalled at the way it has been misinterpreted and has captured the popular imagination.

In this chapter we hope to convey why, in our view, 30 years of research on teacher expectations leads to a series of conclusions that falls between these two extremes. Teacher expectations clearly do influence students—at least sometimes. In the first section of this chapter we briefly review the controversy surrounding Rosenthal and Jacobson's (1968) seminal demonstration of self-fulfilling prophecies in the classroom. However, our view is that 30 years and hundreds of studies have rendered that controversy moot. We then review the evidence leading to the conclusion that expectancy effects are real, although typically neither pervasive nor very large.

Nonetheless, it seemed to us that, under some conditions, teacher expectation effects might be not only larger than usual, but dramatic by any standard. Thus, several years ago we embarked on a quest to identify some of those conditions. We review our own and others' research on "moderators" of (factors increasing or decreasing) expectancy effects. We draw on research in other areas of psychology (social, personality, industrial/organizational, developmental) when it has potential relevance to classrooms (we use the terms "perceive" and "target" when referring to the holders and recipients of expectancies in laboratory studies; mostly, however, we focus on teachers and students). In this section we also review the limited evidence regarding whether expectancy effects accumulate or dissipate with time.

Most research on teacher expectations focuses on individual students. However, expectancies and standards are related constructs, and teachers' standards may often operate at the level of a whole class. Thus, we briefly present an expectancy perspective on the likely effects of standards on whole classes. In addition, we summarize the results of a just-completed study in which we examined the extent to which teachers' expectations for whole classes were self-fulfilling and the extent to which ability grouping practices moderated these expectancy effects.

Whether at the group or individual level, expectancy effects occur neither by magic nor ESP. Many of the interpersonal processes by which teacher expectations influence students have been amply documented. We review that research. In addition, because so many reviews have implicated teacher expectations as a major contributor to social problems, we have included a detailed analysis of the extent to which social stereotypes lead teachers to develop inaccurate perceptions of students from different groups. We also highlight some of the process issues that, even after 30 years, remain unexamined.

After all this time and research, have we learned anything that teachers can actually use in the classroom? In the conclusion to this chapter we address this question but only after summarizing the results of the empirical research—which clearly show that teacher expectations typically are not the vehicle for perpetuating social injustices that once was believed.

SELF-FULFILLING PROPHECIES: A BRIEF INTRODUCTION

A self-fulfilling prophecy occurs when a false belief leads to its own fulfillment. Merton (1948) developed the idea of the self-fulfilling prophecy and applied it to phenomena as diverse as test anxiety, bank failures, prejudice, and discrimination. However, self-fulfilling prophecies received no empirical attention until Rosenthal's groundbreaking work on experimenter effects. In this work, Rosenthal (1963; Rosenthal & Fode, 1963; Rosenthal & Lawson, 1963) showed that researchers sometimes acted in ways that evoked behavior from research subjects (animal and human) that confirmed the researchers' hypotheses.

However, it was Rosenthal and Jacobson's (1968) classic and controversial Pygmalion study that launched the self-fulfilling prophecy as a major area of inquiry in the social sciences. Rosenthal and Jacobson led teachers to believe that certain students in their classes were "late bloomers" who would show dramatic increases in IQ by the end of the year. In fact, however, these students had been selected at random. Thus, there were really no differences between the "late bloomers" and the other students in the class. However, by the end of the school year, the "late bloomers" showed greater increases in IQ than did other students. This study demonstrated a self-fulfilling prophecy because the teachers caused the "late bloomers" to achieve more highly, thereby confirming the teachers' originally false beliefs.

CONTROVERSY: DO TEACHERS' EXPECTATIONS REALLY CREATE SELF-FULFILLING PROPHECIES?

Rosenthal and Jacobson's (1968) study created a furor that raged for several years and has reemerged periodically even decades later. There have been at least two types of gross overreactions to the study. First, researchers have often misinterpreted Rosenthal and Jacobson's study as demonstrating a dramatic impact of teacher expectations (e.g., Hofer, 1994; Gilbert, 1995; see Wineburg, 1987 for a review of such claims). For a study yielding an overall effect size of .15 (the correlation between the experimental manipulation and the IQ outcomes) and an average IQ difference of four points between high expectancy students and controls, such claims seem overstated. We can't help but wonder if these same writers who describe Rosenthal and Jacobson's study in melodramatic terms also would believe that, if one child receives an IQ score of 110 and another receives an IQ score of 114, the second child is "dramatically" smarter than the other.

The study has even been cited in support of arguments claiming that, because teacher expectations are based heavily on social stereotypes, they are potentially a powerful force in the creation of social inequalities and injustices (Gilbert, 1995; Hofer, 1994; Jones, 1986, 1990). We address this issue extensively later in this chapter. For now, we just point out that: (1) Rosenthal and Jacobson did not even examine the role of stereotypes in the formation of teachers' expectations; (2) They only manipulated *positive* expectations, leaving as an open, empirical question the effects of negative expectations; (3) The effects they found were not particularly powerful; and (4) Those effects actually became even weaker over time. This does not appear to us to provide much terra firma for claims regarding the power of teachers' expectations to create social injustices.

There was also a second type of gross overreaction—attempts to deny the existence of self-fulfilling prophecies altogether (e.g., Elashoff & Snow, 1971; Jensen, 1969; Rowe, 1995; Thorndike, 1968; Wineburg, 1987). Our view is that many of the complaints leveled against the original Rosenthal and Jacobson (1968) study were more flawed than the study itself. Although a rehashing of all the arguments for and against the paper is beyond the scope of this chapter (see Rosenthal, 1974 for a synopsis), we will briefly discuss some of the most flawed charges against the study.

One such charge was that the measure of IQ was unreliable (e.g., Thorndike, 1968; Roth, 1995) apparently in an attempt to suggest that any results developed using such a measure were meaningless. In fact, however, lack of reliability in a measure makes it *harder* to find differences between groups. Therefore, finding differences between groups with a measure low in reliability attests to the power of those differences.

Two other early criticisms were that: (1) The IQ test used was not appropriately normed for the youngest children; and (2) The scores of the children tested in kindergarten were so low (mean of 58) as to be manifestly invalid. There is validity to both claims. However, the low scores probably occurred precisely because the test was not created for use on younger children. Furthermore, this critique is irrelevant to understanding Rosenthal and Jacobson's results because it cannot possibly explain why they obtained a pattern of significantly greater IQ increases among the high expectancy students.

RESOLUTION TO THE FUROR

The Meta-analyses

Today, arguments about the strengths and flaws of the original Rosenthal and Jacobson (1968) study have become moot. There have been hundreds of follow-up studies of the effects of expectancies in classrooms, the workplace, and laboratories. Even these studies, however, initially evoked considerable controversy. Consistently, only about one-third of the studies attempting to demonstrate a self-fulfilling prophecy succeeded (Rosenthal & Rubin, 1978). This pattern was often interpreted by the critics as demonstrating that the phenomenon did not exist because support was unreliable. It was interpreted by proponents as demonstrating the existence of self-fulfilling prophecies because, if only chance differences were occurring, replications would only succeed about 5 percent of the time.

This controversy was to have an effect that went well beyond self-fulfilling prophecies. In his attempt to refute critics, Rosenthal developed meta-analysis (Harris, 1989)—a statistical technique for summarizing the results of multiple studies. Although meta-analysis, too, was greeted with considerable skepticism by Rosenthal's critics (see the commentaries on Rosenthal and Rubin's, 1978 meta-analysis in *Behavioral and Brain Sciences*), it has subsequently become the dominant method within the social sciences for summarizing the results of large literatures and resolving controversies about the existence and size of effects.

Rosenthal and Rubin's (1978) meta-analysis of the first 345 experiments on interpersonal expectancy effects conclusively demonstrated the existence of self-fulfilling prophecies. The 345 studies were divided into eight categories. Z-scores representing the combined expectancy effect for all studies in each category were computed. The median of the eight combined Z-scores was 6.62, indicating that the probability of finding the observed expectancy effects, if the phenomenon did not exist, was essentially zero.

The Naturalistic Studies

In addition, there have also been numerous naturalistic investigations of teacher expectation effects (see Jussim & Eccles, 1995 for a review). One of the criticisms leveled against Rosenthal and Jacobson in particular, and experimental studies of expectancies in general, is that researchers induce perceivers (teachers, employers, etc.) to adopt false expectations by misleading or lying to them. For example, Rosenthal and Jacobson (1968) induced teachers to develop false positive expectations by claiming that certain students had been identified as late bloomers by a new test, when, in fact, there was no test of late blooming. Therefore, such studies did not and could not address the extent to which teachers typically develop inaccurate expectations. This is important because only inaccurate expectations can produce self-fulfilling prophecies.

Naturalistic studies eliminated this problem by studying relations between naturally occurring teacher expectations and student achievement. Regardless of whether these studies used quasi-experimental techniques or survey/path analytic techniques, they consistently replicated Rosenthal and Jacobson's (1968) original finding that teacher expectations do indeed create self-fulfilling prophecies—usually with effect sizes closely corresponding to those of that study (see Jussim & Eccles, 1995 for a review). So over the last 30 years, whether Rosenthal and Jacobson (1968) is a good or bad study and did or did not find self-fulfilling prophecies has become moot. Hundreds of studies, both naturalistic and experimental, conducted in classrooms, laboratories, and a wide variety of other real-world contexts, have clearly shown that the self-fulfilling prophecy is a real phenomenon (see reviews by Brophy, 1983; Brophy & Good, 1974; Cooper, 1979; Jussim, 1986; Rosenthal, 1974; Snyder, 1984).

MORE CONTROVERSY: WHAT IS THE MAIN REASON TEACHER EXPECTATIONS PREDICT STUDENT ACHIEVEMENT?

Three Sources of Expectancy Confirmation

Teachers' expectations may be confirmed for any of at least three reasons—two that involve influences of expectations on behavior or perceptions and one that does not. First, expectations may lead to self-fulfilling prophecies—teachers' expectations may influence students' actual achievement. Second, expectations may lead to *perceptual biases*—teachers may interpret, remember, and/or explain student achievement and behavior in ways consistent with their expectations. Self-fulfilling prophecies and perceptual biases both imply that teacher expectations create (or "construct") reality—by creating either an objective reality (when self-fulfilling prophecies change students' actual achievement) or a subjective reality (when perceptual biases influence teachers' evaluations of student achievement or behavior).

Teachers' expectations also may be confirmed for a third reason: They may accurately reflect and predict student achievement, without influencing either objective student accomplishments or subjective perceptions of those accomplishments (see Jussim, 1989, 1991; Jussim & Eccles, 1992, 1995 for detailed discussions of these three sources of expectancy confirmation). There are at least two aspects or types of accuracy relevant to the classroom. *Impression accuracy* refers to the extent to which teachers rely on appropriate criteria (grades, standardized test scores, in-class performance, etc.) when arriving at judgments regarding students. *Predictive accuracy* refers to the extent to which teachers' expectations predict without causing student achievement.

Although self-fulfilling prophecies, perceptual biases, and accuracy are conceptually distinct, they are not mutually exclusive. Any combination of the three (including none at all) can characterize relations between teacher expectations and student achievement (Jussim, 1989, 1991; Jussim & Eccles, 1992). Consider a teacher who believes that a student is especially bright. The teacher may be (largely) accurate—this student may indeed have stronger academic aptitudes than most others. Furthermore, highly positive interactions with the teacher may lead this student to achieve even more highly—thus, a self-fulfilling prophecy. Finally, perceptual biases may lead the teacher to evaluate the student even more favorably than is warranted by the student's objective performance.

Although expectations may lead to many combinations of self-fulfilling prophecy, perceptual bias, and accuracy, they may also lead to none—expectations can be both inaccurate and uninfluential. For example, a teacher may expect a student to be a low achiever. Nonetheless, this student successfully completes most homework assignments in a timely and thorough manner, goes on to perform above average on a highly credible standardized achievement test, and receives mostly "A's" on in-class tests. The teacher may simply acknowledge the error (i.e., the original expectation was erroneous, but there is no perceptual bias) and the student may continue to perform highly (no self-fulfilling prophecy).

More Accuracy Than Self-Fulfilling Prophecy

What, then, is the main reason why teacher expectations predict student achievement? Is it primarily because teachers are accurate, or primarily because their expectations are self-fulfilling? Through the 1980s and early 1990s social psychology abounded with testimonies to the power of expectancies to create social reality (e.g., Fiske & Taylor, 1984; Hamilton, Sherman, & Ruvolo, 1990; Jones, 1986, 1990; Snyder, 1984; see Jussim, 1991 for a review). In contrast, most educational and developmental psychologists argued that expectancy effects were generally small (e.g., Brophy, 1983; Brophy & Good, 1974; Cooper, 1979; Eccles & Blumenfeld, 1985; Eccles & Wigfield, 1985; West & Anderson, 1976). Evidence from naturalistic studies consistently failed to support the strong social psychological claims and instead confirmed the perspective of the educational and developmental psychologists—rarely uncovering expectancy effects larger than .1 to .2 in terms of standardized regression coefficients (see Jussim, 1991; Jussim & Eccles, 1995 for reviews). Furthermore, research has repeatedly shown that the main reason teacher expectations predict student achievement is because they are accurate (see Brophy, 1983; Jussim, 1991; Jussim & Eccles, 1995 for reviews). Because two of our studies provided some of the clearest evidence of teacher accuracy to date (Jussim, 1989; Jussim & Eccles, 1992), we describe them below in some detail.

The Jussim (1989) and Jussim and Eccles (1992) Studies: Overview

These studies were based on the Michigan Study of Adolescent Life Transitions (MSALT), which assessed a variety of social, psychological, demographic, and achievement-related variables in a sample that included over 200 teachers and 3,000 students in sixth and seventh grade (see Eccles et al., 1989; Midgley, Feldlaufer, & Eccles, 1989; Wigfield, Eccles, MacIver, Reuman, & Midgley, 1991 for more details about the MSALT project). About 100 teachers and 1,700 students in sixth-grade math classes were the focus of the two studies summarized here. Both studies tested the hypotheses that: (1) Teachers' expectations early in the year are based on students' previous achievement and motivation; and (2) Teachers' expectations, students' motivation, and students' previous achievement influence students' subsequent achievement (for detailed descriptions of the models and analyses, see Jussim, 1989; Jussim & Eccles, 1992).

Three teacher expectation variables were assessed in early October of sixth grade: Teacher perceptions of students' performance, talent, and effort in math. Student motivation measures included: Self-concept of math ability, intrinsic and extrinsic value of math, and self-reports of effort and time spent on math homework. Fall and spring assessments of these motivational variables were included in Jussim; only fall assessments were included in Jussim and Eccles.

There were two measures of previous achievement: Final marks in fifth-grade math classes, and scores on standardized achievement tests taken in late fifth or early sixth grade. There were two outcome measures of achievement: Final grades in sixth-grade math classes, and scores on the math section of the MEAP (Michigan Educational Assessment Program, a standardized test administered to students in Michigan early in seventh grade). All measures were reliable and valid (for more detail, see Eccles-Parsons, Kaczala, & Meece, 1982; Eccles (Parsons), Adler, & Meece, 1984; Jussim, 1987, 1989; Jussim & Eccles, 1992; Parsons, 1980).

Results

Because results reported here are from two studies, they are presented in pairs below. The first refers to Jussim and the second refers to Jussim and Eccles. These two studies were the first to explicitly assess and compare all three expectancy phenomena (self-fulfilling prophecy, perceptual bias, and accuracy). We assessed whether teacher perceptions early in the school year predicted changes in motivation and achievement (by controlling for previous achievement and motivation).

Consistent with the self-fulfilling prophecy hypothesis, teacher perceptions of students' math performance in October of the sixth grade significantly predicted changes in students' self-concept of ability, final grades, and early seventh-grade standardized test scores. Results consistent with the perceptual bias hypothesis showed that teacher perceptions predicted sixth-grade math grades to a larger extent than they predicted seventh-grade standardized test scores. Teachers assigned higher grades to students whom they believed exerted more effort.

This hypothetically could have represented accuracy—if teachers rewarded students who actually worked harder with higher grades. Instead, however, we found no evidence that the students who received the higher grades actually worked any harder than their peers. In fact, the students who received low grades reported spending more time on homework than the other students (Jussim, 1989; Jussim & Eccles, 1992). It seems, therefore, that the teachers erroneously assumed that higher achievers worked harder. Because effort is difficult to observe directly, we speculated that teachers, perhaps influenced by a belief in a just world (Lerner, 1980) or by the Protestant work ethic (Schuman, Walsh, Olson, & Etheridge, 1985; Weber, 1930), simply assume that "hard work pays off." Therefore, high achievers "must" be working harder. A consequence, however, is that the academically "rich" (the high achievers) get richer (teachers assign them grades that are even higher than they deserve).

There was both accuracy and inaccuracy in teacher perceptions. Teacher perceptions were high in impression accuracy because they were linked most strongly to appropriate criteria: previous grades and standardized test scores, teacher perceptions of in-class performance, and student motivation (the multiple correlation of these factors with teacher expectation variables ranged from about .6 to .8).

Results from both studies also provided considerable evidence of predictive accuracy. The zero-order correlation between teacher expectations early in the year and student achievement late in the year equals expectancy effects (self-fulfilling prophecy and perceptual bias) plus predictive accuracy (teachers basing their expectations on factors that influence student achievement). Therefore, one index of predictive accuracy is the difference between the zero-order correlation and the size of the expectancy effects (see Jussim, 1989, 1991, 1993; Jussim & Eccles, 1992 for more detailed explanations).

The path coefficients relating teacher perceptions to MEAP scores accounted for about 20-30 percent of the zero-order correlations between initial teacher perceptions and subsequent MEAP scores; the remaining 70-80 percent represented predictive validity without influence (i.e., accuracy). There was a similar pattern for final grades in sixth-grade math. The path coefficients relating teacher perceptions to grades accounted for about 30-40 percent of the zero-order correlations between initial teacher perceptions and subsequent grades; the remaining 60-70 percent represented accuracy.

Other Naturalistic Studies

When we first completed these initial studies (Jussim, 1989; Jussim & Eccles, 1992), we were quite surprised. At that time within social psychology nearly every review that addressed self-fulfilling prophecies was virtually a testament to the power of expectations to create reality (e.g., Fiske & Taylor, 1984; Jones, 1986, 1990; Miller & Turnbull, 1986; Snyder, 1984). So, in an attempt to understand why our results were so different from those described by these reviewers, over several years we performed an exhausting (we would like to believe exhaustive) search of the literature regarding the relative sizes of accuracy and self-fulfilling prophecy (see, e.g., Jussim, 1990, 1991; Jussim & Eccles, 1995; Jussim, Madon, & Chatman, 1994; Jussim, Eccles, & Madon, 1996). We briefly describe the results of that search.

Only a few other naturalistic studies have provided evidence capable of empirically comparing self-fulfilling prophecies to accuracy. All have shown that student achievement predicts teacher expectations more strongly than teacher expectations predict changes in future achievement (much higher impression accuracy than self-fulfilling prophecy—Brattesani, Weinstein, & Marshall, 1984; West & Anderson, 1976; Williams, 1976). Only Williams, however, provided the information necessary to separate predictive accuracy from self-fulfilling prophecy (both the zero-order correlations and the path coefficients linking teacher expectations to future achievement). And, as we did (Jussim, 1989; Jussim & Eccles, 1992), Williams found that teacher expectations predicted future achievement more because they were accurate than because they led to self-fulfilling prophecies.

Although only a few naturalistic studies provided data capable of comparing accuracy to self-fulfilling prophecy, many naturalistic studies focused on self-fulfilling prophecies (e.g., Palardy, 1969; Seaver, 1973; Sutherland & Goldschmid, 1974; see Jussim & Eccles, 1995 for a comprehensive review). These studies, too, consistently provided evidence of small self-fulfilling prophecies (effect sizes, in terms of correlation or standardized regression coefficients, typically ranged from 0 to about .3).

Rist (1970)

Because Rist's (1970) study has often been cited as evidence of powerful naturally occurring self-fulfilling prophecies (e.g., Miller & Turnbull, 1986; Myers,

1987; Snyder, 1984), we discuss it in some detail. Rist observed that by the eighth day of class, a kindergarten teacher had divided her class into three groups—a supposedly smart, average, and dumb group. Each group sat at its own table (Tables 1, 2, and 3, respectively). However, the main difference between the students was not intelligence—it was social class. In comparison to the other students, the students at Table 1 came from homes that had greater incomes, were less likely to be supported by welfare, were more likely to have both parents present, and the children themselves were cleaner and more likely to dress appropriately. There were comparable differences between the students at Tables 2 and 3. Table 1 was positioned closest to the teacher, and she proceeded to direct a disproportionately large amount of her time and attention to those students. In addition, she was generally friendlier and warmer to the students at Table 1. Consequently, Rist interpreted his study as documenting strong self-fulfilling prophecies.

The differences Rist observed in teacher treatment of middle-class versus poor students would be inappropriate and unjustified even if there were real differences in the intelligence of the children at the different tables. Nonetheless, despite Rist's conclusions, the study provided no evidence of self-fulfilling prophecy. Although Rist provided a wealth of observations concerning teacher treatment, he provided few regarding student performance. The differential treatment alone is not evidence of self-fulfilling prophecies. Differences in student outcome measures are also needed. The one student outcome measure that Rist provided was students' IQ scores. In contrast to the self-fulfilling prophecy hypothesis, there were no IQ differences between the students at the different tables at the end of the school year. Thus, although the teacher may have held very different expectations for middle- versus lower-class students, and even though the teacher may have treated students from different backgrounds very differently, this did not affect students' IQ scores (see Jussim & Eccles, 1995 for a more detailed critique of this case study).

William's (1976) study of over 10,000 high school students provided a much more rigorous analysis of the role of social class in educational self-fulfilling prophecies. Williams found that teachers held higher expectations for students from upper socioeconomic backgrounds. However, differences in teacher expectations for middle- and lower-class students evaporated after controlling for students' previous levels of performance. This means that, rather than student social class biasing teacher expectations, teachers accurately perceived genuine differences in achievement among students from differing socioeconomic backgrounds. Of course, accurate expectations do not create self-fulfilling prophecies.

A colleague once described the Rist (1970) paper as "a real tearjerker" and we can't help but agree. Nonetheless, the less well-known Williams (1976) study is much stronger than Rist's study on almost all important scientific grounds—Rist relied primarily on his own subjective and potentially biased observations whereas Williams relied on school records and questionnaires; Rist focused on 30 students whereas Williams focused on over 10,000 students; Rist claimed to provide strong evidence of self-fulfilling prophecy but actually provided none, whereas Williams rigorously tested for self-fulfilling prophecies and failed to find any. Although social class may sometimes lead to self-fulfilling prophecies, with respect to drawing scientific conclusions based on evidence, Williams deserves dramatically more weight than Rist.

The Experimental Research

Many of the strongest testaments to the supposed power of expectancies to create reality have drawn almost exclusively on the experimental evidence (e.g., Devine, 1995; Fiske & Taylor, 1984; Gilbert, 1995; Jones, 1986, 1990; Miller & Turnbull, 1986; Snyder, 1984). Therefore, it is especially important to critically evaluate and interpret the meaning of the experimental studies.

Although experiments have major strengths, they also have important weaknesses. Following Rosenthal and Jacobson's lead, in most self-fulfilling prophecy studies, experimenters induced teachers to develop false expectations for some students, usually by providing bogus background information (e.g., false standardized test scores). Thus, the experiments typically only demonstrated that inaccurate teacher expectations could lead to self-fulfilling prophecies; they could not demonstrate that teacher expectations typically do lead to self-fulfilling prophecies. Second, they did not examine accuracy. Therefore, even when such experiments provide evidence of self-fulfilling prophecies, they leave as an open, unanswered, empirical question the extent to which naturally developed teacher expectations predict student achievement because they are accurate.

In fact, many of the experimental studies demonstrating self-fulfilling prophecy can be readily interpreted as suggesting that under naturalistic conditions, teachers are likely to be accurate. By randomly assigning students to late-blooming versus control conditions, Rosenthal and Jacobson rendered standardized test information independent of students' actual achievement. However, standardized tests are usually excellent predictors of student achievement. In general, the more that teachers use standardized test information, the more accurate their expectations will be. Therefore, the teachers in Rosenthal and Jacobson's study developed erroneous expectations *by doing something that in nearly all other circumstances would lead them to be highly accurate*—that is, utilizing standardized test information as a basis for expectations. Even Rosenthal and Jacobson's study strongly suggests that under naturalistic conditions, in which standardized tests generally provide a sound basis for teacher expectations, those expectations are likely to be highly accurate.

Meta-analyses of the Expectancy Literature

Although several meta-analyses have clearly shown that self-fulfilling prophecies occur in classrooms, laboratories, the workplace, and other settings, the average effect sizes (in terms of correlations between expectations and outcomes) are quite modest—typically around .2 (e.g., Cooper & Hazelrigg, 1988; Harris & Rosenthal, 1985; Rosenthal & Rubin, 1978; Smith, 1980). Furthermore, self-fulfilling prophecies have only occurred in slightly over a third of the studies testing for them. Clearly, notwithstanding the claims in the social psychological reviews (e.g., Devine, 1995; Gilbert, 1995; Jones, 1986; Snyder, 1992) self-fulfilling prophecies are typically neither pervasive nor powerful.

ARE EXPECTANCY EFFECTS EVER POWERFUL?

Educational psychologists working in the expectancy area have long argued that self-fulfilling prophecy effects are generally small and that many teachers are highly accurate (e.g., Brophy, 1983; Brophy & Good, 1974; Cooper, 1979; West & Anderson, 1976). Although the general evidence supports this view, we strongly suspected that there were conditions under which expectancy effects were substantially larger. Thus, we embarked on a quest to identify conditions under which powerful self-fulfilling prophecies actually did occur.

Demographic Moderators of Expectancy Effects

For several reasons we suspected that students from stigmatized social groups might be more vulnerable to expectancy effects (see Jussim, Eccles, & Madon, 1996 for more details on both the conceptual analysis and the empirical results). For African-American students, students from lower social class backgrounds, and girls (at least in math classes), school may not be a particularly friendly place (AAUW, 1992; Lareau, 1987; Steele, 1992). When faced with negative teacher expectations, such students may be particularly likely to devalue the importance they place on achievement. Similarly, when faced with a supportive and demanding (high expectancy) teacher, it may feel like a breath of fresh air, thereby inspiring these students to new heights. This perspective may not be as pollyannish as it sounds. In his influential article on black disidentification with school, Steele (1992) described academic programs in which previously low-performing students (e.g., some with SATs in the 300s) took on difficult honors-level work and outperformed their white and Asian classmates.

Therefore, we (Jussim, Eccles, & Madon, 1996) assessed whether student demographics moderated relations between teacher expectations (assessed early in the school year) and students' subsequent achievement, in the context of a model that included many controls (previous achievement, motivation, etc.). We found virtually no evidence that student sex moderated expectancy effects—for both boys and girls the standardized regression coefficients relating teacher expectations to future achievement were typically small (.1 to .2).

However, we found considerable evidence of moderation by ethnicity and by social class. Standardized coefficients relating teacher perceptions to future standardized test scores were .14 for white students, .37 for African-American students, .11

for students from higher SES backgrounds, and .25 for students from lower SES backgrounds. These findings mean that for white and middle-class students, the difference in achievement outcomes predicted by the highest and lowest teacher expectations was about 20 percentile points. But for African-American students and students from lower SES backgrounds, the difference in achievement outcomes predicted by the highest and lowest teacher expectations was about 50 percentile points.

That is a dramatic difference and we found the same pattern when using grades as the achievement outcome. In addition, the effects of ethnicity and social class were independent—excluding the African-American students from analyses did not change the social class effects and expectancy effects were just as powerful among middle-class African-American students as among less well off African-American students.

Prior Achievement as a Moderator

We wondered whether this pattern of greater susceptibility among students from stigmatized groups might also extend to students who have histories of poor academic performance. Teachers often communicate negative expectations to lower achieving students by criticizing them more often, interacting with them less often, and demanding less from them (Brophy, 1983). These types of behaviors may undermine these students' motivation, rendering them more susceptible to confirming negative expectations. However, when teachers communicate positive expectations to lower achieving students, their motivation for school may rise dramatically and inspire them to make substantial gains in their achievement (Eccles & Wigfield, 1985).

We know of only one study that has empirically examined whether students' prior achievement moderates expectancy effects (Madon, Jussim, & Eccles, 1997). Our results mirrored the patterns that we found among African-American and lower social class students. Teachers' expectations predicted students' future achievement more strongly among students with lower previous standardized test scores than for students with higher previous standardized test scores. Thus, it seems that students who are stigmatized, whether because of their demographic group membership or because of their history of achievement, are more vulnerable to expectancy effects.

Multiple Vulnerabilities

We wondered whether the expectancy moderators we discovered were additive—for example, did teacher expectations predict students' future achievement more strongly among low performing students from lower class backgrounds? They did. The relation (standardized regression coefficient, obtained in context of a model with all the controls) of teacher expectations to future achievement among low performing students from lower social class backgrounds was .62. This is one of the strongest relations between teacher expectations and student achievement ever obtained, especially in a naturalistic study. Unfortunately, however, because of the small number of African-American students in the sample, we could not examine whether teacher expectations also more strongly predicted the future achievement of low performing African-American students.

The research described thus far did not address whether self-fulfilling prophecies mostly helped or harmed students. In a subsequent section, however, we describe other research that specifically examined whether positive or negative teacher expectations more strongly predicted students' future achievement.

Other Student Characteristics

Goals

Students may become more or less susceptible to self-fulfilling prophecies, depending upon their goals. When perceivers' have something targets want (such as approval), and when targets are aware of the perceiver's beliefs, they often confirm those beliefs in order to create a favorable impression (Zanna & Pack, 1975; von Baeyer, Sherk, & Zanna, 1981). Similarly, when targets desire to facilitate smooth social interactions, they are also more likely to confirm perceivers' expectations (Snyder, 1992).

Although these findings were all obtained in laboratory studies of college students, similar processes seem likely to occur at times in classrooms. Teachers have control over many things students want (approval, higher grades, free time, etc.), and students who want these enough may be particularly vulnerable to expectancy effects. Similarly, because of teachers' greater power, the onus will more often be on the student than on the teacher to smooth interactions (at least among older students). This, too, may render some students particularly vulnerable to expectancy effects. Of course, when teachers' expectations are positive, this may be quite beneficial to the student.

In contrast, when targets believe that perceivers hold a negative belief about them, they often act to disconfirm that belief (Hilton & Darley, 1985). That is, targets may resist negative expectations far more than they resist positive ones, especially when they are confident that those negative expectations are unjustified (e.g., Swann & Ely, 1984).

Age

Self-fulfilling prophecies were strongest among the youngest students in the original Rosenthal and Jacobson (1968) study, suggesting that younger children may be more vulnerable to expectancy effects than are older children and adults. However, a meta-analysis has shown that the strongest teacher expectation effects occurred in first, second, *and* seventh grade (Raudenbush, 1984). Furthermore, the largest self-fulfilling prophecy effects yet reported were obtained in a study of adult Israeli military trainees (Eden & Shani, 1982). Although these findings do not deny a moderating role for age, they do suggest that situational factors may also influence students' susceptibility to self-fulfilling prophecies.

Situational Factors

New Situations

People may be more susceptible to confirming others' expectations when they enter new situations. Whenever people engage in major life transitions, such as entering a new school or starting a new job, they may be less clear and confident in their self-perceptions. Unclear self-perceptions render targets more susceptible to confirming perceivers' expectations (Jussim, 1986; Swann & Ely, 1984).

This analysis may help explain the seemingly inconsistent findings regarding age. Students in first, second, and seventh grade, and new military inductees, are all in relatively unfamiliar situations. Therefore, all may be more susceptible to self-fulfilling prophecies.

Class Size and Resources

Expectancy effects also may be more likely to occur in classrooms with large numbers of students than they are in smaller classrooms. People are more susceptible to biases when more of their "cognitive capacity" is being used—when they are trying to do several things at once (Gilbert & Osborne, 1989). The more students in a class, the more "cognitively busy" the teacher is likely to be, and, therefore, the more susceptible to biases and expectancy effects.

A related moderator may be class and school resources. Not only do resources (access to books, computers, laboratories, indoor and outdoor athletic facilities, fine arts, etc.) create a more pleasant learning environment, they probably make it easier for teachers to manage the students in their classes. Consequently, teachers from schools with many resources, too, may typically be less cognitively overloaded, and, therefore, less susceptible to self-fulfilling prophecies.

At least one study (Finn, 1972) found results consistent with this perspective. Finn found that teacher expectations influenced the grades teachers assigned, but only in urban schools (not in suburban schools). Although urban and suburban schools differ on many dimensions, two key areas are class size (suburban schools often have smaller class sizes) and resources (suburban schools are often wealthier).

Tracking

School tracking refers to the policy of segregating students into different classes according to their ability. For example, smart students may be assigned to one class, average students to another, and slow students to a third. Tracking may be intended as a helpful intervention. By putting students with similar capacities together, teachers have the opportunity to tailor their lessons in such a way as to maximize those students' learning and achievement.

However, researchers have long speculated that tracking may also moderate expectancy effects. Tracking represents institutional justification for believing that some students are smarter than others. Thus, it may lead to the type of rigid teacher expectations that are most likely to evoke self-fulfilling prophecies and perceptual biases (Eccles & Wigfield, 1985; Jussim, 1986, 1990; Oakes, 1987).

In addition, poor quality instruction may occur in at least some low-track courses. These classes are harder to manage and traditional teaching techniques are not likely to be successful. But, at least sometimes, teachers' expectations may exacerbate the poor environment. Lower teacher expectations for low-track classes (Brophy & Good, 1974; Oakes, 1985), and the fact that less experienced teachers often teach low-ability groups, can limit students' opportunity to learn (Slavin, 1993). Furthermore, if teachers think that low-skill children cannot learn, or do not want to learn, they may reduce their teaching efforts (Allington, 1980; Evertson, 1982)—which is one of the behaviors that often leads to self-fulfilling prophecies (Harris & Rosenthal, 1985).

However, despite the harsh criticisms against tracking, several recent meta-analyses have demonstrated that tracking can have positive effects on student achievement and attitudes toward school (Kulik & Kulik, 1982, 1987, 1992). In addition, tracking may increase the accuracy of teachers' expectations. If valid factors (e.g., previous grades and standardized test scores) influence grouping decisions, and if teachers rely on grouping as a basis for expectations, those expectations will be largely accurate.

In addition, large differences in student ability are reduced by grouping together students of similar ability. Thus, it may be easier for teachers to assess student achievement accurately because (1) there is lowered variability within ability groups; and (2) the approximate level of students' achievement is clearly communicated by their assignment to a high, average, or low group. Thus, teachers may be less likely to either overestimate (leading to frustrated students) or underestimate (leading to bored students) student ability. If ability grouping increases accuracy, it could decrease the strength of teacher expectation effects.

We recently completed a study that empirically examined whether ability grouping practices (both between-class grouping and within-class grouping) moderate expectancy effects (Smith, Jussim, Eccles, Vannoy, Madon, & Palumbo, 1997). Self-fulfilling prophecies were strongest among students in the low groups when teachers used *within-class* grouping. Ability differences and group labels may be more salient to teachers who use within-class grouping. This may increase differential treatment (Weinstein, 1985), leading to greater self-fulfilling prophecies. However, these self-fulfilling prophecies may be more likely to help students than to harm them, because research based on the same data has shown that, in general, positive self-fulfilling prophecies were more powerful than negative ones (Madon et al., 1997).

In addition, we (Smith et al., 1997) found that self-fulfilling prophecies were no more powerful among students grouped by ability between classes than among students in heterogenous classes. Furthermore, we only found evidence of perceptual biases when *no* grouping was used. Thus, between-class grouping, rather than strengthening expectancy effects, actually seemed to reduce them. Altogether, the results from our study and the meta-analyses provide virtually no support for sweeping claims that ability grouping harms students.

Teachers' Characteristics

Goals

Goals moderate the influence of expectations (Hilton & Darley, 1991). Self-fulfilling prophecies are more likely to occur when perceivers desire to arrive at a stable and predictable (although not necessarily accurate) impression of a target (Snyder, 1992), when perceivers are more confident in the validity of their expectations (Jussim, 1986; Swann & Ely, 1984), and when they have an incentive for confirming their beliefs (Cooper & Hazelrigg, 1988). Self-fulfilling prophecies and perceptual biases are less likely when perceivers are motivated to develop an accurate impression of a target (Neuberg, 1989), when perceivers' outcomes depend on the target (Neuberg, 1994), and when perceivers' main goal is to get along in a friendly manner with targets (Snyder, 1992). Perceptual biases are more likely when perceivers strive to reach a particular conclusion rapidly (Kunda, 1990; Neuberg, 1994; Pyszczynski & Greenberg, 1987). These findings raise the following question: When are perceivers likely to be motivated by accuracy or a desire to get along in a friendly manner, and when are they likely to be overconfident in their beliefs or motivated by desires to reach a particular conclusion?

Prejudice, Cognitive Rigidity, and Belief Certainty

Prejudiced individuals seem especially unlikely to be motivated by either accuracy concerns or the desire to get along with members of the group they dislike. Instead, they seem likely to desire to reach the particular conclusion that members of the stigmatized group have negative, enduring attributes (Pettigrew, 1979). People high in cognitive rigidity or belief certainty also may not be motivated to consider different viewpoints. Cognitive rigidity, which is usually construed as an individual difference factor (e.g., Adorno, Frenkel-Brunswick, Levinson, & Sanford, 1950; Allport, 1954; Harris, 1989), and belief certainty, which is usually construed as a situational factor (Jussim, 1986; Swann & Ely, 1984), are similar in that they describe people who may be unlikely to alter their beliefs when confronted with disconfirming evidence. Whether the source is prejudice, cognitive rigidity, or belief certainty (which may tend to co-occur within individuals—see Adorno et al., 1950), people who are overly confident in their expectations may be most likely to maintain biased perceptions of individuals and to create self-fulfilling prophecies (Babad, Inbar, & Rosenthal, 1982; Harris, 1989; Swann & Ely, 1984).

Other Individual Differences

Experienced teachers may be less likely to create self-fulfilling prophecies. We use the term "experienced" here in two different but related senses. One aspect of experience refers to time on the job or in one's role. Thus, for example, more experienced teachers, therapists, doctors, and so on have probably developed considerably more competence and expertise at appraising students, clients, patients, and so on. If so, then their impressions may be more accurate and less self-fulfilling.

The second sense in which we use "experience" involves teachers' experience with students. Teachers who have greater information and more opportunities to interact with students are more likely to develop accurate expectations, and such accuracy reduces the potential for self-fulfilling prophecies.

Professional efficacy may also moderate expectancy effects. In general, efficacy refers to beliefs concerning one's ability to engage in the behaviors necessary for accomplishing a particular goal (Bandura, 1977). Professional efficacy, therefore, refers to beliefs regarding one's ability to engage in the behaviors necessary for accomplishing the essential work of one's profession. For example, teaching efficacy would refer to beliefs regarding one's ability to teach. When teachers are less confident in their teaching ability (low teaching efficacy), they may be more likely to create expectancy effects. Teachers low in teaching efficacy may feel less able to improve the skills of low expectancy students; consequently, they may spend less time and effort with lows than do teachers high in teaching efficacy. By virtue of spending less time with low expectancy students (and perhaps more time with high expectancy students), teachers low in teaching efficacy may exacerbate differences among high and low expectancy students to a greater extent than do teachers high in teaching efficacy (Midgley, Feldlaufer, & Eccles, 1989).

A need to control others may also moderate expectancy effects. For example, the more that teachers strive to control students, the more likely it may be that their expectancies will be self-fulfilling and biasing. A high emphasis on control may include a particularly strong preference for having one's expectations confirmed. Control implies predictability, so that unpredictable situations (or students) may be perceived as implying a lack of control. When students disconfirm expectations, therefore, teachers who emphasize control may feel threatened. These teachers may be most motivated to "insure" that students confirm their expectations. This analysis is consistent with a less well-known finding of the original Rosenthal and Jacobson (1968) study—some teachers responded especially negatively to the successes of students not specifically designated as late bloomers.

Student Perceptions of Differential Treatment

Another condition under which self-fulfilling prophecies are more powerful is when students believe teachers treat high expectancy students differently than they treat low expectancy students (Brattesani, Weinstein, & Marshall, 1984). Students in high differential treatment classrooms perceive greater differences in how teachers treat high expectancy and low expectancy students than do students in low differential treatment classrooms. In high differential treatment classes students report that teachers are more demanding of high expectancy students and grant them special privileges. Students in such classes also report that lows are treated with more concern and vigilance, and receive more negative feedback (Weinstein, 1985).

Students' perceptions of differential treatment correspond to actual differences in how teachers treat high and low expectancy students (Cooper & Good, 1983, Weinstein, 1985). Thus, in high differential treatment classrooms, teachers clearly convey their expectations about which students are going to succeed, and students pick up on these verbal and nonverbal cues (Brophy, 1983). As a result, teacher expectations more strongly influence both student self-concepts and student achievement in high differential treatment classrooms than in low differential treatment classrooms (Brattesani, Weinstein, & Marshall, 1984).

Positive versus Negative Expectancy Effects

Many discussions of expectancy effects assume or imply that they are mostly bad (e.g., Devine, 1995; Fiske & Taylor, 1984, 1991; Gilbert, 1995; Jones, 1986, 1990) or that negative self-fulfilling prophecies (which undermine students' achievement gains) are likely to be more powerful than positive self-fulfilling prophecies (which increase students' achievement gains—e.g., Brophy, 1983; Brophy & Good, 1974; Eccles & Wigfield, 1985; Weinstein, 1985). In fact, however, we know of only three studies that seem to have addressed the issue of whether self-fulfilling prophecies are more likely to hurt students than to help them.

The first (Babad, Inbar, & Rosenthal, 1982) examined the power of negative and positive self-fulfilling prophecies among students in gym classes who had either low bias or high bias teachers. Although there were no differences between high and low expectancy students' athletic performance among low bias teachers, the high expectancy students consistently performed more highly than did the low expectancy students among high bias teachers (a result demonstrating the occurrence of a self-fulfilling prophecy). However, a simple difference between high and low expectancy students is insufficient to determine whether self-fulfilling prophecies primarily helped or hurt students. A difference between highs and lows could occur if: (1) high expectations helped students and low expectations had no effect; (2) low expectations harmed students and high expectations had no effect; or (3) high expectations helped students *and* low expectations hurt students.

Because there was no evidence of self-fulfilling prophecies among low bias teachers, students' performance among low bias teachers could be used as a sort of control group for determining whether self-fulfilling prophecies primarily helped or hurt students with high bias teachers. Among students with high bias teachers, if negative self-fulfilling prophecies were more powerful than positive self-fulfilling prophecies, then: (1) lows with high bias teachers should have consistently performed worse than lows with no bias teachers; and (2) there should be little difference between the performance of highs with high or no bias teachers.

This was not the case. There were three performance measures: sit-ups (for girls) and push-ups (for boys); distance jump; and speed. For sit-ups/push-ups, the power of negative and positive self-fulfilling prophecies was similar. The performance differences between lows with no and high bias teachers were similar in magnitude to the performance differences between highs with no and high bias teachers.

For the distance jump, negative self-fulfilling prophecies were more powerful than positive ones. Lows with no bias teachers jumped farther than lows with high bias teachers, whereas highs with high bias teachers jumped the same distance as highs with low bias teachers. For the speed measure, there was no evidence of self-fulfilling prophecies. The performance of lows with no and high bias teachers were similar, indicating that negative self-fulfilling prophecies did not occur. Highs with high bias teachers actually performed worse than highs with no bias teachers, which may be an interesting effect of teacher bias, but does not represent a self-fulfilling prophecy.

Overall, therefore, Babad, Inbar, and Rosenthal's results provided a decidedly mixed picture. Although they clearly did find evidence of negative self-fulfilling prophecies, they also found evidence of positive self-fulfilling prophecies. Their research provided no evidence that negative self-fulfilling prophecies are stronger than positive self-fulfilling prophecies.

There is a less frequently cited study that did test this hypothesis (Sutherland & Goldschmid, 1974). Their results indicated that negative expectations undermined the future IQ scores of high IQ students, whereas positive expectations had no significant effects on the future IQ scores of low IQ students. Although these results suggest that negative self-fulfilling prophecies are more powerful than positive ones, this study suffers from a serious methodological weakness. Specifically, negative expectations underestimated students more than positive expectations overestimated them. Thus, the greater power of negative versus positive self-fulfilling prophecies that emerged may have reflected the greater inaccuracy of negative expectations in their particular sample, rather than any generally greater power of negative expectations.

We know of only one other study that compared the power of positive and negative self-fulfilling prophecies (Madon et al., 1997). We examined whether positive or negative self-fulfilling prophecies are more powerful among students in general, and whether students' prior level of achievement moderated the power of positive and negative self-fulfilling prophecies. Our results were consistent with the conclusion that positive self-fulfilling prophecies are more powerful than negative self-fulfilling prophecies. Although the standardized test scores of high expectancy students increased (relative to their classmates) and the standardized test scores of low expectancy students decreased (relative to their classmates), highs increased more than lows decreased.[1]

However, both positive and negative self-fulfilling prophecies were also more powerful among low achievers than among high achievers (and the effects of moderation by achievement and by positive/negative expectancy were additive). Consequently, teacher expectations most strongly predicted students' future standardized test scores when they overestimated low achievers. The effect size associated with strong overestimates of low achievers was about .4—which is one of the largest found in naturalistic studies of self-fulfilling prophecies.

Clearly, additional research on the relations between positive and negative expectations and students' future achievement is sorely needed. However, our results suggest that among underachieving students, self-fulfilling prophecies are more likely to increase the rate of achievement gains than they are to slow them down. And our results certainly provided no evidence that self-fulfilling prophecies maintain a caste-like system by limiting the achievement gains of students from the wrong side of the tracks (see, e.g., Gilbert, 1995; Hofer, 1994; Rist, 1970).

Do Expectancy Effects Accumulate?

Even if self-fulfilling prophecies tend to be small over a single school year, if they accumulate over several years, they might eventually lead to large differences between high and low expectancy students. Accumulation means that, over time, the effects of self-fulfilling prophecies on student achievement would be greater and greater. For example, suppose that two sixth-grade students of equal ability start out with IQs of 100 and that one is the target of high teacher expectations and the other is the target of low teacher expectations. If teachers' expectations change student IQ by three points a year, by the end of sixth grade, the high expectancy student would have an IQ of 103, and the low expectancy student would have an IQ of 97. If this trend continued, by the end of high school, the high expectancy student would have an IQ of 118 and the low expectancy student would have an IQ of 82. Thus, small effects could translate into dramatic differences if self-fulfilling prophecies accumulated.

However, self-fulfilling prophecies could also dissipate. Dissipation means that, over time, the effects of self-fulfilling prophecies become smaller and smaller. For example, even if a teacher increases a student's IQ score by three points, unless some process serves to sustain or increase that effect next year, the student's IQ may fall back toward its original level. Thus, the expectancy effect may dissipate. To date, five studies have empirically examined the issue of accumulation. We review them next.

Rosenthal and Jacobson (1968)

Rosenthal and Jacobson followed their students for two years. If self-fulfilling prophecies accumulated, then the differences between the "late bloomers" and the control students should have been even greater in the second year. This was not the case. "Late bloomers" gained 3.18 more IQ points, on average, compared to other students in the first year of the study, but were only 2.67 points higher in the second year. These results are consistent with dissipation—over time, teachers' expectations had less, rather than more of an effect on student achievement.

Rist (1970)

Rist followed a group of inner-city students from kindergarten to second grade. Because there were no quantitative measures of achievement after kindergarten, the evidence regarding accumulation/dissipation consisted entirely of Rist's observations regarding seating table assignment.

As in kindergarten, in first grade the students were again placed at tables supposedly reflecting achievement. All of the students from kindergarten Table 1 (the high table) were placed at the first-grade Table A (high group). Nearly all of the students from kindergarten Tables 2 (middle) and 3 (low) were placed at first-grade Table B (middle group). Only one of the students from the kindergarten class was placed at the lowest first-grade table, Table C. Although students from the high-ability table remained at a high-ability table, the students from the middle- and low-ability tables in kindergarten were combined into one middle-ability table in first grade. Thus, if seating assignment is the criterion, at this first transition, differences among students based on reading-table assignment had declined.

By second grade, the students from Table A were assigned to the "Tigers" (high group) and students from Tables B and C were assigned to the "Cardinals" (middle group). None of the students from the first-grade class were assigned to the "Clowns" (low group). In addition, that year, two students from the "Tigers" were moved down to the "Cardinals" and two students from the "Cardinals" were moved up to the "Tigers." Although the groups created by the kindergarten teacher did remain somewhat intact from year to year, by the end of second grade, initial differences (as indicated by seating assignments) between students had decreased. Thus, although Rist interpreted his study as demonstrating that expectancies contribute to a caste-like system based on social class, his actual results are consistent with dissipation.

West and Anderson (1976)

West and Anderson examined data from 3,000 male students in their freshman, sophomore, and senior years of high school that included information on both teachers' expectations and student achievement. Results from this study would have supported accumulation if the path coefficients relating freshman year teachers' expectations to senior year achievement were larger than the path coefficients relating freshman year teachers' expectations to sophomore year achievement. This was not the case. The path coefficient relating teachers' expectations to sophomore year achievement was .12, whereas the path coefficient relating teachers' expectations to senior year achievement was .06. These results support dissipation. Teachers' expectations from freshman year predicted senior year achievement less strongly than they predicted sophomore year achievement.

Frieze, Olson, and Russell (1991)

Frieze, Olson, and Russell examined the relationship between facial attractiveness and salaries of MBA graduates. Results from this study indicated that attractive MBA graduates, both men and women, made more money over time than less attractive MBA graduates. These results support accumulation. That is, over time, the effect of facial attractiveness on later salary increased.

However, it is not clear that the results from this study reflect self-fulfilling prophecies for several reasons. First, expectancies were never measured, so that it is impossible to determine whether expectancies were self-fulfilling at all, let alone that they accumulated. Second, social skill correlates with attractiveness (Eagly, Makhijani, Ashmore, & Longo, 1991; Feingold, 1992). Therefore, attractive MBAs' earning higher salaries over time could be due to better interpersonal skills on the job rather than to false expectations on the part of their employers.

Smith, Jussim, and Eccles (1997)

We have just completed a study addressing the accumulation issue. We examined the relationship between teachers' expectations in sixth and seventh grade and student achievement through high school. This study employed more extensive control variables than have previous studies, including two measures of previous achievement (final marks and standardized test scores) and five measures of student motivation (self-concept of ability, self-perception of effort, time spent on homework, and intrinsic and extrinsic value of math). The results were consistent with the dissipation hypothesis for both final marks and standardized test scores across school years (through twelfth grade) and even within the same school year (using semester marks). Over time, teacher expectation effects became smaller and smaller. These results corroborate findings from the three other teacher expectation studies on accumulation, suggesting that self-fulfilling prophecies in the classroom generally dissipate over time.

Do Expectancy Effects Accumulate? Conclusion

The idea that expectancy effects accumulate has been very popular among many social scientists (e.g., Hofer, 1994; Jones, 1990; Snyder, 1984). Unfortunately, however, there is almost no evidence in support of the accumulation hypothesis and much evidence that contradicts it. The one study that yielded results consistent with the accumulation hypothesis (Frieze, Olson, & Russell, 1991) was a workplace rather than a classroom study, it did not measure expectancies, and it failed to rule out plausible alternative explanations. All classroom studies that have examined the accumulation issue have instead found support for dissipation. At least in the classroom, therefore, we think it is now appropriate for researchers to assume that dissipation, rather than accumulation, is the norm, until new data comes in that clearly and consistently documents accumulation of expectancy effects in the classroom.

GENERALIZED TEACHER EXPECTATIONS

Although a review of how teachers' general beliefs influence teaching and learning is beyond the scope of this chapter, we do address two specific aspects of teacher beliefs that fit well into an expectancy perspective: (1) teachers' general beliefs about the appropriate standards to which students should be held; and (2) teachers' expectations for whole classes.

Standards

The public discourse on education often involves issues of standards. Many people seem to believe that standards are too low in many schools and need to be raised. The federal government and many states periodically proclaim new lists of goals or standards that they expect all or most students to meet. Standards are closely related to expectations. Teachers with higher standards expect students, in general, to learn a greater amount of and more difficult material than do teachers with lower standards. Given any tendency for self-fulfilling prophecies to occur, one might conclude that teachers should hold much higher standards for all of their students. Next we discuss the evidence suggesting that this may be largely true, and factors likely to limit the potential benefits of raising standards.

The Expectancy Case for Higher Standards

In school, business, and athletic settings, hard, specific goals often increase performance, achievement, and productivity (Locke & Latham, 1990). In general higher goals probably go hand in glove with higher standards and high expectations. Perhaps the most dramatic example of this are minority students with histories of low achievement who have done as well as other students when admitted into honors programs (Steele, 1992).

Several different aspects of the expectancy literature, too, are consistent with the idea that standards can probably be raised in many classrooms. First, our own research has shown that positive expectations were more powerful than negative ex pectations—even among high achievers (Madon et al., 1997). This implies it may often be easier to raise than to undermine achievement gains and, perhaps, that higher standards/expectations will often have some beneficial effect on achievement.

Second, an experimental study of Israeli military inductees (Eden, 1990) showed that inducing trainers to develop higher expectations for their whole group of recruits increased (compared to no expectancy controls) the performance of those groups. Third, the process research clearly shows that high expectations increase achievement because they lead teachers to provide a demanding and supportive environment. Such an environment is likely to maximize students' academic potential.

Caveats about and Limitations to Raising Standards

Despite the many beneficial effects of hard goals, high standards, and high expectations, there are important limitations to their effects. First, there are limits to how much students can learn. First graders who have not yet learned how to add cannot be taught calculus. Even the expectancy literature is consistent with this perspective—expectancy effects tend to be small, implying that there are indeed limits to how much students can meet even high expectations.

The goal-setting literature, too, is consistent with this perspective. Hard goals raise performance, but generally only when people accept those goals and believe they can attain them (Locke & Latham, 1990). People will not usually accept goals that seem to be unreachable, so that goals, standards, and expectations that are too high may boomerang, thereby undermining performance. Excessively high standards have considerable potential to have negative side effects—frustration and loss of confidence among students ("Why can't I do this?") and frustration on the part of the teacher ("Is something wrong with them or with me?"). Similarly, excessively high standards probably lead to excessively difficult assignments, which would likely lead students to feel that school is an oppressive place and learning an unpleasant experience.

Another important caveat to the assumption that high standards are likely to be beneficial is that what constitutes a "high" expectation/standard will vary between students. Encouraging a student with a history of Cs and Ds to perform at a B level is probably a pretty high standard/expectation; encouraging a student with a history of As to perform at a B level is clearly an expectation that is too low (see Madon et al., 1997 for a more detailed discussion of this issue).

Last, raising standards without providing the resources necessary to increase achievement is not likely to be of much benefit. When schools are literally falling apart, when there are 40-50 students in a classroom, when classes are held in large closets and bathrooms, when teachers are so grossly underpaid that they rarely teach for more than a month before finding another job, when books are in such short supply that students must share them and cannot bring them home for homework or study, attempts to raise standards are meaningless. Although such conditions are not typical in most of America's schools, they are common in some of the poorest school districts (Kozol, 1991). These often are the districts in which students perform most poorly and would appear to be ripe targets for "raising standards." In these poorest districts, however, attempts to raise standards will be little more than bad, sad jokes until those districts are equipped with minimally adequate resources for providing students with an education.

Whole-Class Effects

Some researchers have suggested that self-fulfilling prophecies may be more powerful for whole classes than for individual students (Brophy, 1983; Jussim & Fleming, 1996). The apparent logic of this argument is compelling. Teachers' expectations for their classes may be conveyed more powerfully because teachers spend much more time addressing their classes as a whole than addressing individual students. In addition, teachers may teach more or less difficult material to different classes, depending on their expectations for those classes. This seems especially likely when classes are grouped by ability (which was previously discussed in detail at the individual level). Alternatively, however, (as previously discussed) ability grouping may, under many conditions, have beneficial effects.

We have just completed a study that empirically examined: (1) the relation between teachers' expectations for their whole classes and classes' mean achievement level; and (2) whether expectancy effects were stronger among classes that were grouped by ability level (Smith et al., 1997). Although we found no evidence of self-fulfilling prophecy (whole-class teacher expectations did not predict a change in mean standardized test scores) there was evidence consistent with small perceptual bias (whole-class expectations did predict a change in mean grades). Accuracy fully accounted for the relation of whole-class expectations to future standardized test scores; and accuracy accounted for about half of the relation of whole-class expectations to future grades.

When we examined moderation by ability grouping, we found that perceptual biases were stronger when *no* grouping was used than when between-class grouping was used. This finding runs contrary to claims that grouping leads teachers to develop rigid and biased expectations for their classes. This finding is also consistent with our other results focusing on teacher expectations for individual students demonstrating no evidence at all of perceptual bias when students are in between-class ability groups (Smith et al., 1997). However, we did find evidence of self-fulfilling prophecies for within-grouped classes. This, too, is consistent with our results for individual students.

Although one of the current authors has been among those who have speculated that self-fulfilling prophecies for whole classes could be stronger than those for individual students (Brophy, 1983; Jussim & Fleming, 1996), this was not the case. Relations of whole-class teacher expectations to mean future achievement were never particularly powerful. Consistent with naturalistic research on teacher expectations for individual students (e.g., Jussim, 1989; Jussim & Eccles, 1992; West & Anderson, 1976; Williams, 1976), accuracy was the main reason whole-class expectations predicted future achievement.

SELF-FULFILLING PROPHECY PROCESSES

Starting in the early 1970s many researchers sought to discover the social and psychological processes by which expectations became self-fulfilling (e.g., Brophy & Good, 1970; Brophy & Good, 1974; Rosenthal, 1973). Although many researchers have proposed models of the self-fulfilling prophecy process (e.g., Brophy & Good, 1974; Cooper, 1979; Darley & Fazio, 1980; Jussim, 1986; West & Anderson, 1976), all agree on three broad stages. First, teachers must develop expectations; second they must treat high expectancy students differently than low expectancy students; third, students must react in ways that confirm the teachers' expectations. Next we review evidence on each of these three stages.

Development of Expectations

Past Achievement

Teachers often develop expectations for students early in the year (see reviews by Brophy & Good, 1974; Jussim, 1986). By far, the strongest influences on teachers' expectations are usually students' past performance and motivation. Naturalistic studies consistently find that the influence of previous grades, IQ, standardized test scores, and student motivation on teacher expectations is three to five times that of all other influences combined (e.g., Jussim, Eccles, & Madon, 1996; Williams, 1976; see Jussim & Eccles, 1995 for a review). Similarly, experimental studies find that teacher perceptions are influenced far more by performance than by other information (Reschly & Lamprecht, 1979; Ross & Jackson, 1991).

Stereotypes

However, teachers may also be influenced to some degree by social stereotypes associated with physical attractiveness, social class, ethnicity, gender, speech style, or diagnostic label (learning disabled, emotionally disturbed, etc.) (see reviews by Braun, 1976; Brophy, 1983; Dusek, 1975; Jussim, 1986; Jussim, Eccles, & Madon, 1996; West & Anderson, 1976). Because of the potential for stereotype-based beliefs to contribute to social problems and educational injustices, we next discuss their role in the development of teachers' expectations in detail.

In lay parlance, and even among some social scientists, the term "stereotype" is often used to refer to *someone else's* beliefs about members of some social group (often, but not always, a demographic group) as a way of dismissing those beliefs as false, rigid, and irrational (Brigham, 1971; Fox, 1991). However, there has been a slow revolution in social scientific perspectives on stereotypes which has expunged the inaccurate and irrational component of the definition of "stereotype" (e.g., Ashmore & Del Boca, 1981; Judd & Park, 1993; Jussim, McCauley, & Lee, 1995; Oakes, Haslam, & Turner, 1994). In this chapter, therefore, we define stereotypes as people's beliefs about social groups and their individual members. This definition is completely neutral on the inaccuracy and irrationality issue—it neither assumes inaccuracy nor denies it. Determining whether stereotypes are accurate or inaccurate, rational or irrational, rigid or flexible is an open, empirical question (Judd & Park, 1993; Jussim, 1991; Jussim, McCauley, & Lee, 1995; Jussim, Eccles, & Madon, 1996; McCauley, Stitt, & Segal, 1980).

We frame our analysis of the role of stereotypes as bases for teachers' expectations in terms of the following three questions: 1. Do teachers perceive differences between students from different social groups? 2. Are stereotype-based expectations inaccurate? and 3. Do stereotypes lead teachers to ignore individual differences among students (this is a variation on the "stereotypes are rigid" hypothesis, and one of the classic accusations leveled against stereotypes)?

Do Teachers Perceive Differences Between Students from Different Social Groups?

One of the clearest and most comprehensive analyses of the role of stereotypes in the development of teachers' expectations was a meta-analysis addressing the influence of students' physical attractiveness, gender, social class, and ethnicity (Dusek & Joseph, 1983). This meta-analysis showed that, of these categories, teachers relied most heavily on social class when developing their expectations. They held significantly higher expectations for middle- versus lower-class students, and the difference was of moderate size ($d = .47$—this means that teacher expectations were .47 standard deviations higher for middle-class students than for lower-class students). Although most of the studies included in Dusek and Joseph's meta-analysis were experiments, naturalistic research has found substantially similar patterns (Jussim, Eccles, & Madon, 1996; Williams, 1976).

This meta-analysis also found that teachers held higher expectations for more physically attractive students ($d = .30$). However, these expectation differences usually disappeared after teachers had become familiar with their students. Thus, Dusek and Joseph concluded that physical attractiveness is a basis only for teachers' initial expectations. This result is also consistent with a meta-analysis of the general literature on the physical attractiveness stereotype (Eagly et al., 1991), which found that the more personal information perceivers had about targets, the less likely they were to rely on physical attractiveness.

Dusek and Joseph's meta-analysis failed to find a significant effect of student gender on teachers' expectations. This contrasts with our own recent research showing that in sixth-grade math classes, teachers believe girls performed at slightly higher levels than boys and that girls tried harder at math than did boys (Jussim, Eccles, & Madon, 1996; Madon et al., 1997). Because the studies in Dusek and Joseph primarily focused on performance, there was no reason for them to have uncovered sex stereotype effects regarding effort. However, there is considerable evidence that girls are more pleasant and cooperative, and less disruptive than boys (see, e.g., Bye, 1994; Bye & Jussim, 1993 for reviews). To many teachers, this may translate into a belief that girls care more about, and try harder in, school.

Of course, this does not explain the discrepancy between their results and our findings regarding perceptions of performance. However, the perceived sex differences in performance that we (Jussim, Eccles, & Madon, 1996; Madon et al., 1997) have found have been very small (statistically significant because of the large sample sizes, but below .1 in both studies) and, especially in math classes, is in a counter-stereotypic direction (teachers favor girls). Therefore, it will be up to future research to determine whether this pattern holds up.

The relationship between teacher expectations and ethnicity/race has also received considerable attention (e.g., see reviews by Jussim, 1986, 1990; Jussim, Madon, & Chatman, 1994). The studies on ethnicity included in Dusek and Joseph's meta-analysis overwhelmingly focused on comparing expectations for African-American and white students (too few other groups were examined to perform a meaningful meta-analysis regarding the role of other ethnic stereotypes). There was a small but statistically significant tendency for teachers to develop more favorable expectations for white students ($d = .11$). This result suggests that ethnicity was only a weak basis of teacher expectations.

In fact, however, the role of ethnicity in teacher expectations is considerably more complicated than the experimental research suggests. This research is probably most relevant to understanding teacher expectations in integrated schools, where variations in student ethnicity may contribute to variations in teacher expectations. However, residential patterns in the United States remain highly segregated so that most students in the United States attend segregated schools (Marger, 1991). Even when government policies and court mandates call for desegregating schools, white flight leaves many schools ethnically homogeneous (Armor, 1988). And even when the overall student population of a school appears ethnically diverse, resegregation may occur, such that the students in some classes are predominantly white, and students in other classes are predominantly African American (Epstein, 1985). In ethnically homogeneous schools and classes there are no variations in ethnicity to contribute to variations in teacher expectations for individual students.

It is possible that the general level of expectations held by teachers is higher in some segregated schools or districts than in others. Perhaps, for example, teachers in predominantly white school districts hold higher expectations than teachers in predominantly African-American school districts. This, however, would be a school- or district-level phenomenon, rather than a dyadic-level phenomenon. Furthermore, the only study of which we are aware that has compared teachers' expectations among white and African-American segregated districts has found little difference (Jussim, Eccles, & Madon, 1996). This study, however, only focused on sixth-grade math classes in districts in southeastern Michigan. Whether this pattern holds in other grade levels, geographic areas, in integrated districts, or among different ethnic groups is unknown.

Nonetheless, current evidence (Dusek & Joseph, 1983; Jussim, Eccles, & Madon, 1996) suggests that teachers generally perceive smaller differences between white and African-American students than is indicated by most measures of achievement (e.g., Helms, 1992; Scarr, 1981). This pattern is consistent with research showing that people generally *underestimate* objective differences (as indicated by census data) between African Americans and other Americans in areas such as likelihood of completing high school, being on welfare, and so on (McCauley & Stitt, 1978). Additional research on the role of student ethnicity in teacher expectations is clearly needed.

One major source of teacher expectations that was not directly addressed in the Dusek and Joseph meta-analysis was children's diagnostic labels (emotionally disturbed, neurologically impaired, learning disabled, etc.). In contrast to social stereotypes involving gender, ethnicity, social class, and so on, diagnostic labels have the stamp of approval of highly credible experts. Consequently, they are a major basis of teacher expectations (see Rist & Harrell, 1982 for a review). Teachers generally expect lower achievement and greater amounts of behavioral problems from students labeled learning disabled and emotionally disturbed (Thurlow, Christensen, & Ysseldyke, 1983; Ysseldyke & Foster, 1978). These labels are apparently so powerful that teachers even expect less of the *siblings* of such labeled students (Thurlow, Christensen, & Ysseldyke, 1983).

In sum, it appears that social class and diagnostic labels serve as major bases of teacher expectations; physical attractiveness, ethnicity, and perhaps gender serve as minor bases. It is hypothetically possible that research will uncover powerful influences of other stereotypes on teacher expectations. For now, however, there is little evidence that other stereotypes are major influences on teacher expectations.

Are Stereotype-based Expectancies Inaccurate?

Showing that teachers perceive differences between students from different groups provides no information about the appropriateness or accuracy of those perceptions. Determining whether stereotypes lead teachers to see differences that do not exist, or are greater than really exist, involves understanding two issues: (1) The validity of teachers' beliefs that students from different groups differ; and (2) Whether teachers' impressions of individual students are responsive to evidence that disconfirms the stereotype.

The belief that students from higher socioeconomic backgrounds (SES) will achieve more highly in school, on average, than their lower SES classmates is accurate. In comparison to upper and middle SES students, low SES students do not receive as high grades or score as highly on standardized tests, they are less likely to complete high school or attend college, their parents are less likely to be actively involved in their education, and they have lower educational aspirations (Gordon, 1976; Hallinan & Williams, 1990; Lareau, 1987; Rumberger, Ghatak, Poulos, Ritter, & Dornbusch, 1990; Scarr, 1981; Wright & Wright, 1976).

Our own recent research has directly examined the accuracy of teachers' perceptions of social class differences (Jussim, Eccles, & Madon, 1996; Madon et al., 1997). This research shows that the differences teachers perceive between students from differing social class backgrounds closely correspond to real differences in those students' grades and standardized test scores. Across two studies, student social class consistently correlated .2 to .3 with teacher expectations (indicating that teachers held higher expectations for students from higher social class backgrounds); and that grades and standardized test scores (obtained *prior* to interaction with the teacher) consistently correlated about .3 with student social class. So teachers' perceptions of differences between students from differing social class backgrounds were largely accurate; when inaccurate, teachers tended to slightly underestimate the real differences.

Evidence regarding the accuracy of diagnostic labels (learning disabled, emotionally disturbed, neurologically impaired, etc.) is more mixed. Undoubtedly, many children are labeled appropriately, and this facilitates their receipt of appropriate attention and special programs. There are real differences, for example, between children diagnosed with attention deficit-hyperactivity disorder and normal children—differences that are readily apparent even to other children (Harris, Milich, Corbitt, Hoover, & Brady, 1992). However, as many as 40 percent of the children who receive some label are misclassified (Ysseldyke, Algozzine, Shinn, & McGue, 1982). In addition, teachers who are more competent and more self-confident are *less* likely to refer children for the type of psychological evaluation that might lead to a label (Gersten, Walker, & Darch, 1988; Itskowitz, Abend, & Dimitrovsky, 1986; Meijer & Foster, 1988). Thus, at least sometimes, the label reflects characteristics of the labelers as much or more than characteristics of the labelee (see Rist & Harrell, 1982 for an extended discussion of this issue).

Teachers' belief that physically attractive students are smarter, however, is clearly inaccurate. Attractive students are no more successful in school than their less attractive peers (Clifford, 1975; Feingold, 1982; Maruyama & Miller, 1981). Of course, this also means that, in general, this stereotype *does not* lead to educational self-fulfilling prophecies. If it did, there would be more evidence of physically attractive students actually achieving more highly.

In our own research (Jussim, Eccles, & Madon, 1996; Madon et al. 1997), we have found both accuracy and inaccuracy in teachers' perceptions of similarities and differences between boys and girls. On the accuracy side, teachers perceived girls as performing slightly higher than boys, and girls did indeed receive slightly higher grades the previous year. Teachers also perceived boys and girls as equally talented, and, consistent with this perception, there was no difference in the standardized test scores of boys and girls. On the other hand, teachers also perceived girls as trying harder than boys, but boys and girls claimed to be exerting similar amounts of effort, spending similar amounts of time on math homework, and boys had slightly higher self-concepts of ability (in many theoretical perspectives this is a major source of motivation—e.g., Bandura, 1977; Eccles & Wigfield, 1985). Thus, there was no evidence to support the belief that girls try harder than boys.

Overall, even though there is clear evidence that teachers often perceive differences between different groups of students, there is little evidence that teachers consistently use grossly inaccurate stereotypes as a basis for their expectations. Social class and diagnostic labels have the largest influence on teacher expectations, and although there is room for error, both factors often predict student achievement. Existing evidence suggests that teachers do rely on inaccurate stereotypes regarding gender and attractiveness, but only to a relatively modest extent. Neither the experimental nor naturalistic research provides evidence of teachers' expectations inventing or exaggerating achievement differences between ethnic groups.

Do Stereotypes Lead Teachers to Ignore Individual Differences among Students?

One of the classic accusations against stereotypes is that they lead people to ignore differences among individual members of the stereotyped group (see, e.g., Jussim, McCauley, & Lee, 1995 for a review). This is important because even stereotypes that accurately describe group differences do not perfectly fit all individual group members. Therefore, stereotypes may lead to errors and biases in evaluations of students if they are applied rigidly. Similarly, even inaccurate stereotypes may not lead to errors and biases in judgments of individual students, if teachers readily revise their impressions when faced with disconfirmation. Thus, the accuracy of teachers' stereotype-based expectations also depends upon how readily teachers revise their impressions when students perform at levels inconsistent with those expectations.

How flexible are stereotype-based expectations regarding individuals? Abundant evidence within social psychology shows that perceivers generally judge individual targets far more on the basis of targets' personal characteristics than on their membership in social groups (see reviews by Jussim, 1990, 1991; Jussim, McCauley, & Lee, 1995; Jussim, Eccles, & Madon, 1996; Kunda & Thagard, 1996). Even small amounts of information inconsistent with the stereotype often lead to revision of perceivers' impression of a particular person (Locksley, Borgida, Brekke, & Hepburn, 1980). In general, research in education and social psychology shows that the more personal information available, the less stereotypes bias judgments (Dusek & Joseph, 1983; Eagly et al., 1991; Krueger & Rothbart, 1988; Reschly & Lamprecht, 1979). The general finding that teachers are heavily influenced by objective performance also implies considerable flexibility in teachers' expectations—that is, they judge students primarily on the basis of their achievement.

However, the news is not all good. The claim that teachers use information about individual students more than stereotypes still allows room for stereotypes to bias judgments (see Jussim, 1991). Even if teachers evaluate students who score one standard deviation above the mean on a standardized test more favorably than students who score at the mean regardless of their group membership, they may still evaluate one group more favorably than another among students with similar scores. For example, elementary and junior high school teachers often assign higher math grades to girls than boys, even though they perform similarly on standardized tests (Jussim, Eccles, & Madon, 1996; Kimball, 1989). However, research suggests that girls may earn higher grades in comparison to boys (although boys and girls perform similarly on standardized tests) because girls may turn in more assignments, do better on in-class tests, and may be more pleasant and cooperative in class (Brophy & Good, 1974; Bye, 1994; Bye & Jussim, 1993; Kimball, 1989). This suggests that differential grading of boys and girls may not necessarily reflect bias, but simply differential accomplishment in class.

Apparently, there is even room for bias when students' achievement is obviously high. Minner, Prater, Bloodworth, and Walker (1987) directly examined whether providing evidence that clearly disconfirms a negative stereotype could overcome bias against children with stigmatizing labels. They found that, in comparison to a high performing child without any stigmatizing labels, teachers are less likely to recommend that an equally high performing child who is handicapped or learning disabled be evaluated for entry into a program for intellectually gifted children. This particular bias may be especially important because it denies stigmatized children a major opportunity to enrich their educational experience and demonstrate high achievement.

How Do Teachers Act on Their Expectations?

Rosenthal's Four-factor Theory

Rosenthal (1974) identified four broad ways in which teachers treat high expectancy students more favorably than low expectancy students (see Harris & Rosenthal, 1985 for a meta-analysis; see Brophy, 1983; Brophy & Good, 1974; Jussim, 1986 for reviews). These types of treatment are generally referred to as climate, feedback, input, and output. First, teachers provide a more supportive emotional climate for high expectancy students. They are warmer, smile more, and offer them more encouragement.

Second, teachers provide clearer and more favorable feedback to highs. Feedback (positive or negative) received by highs tends to focus on performance. In contrast, lows receive considerably more feedback that is unrelated to achievement—instead, it more likely focuses on behavior, cooperativeness, aggression, and so on. In addition, highs are praised more and criticized less than are lows.

Third, teachers often provide greater input into highs' education. They spend more time with and provide more attention to highs. They also may teach more material to highs. Fourth, teachers often provide highs with more opportunities for output. They call on highs more often; give highs more hints and prompts when they seem hesitant and unsure; provide highs with more time to respond to verbal questions; teach more difficult material to highs; and give highs more challenging classwork and homework assignments.

Why Differential Treatment?

Why do teachers sometimes treat high expectancy students so much more favorably than lows? The four classes of differential treatment are broadly consistent with affect-effort theory (Harris et al., 1992; Rosenthal, 1989). The main idea of affect-effort theory is quite simple: perceivers (including teachers) often like highs more than lows (accounting for climate differences, and perhaps some of the feedback, input, and output differences) and often exert more effort in teaching highs (accounting for many of the input and output differences).

Why do teachers like highs more and work harder teaching them? Although there may be many reasons, some of the strongest contenders include similarity of the students to the teacher, cooperativeness of the students, and how rewarding it is to teach highs and lows. First, abundant evidence within social psychology has documented that the more similar others are to us, the more we like them (Rubin, 1973). Similar beliefs and values are especially powerful influences on attraction (Byrne, 1971; Rokeach, 1960). It seems reasonable to assume that most teachers consider education important. Teachers often believe that students doing well hold positive attitudes toward school (Jackson, 1968). They erroneously assume that highs try harder than lows, when, in fact, lows usually need to (and claim that they do) spend more time on homework (Jussim, 1989; Jussim & Eccles, 1992). It seems particularly ironic that highs need to exert less effort, and yet receive the benefits (in terms of teacher good will, positive affect, and climate) of the erroneous belief that they try harder.

Independent of students' objective achievement, teachers often assume that pleasant, cooperative students are brighter and they reward those students with a more positive classroom environment (e.g., Brophy & Good, 1974). In general, students' attempts to comply with behavioral standards in the classroom are positively and significantly correlated with grade point average but not with SAT scores (Wentzel, 1989). Even if they scored relatively high on SATs, students with low grades did not conform to what Wentzel referred to as the "normative standards of the classroom." That is, they were uncooperative and/or did not place much value on school. Many teachers manifest a strong preference for interacting with more pleasant students, and assume, at least sometimes erroneously, that the less pleasant students are not as smart as the others.

Many teachers probably feel that it is more rewarding to teach highs than lows. As a result of highs' more positive attitude and greater cooperativeness (illusory or real), and their (perceived) greater ability to learn material quickly and efficiently, many teachers may feel a greater sense of accomplishment teaching them than in teaching lows. If they believe their efforts are more productive among highs, they may spend more time teaching them.

Some Caveats about Differential Treatment

Thus far, we have emphasized how differential treatment may mediate the self-fulfilling effects of high and low teacher expectations. This discussion could implicitly convey the related ideas that all differential treatment leads to self-fulfilling prophecies and that, therefore, all students should be treated identically. This would be a gross misinterpretation of the meaning of research on self-fulfilling prophecies.

There are many ways teachers may treat highs and lows differently that are completely justified and appropriate for maximizing the learning of both types of students. Students who are struggling may often need to spend more time working on basic material, whereas it is appropriate to provide extra and more challenging assignments to students who might otherwise lose interest because they master the basics quickly.

This is especially important because teachers' expectations are so often highly accurate. Although we have followed tradition in this area by referring to students as "high" or "low" expectancy students, in general, a high expectancy student and a student with a history of high achievement are the same person; as is a low expectancy student and a student with a history of low achievement. And the idea that teachers should treat a student with straight As who scores in the top 5 percent on standardized tests the same as a student with Cs and Ds who scores in bottom 20 percent of standardized tests is untenable. Although a thorough review of appropriate teaching techniques for maximizing the learning of high and low achievers is beyond the scope of this chapter, it is clear that, for some very good reasons, much of modern educational philosophy is directed toward tailoring teaching, at least to some degree, to students' individual needs. And that means some (appropriate and justified) differential treatment.

Students' Reactions

It may appear obvious that many of the types of differential treatment described in Rosenthal's (1973) four-factor theory lead to self-fulfilling prophecies. However, it is not necessarily true. For a self-fulfilling prophecy to occur, students must react in specific ways to such differential treatment. Therefore, we next discuss students' reactions to differential treatment. Three broad types of student reactions to differential treatment seem to be involved in self-fulfilling prophecies: student skills, self-concept, and motivation.

Skills

Aspects of differential treatment may directly influence students' academic skills. Clear performance feedback, received more by highs than lows, provides useful information for distinguishing between high- and low-quality work. In contrast, feedback focusing more on normative behavior, received more by lows, provides little information about quality of work, and instead focuses lows' attention on conforming to classroom behavioral norms. Thus, high expectancy students may develop academic skills more quickly than low expectancy students because teachers give highs clearer performance-related feedback.

Differential input and output also seem likely to directly enhance highs' skills. By spending more time with highs, teachers provide them with more knowledge and more strategies for problem solving than lows (Eccles & Wigfield, 1985). Highs' greater opportunity to perform in class, by being called on more often and given more time to answer questions, also increases their chances to think spontaneously, articulate ideas, become aware of their mistakes and attempt corrections. Thus, higher teacher expectation for highs results in highs having more skills than lows.

Self-concept and Motivation

Differential treatment may also influence students' self-concept and motivation. Teacher expectations early in the year influence students' self-concept of ability and expectations late in the year (Jussim, 1989; Parsons, Kaczala, & Meece, 1982), and these types of effects are stronger in classrooms where students perceive greater differential treatment (Brattesani et al., 1984). Receiving more positive and less negative performance feedback, and higher grades, may enhance highs' self-concept of ability, self-efficacy, self-esteem, and so on. These factors, in turn, influence actual achievement (see reviews by Bandura, 1977; Eccles & Wigfield, 1985; Jussim, 1986).

Differential treatment seems especially likely to influence students' intrinsic motivation (the enjoyment or pleasure one receives from simply engaging in an activity regardless of outcomes and evaluations). Cognitive evaluation theory has identified three types of feedback that undermine intrinsic motivation: (1) feedback intended to control behavior; (2) feedback not contingent on performance; and (3) negative feedback (e.g., Deci & Ryan, 1980; Jussim, Soffin, Brown, Ley, & Kohlhepp, 1992). These are exactly the types of feedback more likely to be experienced by lows. Thus, differential treatment due to lowered teacher expectancies (especially climate and feedback), may influence intrinsic motivation in ways likely to result in a self-fulfilling prophecy.

The emotional climate teachers provide may also influence motivation in a different way. School may simply be a more pleasant and positive place for high expectancy students who typically are treated supportively. However, for students who are the target of low expectations, school may be an unfriendly environment. Under such conditions, lows may not want to exert the effort required to perform well. For lows, avoiding school by missing classes, withdrawing from classroom activities, and so on, may become more appealing than trying to conform and learn.

Un(der)-explored Process Issues

There has been relatively little research on the processes mediating self-fulfilling prophecies in the last 10 years, probably because the early research so clearly supported Rosenthal's four-factor theory (e.g., Harris & Rosenthal, 1985; Rosenthal, 1973). However, other types of mediators have been under-explored. In performance situations abundant research attests to the power of setting high goals for students, employees, athletes, and so on (Locke & Latham, 1990). However, whether high expectations often lead teachers to explicitly set higher goals for students is not known. But even if teachers do not set explicit goals for individual students, they may sometimes explicitly convey high expectations—which may have an effect much like setting high goals. However, both the extent to which teachers do this, and its effect on students, are currently unknown.

Two recent studies suggest that the role of affect in driving "expectancy" effects has been under-explored. The first found that children were less warm, friendly, and involved when playing with other children who were stigmatized (Harris et al., 1992). The second found that perceivers' liking or disliking of (prejudice toward) a target's group was a more potent source of biases in judgments of that target's sanity than were perceivers' beliefs (stereotypes) about that group (Jussim, McCauley, & Lee, 1995). Although neither of these studies examined teachers' expectations, they are broadly consistent with Rosenthal's (1989) affect-effort theory of the mediation of expectancy effects. This research raises the possibility that teachers' liking and disliking of students may drive some expectancy effects.

Another under-explored mediator is students' beliefs about teachers' beliefs. A few experiments have shown that targets sometimes confirm the beliefs that they (erroneously) think perceivers hold (Farina, Allen, & Saul, 1968; Farina, Gliha, Boudreau, Allen, & Sherman 1971; von Baeyer, Sherk, & Zanna, 1981; Zanna & Pack, 1975). The general question here is: How important is students' awareness (accurate or not) of teachers' expectations? We speculate that although awareness is not a necessary mediator of self-fulfilling prophecies (i.e, self-fulfilling prophecies may occur without student awareness of the teachers' expectancies), awareness will often tend to enhance the power of self-fulfilling prophecies, especially among children and people in new situations. Of course, targets may sometimes intentionally resist confirming expectations when they believe that a perceiver holds inappropriate expectations (Hilton & Darley, 1985; Swann & Ely, 1984). Understanding the role of target awareness in self-fulfilling prophecies, then, is an important focus for future research.

CONCLUSION: IMPLICATIONS FOR TEACHING

Improving Teaching

The expectancy literature does provide some valuable information regarding how to improve teaching. Before discussing how teachers (and administrators) might make use of some of the ideas and findings from expectancy research, we raise two caveats.

First, we make no assumption regarding whether teachers generally engage in these behaviors; probably, many teachers do indeed already engage in the behaviors we recommend. On the other hand, people in general (including teachers), are clearly subject to self-serving biases (e.g., Pyszczynski & Greenberg. 1987; see also Good & Brophy, 1997), so that many teachers may believe they engage in these behaviors more frequently than they actually do. Inasmuch as beliefs that are salient are more likely to influence behavior (Sedikedes & Skowronski, 1991), it may be useful to review these behaviors, even for teachers who believe they already engage in them.

Another caveat involves motivation on the part of teachers and administrators. Instituting the recommendations described below will often require some additional effort. Therefore, a genuine commitment to increasing student achievement is a necessary condition for any of these recommendations to have much potential for success.

Teachers

High teacher expectations can increase students' achievement and unduly low expectations can undermine students' achievement. Furthermore, as discussed previously, we suspect that many teachers could successfully raise their expectations and standards for many of their students. Thus, the first question is: How can teachers know when their expectations are appropriately high or inappropriately low? It might often be easier and more productive to start with one's expectations for a whole class; easier because only a single expectation or standard is involved (in contrast to 20, 30, or more for each individual student in a class) and more productive because, by definition, a general shift upward in the achievement of a whole class likely means that, on average, most students are performing more highly.

Good and Brophy (1997, chap. 3) describe many expectancy research-inspired techniques teachers can use to improve achievement. These include setting minimum acceptable standards, providing clear instructions for how to attain objectives, and teaching difficult material in several different ways. In addition, teachers should generally strive to set hard, specific, ambitious goals for their classes. If they can make learning fun and interesting, especially by incorporating students' own ideas, students might accomplish goals that did not seem possible.

Time spent preparing for teaching each class is clearly important. Teachers sometimes spend more time preparing for high track/high expectancy classes than for low track/low expectancy classes, and, at least sometimes, this contributes to achievement differences among students in different tracks (Brophy, 1983; Jussim, 1986). When teachers are motivated to improve students' achievement, however, they clearly should spend at least as much time preparing for their low expectancy classes.

The expectancy literature also provides some important insights into how to harness expectancy effects for individual students (see Good & Brophy, 1997 for a comprehensive review of these ideas). As much as possible, teachers should develop practices that fit the positive side of Rosenthal's (1973) four factors (climate, feedback, input, output). In short, this means being challenging and supportive. "Challenging" means that teachers should try, at least sometimes, to provide students with work that will cause them to stretch their academic abilities. For example, in addition to dittos and seat work, teachers can require that students critically evaluate ideas or problems, and apply simple skills in more complex settings. As Good & Brophy (1997) note, students are not likely to maximize their skills if they are spending lots of time on phonics and penmanship, but little time reading and analyzing stories, and little time actually writing. Of course, what is challenging for one student may not be challenging for another student—thus, certain types of differential treatment, if appropriately tailored to different students, may enhance all students' achievement.

What does "supportive" mean? To some extent, this goes hand in glove with challenging. In general, challenging work is more interesting and fun than remedial and repetitive work (Deci & Ryan, 1980; Good & Brophy, 1997; Steele, 1992). Making schoolwork more fun is clearly a part of what it means to be supportive. Consider the following concrete example (from one of the authors' daughter's experience). In a fourth-grade math class, rather than assign kids dittos with problems, the teacher started off each class with "minute math." The teacher handed out a sheet with 100 fairly simple math problems, and the students' job was to solve as many as possible in one minute. The students loved this "game"—especially since the teacher emphasized student *improvement* in performance compared to previous games, and students could see for themselves how much better they became as they year went on. Within a few months most students in the class could do basic and some intermediate computations in their heads in a snap.

Supportive also means engaging in behaviors that fit the "climate" factor of differential treatment (warmth, smiling, etc.). In particular, teachers could make sure to acknowledge the improvements of low achievers, even if their absolute level of achievement is not that high. Similarly, when students indicate they do not understand something difficult, a more positive classroom environment will be created if the teacher explains it calmly and in a variety of ways, rather than if the teacher becomes manifestly frustrated.

Often, however, teachers may be unaware of differences in the ways they treat high and low achievers (e.g., Good & Brophy, 1997—they also describe detailed observational techniques for heightening teachers' awareness of differences in the ways they treat high and low expectancy students). Consider one example: When high expectancy students are called on but have trouble answering, teachers often provide clues; when low expectancy students have similar difficulties, teachers often move on or provide the answer (Brophy, 1983; Brophy & Good, 1974). Providing struggling students with clues is clearly a specific behavioral technique that: (1) should be administered at least nearly equally to high and low expectancy students; and (2) could be explicitly incorporated into teacher training programs so that teachers enter classrooms with this technique already in their teaching repertoire.

In general, it may often be far easier to be warm and pleasant when one's students are calm, obedient, and under control. In contrast, it may be much more difficult to be warm and pleasant when faced with a group of wild or disobedient students. Thus, it seems that good classroom management clearly has its place in creating a high achieving class.

Administration

The expectancy literature also has clear implications for school administration (see, e.g., Eden, 1984, 1986 for discussions of harnessing expectancy effects to increase organizational effectiveness). Superintendents, school boards, and school principals can use their positions to set a general tone (climate) and standards (expectancies) which, if followed through with specific behaviors, can improve education and achievement. First, administrators should develop clear, hard, specific goals for their schools (Locke & Latham, 1991). This implicitly conveys high expectations and, if combined with a clear plan, is likely to have beneficial effects.

For example, "improving standardized test scores" is not a clear, hard, specific goal. However, in a low achieving school, "decreasing the proportion of our students who score in the bottom tenth on standardized tests from 25 percent to 15 percent" is a clear and specific goal, and probably a hard one, too. This should be accompanied by some sort of clear plan for doing so. For example, educators could try to meet with struggling students and their parents, discuss their strengths and weaknesses, and cooperatively develop a plan to improve students' skills.

In general, any such plan should be arrived at cooperatively with teachers—who often have more in-class experience than administrators and who must consider the feedback credible and the goals appropriate if they are to succeed. Just as teachers must be fair with students, so, too, administrators must be fair with teachers. We think it should be clear (but doubt that it often is) that an eleventh-grade teacher whose students average 850 on the SAT's could have successfully taught much more than one whose students average 1000. How could this be? It is possible that the students in the first teacher's class came into the class functioning at a dramatically lower level than did the students in the second teacher's class. Improvement, much more than absolute performance, is what education is all about. And teachers who can harness positive expectancy effects for the benefit of their students are priceless.

A Series of Liberating Implications

Because of the alleged power of expectancy effects to create social problems, teachers have sometimes been beaten over the head for being perpetrators of injustices based on race, class, sex, and other demographic categories (e.g., Gilbert, 1995; Hofer, 1994; Rist, 1970). However, the naturalistic research on teacher expectations makes it abundantly clear that this accusation is not justified. Teacher expectations predict student achievement primarily because those expectations are accurate. Even teacher perceptions of differences between students from different demographic groups are mostly accurate. Furthermore, even when inaccurate, teacher expectations do not usually influence students all that much. Even when they do influence students, they may be more likely to help than to harm students (especially low achievers) and the effects typically dissipate with time anyway.

Nonetheless, self-fulfilling prophecy research has some clear practical implications for teaching and learning. Research showing that self-fulfilling prophecies help more than harm students (Madon et al., 1997), and that some of the most powerful self-fulfilling prophecies occur among low achievers and students from historically stigmatized or disadvantaged groups (lower social class, African Americans—Jussim, Eccles, & Madon, 1996), strongly implies that teachers can make a major positive difference in the lives of students who need it most. The self-fulfilling prophecy process research also provides some clear information regarding behaviors and strategies teachers and administrators can use to enhance students' achievement. Although teacher expectation effects may not provide a particularly strong explanation for the origins of many social injustices, harnessing those effects may indeed contribute, in at least a small way, to redressing some of those injustices.

NOTE

1. In this study the schools gave different tests in fifth, sixth, and seventh grade. Therefore, we did not assess the amount of absolute achievement gain or loss. Few, if any, students probably showed an absolute decline in knowledge and achievement from fifth through seventh grade. Thus, the terms "increase" and "decrease" refer to achievement relative to students' classmates. For example, a student whose performance increased may have performed near the mean on a standardized test in early sixth grade, and above the mean on a standardized test in seventh grade. A student whose performance decreased may have performed near the mean in sixth grade and below the mean in seventh grade. Even this latter student, however, probably had considerably more knowledge in seventh grade than in early sixth grade.

Various aspects of the achievement test score distributions could have provided alternative explanations for our results suggesting that positive expectancy effects were stronger than negative ones (ceiling and floor effects, non-normal test score distributions, etc.). When such distributional explanations were examined in depth in our original empirical investigation, they were unable to account for the results (see Madon et al., 1997, for more details).

REFERENCES

Adorno, T., Frenkel-Brunswick, E., Levinson, D., & Sanford, R. N. (1950). *The authoritarian personality.* New York: Harper.
Allington, R. (1980). Teacher interruption behaviors during primary grade oral reading. *Journal of Educational Psychology, 72,* 371-377.
Allport, G. (1954). *The nature of prejudice.* Cambridge, MA: Addison-Wesley.
American Association of University Women (1992). *How schools shortchange girls.* Washington, DC: American Association of University Women Education Foundation.
Armor, D. J. (1988). School busing: A time for a change. In P. A. Katz & D. A. Taylor (Eds.), *Eliminating racism.* New York: Plenum Press.
Ashmore, R. D., & Del Boca, F. K. (1981). Conceptual approaches to stereotypes and stereotyping. In D. L. Hamilton (Ed.), *Cognitive processes in stereotyping and intergroup behavior* (pp. 1-35). Hillsdale, NJ: Erlbaum.
Babad, E., Inbar, J., & Rosenthal, R. (1982). Pygmalion, Galatea, and the Golem: Investigations of biased and unbiased teachers. *Journal of Educational Psychology, 74,* 459-474.
Bandura, A. (1977). Self-efficacy: Toward a unifying theory of behavioral change. *Psychological Bulletin, 84,* 191-215.
Brattesani, K. A., Weinstein, R. S., & Marshall, H. H. (1984). Student perceptions of differential teacher treatment as moderators of teacher expectation effects. *Journal of Educational Psychology, 76,* 236-247.
Braun, C. (1976). Teacher expectation: Sociopsychological dynamics. *Review of Educational Research, 46,* 185-213.
Brigham, J. C. (1971). Ethnic stereotypes. *Psychological Bulletin, 76,* 15-38.
Brophy, J. (1983). Research on the self-fulfilling prophecy and teacher expectations. *Journal of Educational Psychology, 75,* 631-661.
Brophy, J., & Good, T. (1970). Teachers' communications of differential expectations for children's classroom performance. *Journal of Educational Psychology, 61,* 365-374.
Brophy, J., & Good, T. (1974). *Teacher-student relationships: Causes and consequences.* New York: Holt, Rinehart, and Winston.
Bye, L. (1994). *Referral of elementary age students for social skill training: Student and teacher characteristics.* Doctoral dissertation, Rutgers University, New Brunswick, NJ.
Bye, L., & Jussim, L. (1993). A proposed model for the acquisition of social knowledge and social competence. *Psychology in the Schools, 30,* 143-161.
Byrne, D. (1971). *The attraction paradigm.* New York: Academic Press.
Clifford, M. M. (1975). Physical attractiveness and academic performance. *Child Study Journal, 5,* 201-209.
Cooper, H. (1979). Pygmalion grows up: A model for teacher expectation communication and performance influence. *Review of Educational Research, 49,* 389-410.
Cooper, H., & Good, T. (1983). *Pygmalion grows up: Studies in the expectation communication process.* New York: Longman.
Cooper, H., & Hazelrigg, P. (1988). Personality moderators of interpersonal expectancy effects: An integrative research review. *Journal of Personality and Social Psychology, 55,* 937-949.
Darley, J. M., & Fazio, R. H. (1980). Expectancy-confirmation processes arising in the social interaction sequence. *American Psychologist, 35,* 867-881.
Deci, E. L., & Ryan, R. M. (1980). The empirical exploration of intrinsic motivation processes. *Advances in Experimental Social Psychology, 13,* 39-80.

Devine, P. G. (1995). Prejudice and out-group perception. In A. Tesser (Ed.), *Advanced social psychology* (pp. 467-524). New York: McGraw-Hill.
Dornbusch, S. (1994, February). Off the track. Presidential address at the biennial meeting of the Society for Research on Adolescence, San Diego, CA.
Dusek, J. (1975). Do teachers bias children's learning? *Review of Educational Research, 45*, 661-684.
Dusek, J., & Joseph, G. (1983). The bases of teacher expectancies: A meta-analysis. *Journal of Educational Psychology, 75*, 327-346.
Eagly, A. H., Makhijani, M. G., Ashmore, R. D., & Longo, L. C. (1991). What is beautiful is good, but...: A meta-analytic review of research on the physical attractiveness stereotype. *Psychological Bulletin, 110*, 109-128.
Eccles (Parsons), J., Adler, T., & Meece, J. L. (1984). Sex differences in achievement: A test of alternate theories. *Journal of Personality and Social Psychology, 46*, 26-43.
Eccles, J., & Blumenfeld, P. (1985). Classroom experiences and student gender: Are there differences and do they matter? In L. C. Wilkinson & C. Marrett (Ed.), *Gender influences in classroom interaction* (pp.79-114). Hillsdale, NJ: Lawrence Erlbaum Associates.
Eccles, J., & Jacobs, J. E. (1986). Social forces shape math attitudes and performance. *Signs, 11*, 367-380.
Eccles, J., & Wigfield, A. (1985). Teacher expectations and student motivation. In J. Dusek (Ed.), *Teacher expectancies*. Hillsdale, NJ: Erlbaum.
Eccles, J., & Wigfield, A., Flanagan, C. A., Miller, C., Reuman, D. A., & Yee, D. (1989). Self concepts, domain values, and self-esteem: Relations and changes at early adolescence. *Journal of Personality, 57*, 283-310.
Eccles-Parsons, J., Kaczala, C. M., & Meece, J. L. (1982). Socialization of achievement attitudes and beliefs: Classroom influences. *Child Development, 53*, 322-339.
Eden, D. (1984). Self-fulfilling prophecy as a management tool: Harnessing Pygmalion. *Academy of Management Review, 9*, 64-73.
Eden, D. (1986). OD and self-fulfilling prophecy: Boosting productivity by raising expectations. *Journal of Applied Behavioral Science, 22*, 1-13.
Eden, D. (1992). Interpersonal expectations in organizations. In P. D. Blanck (Ed.), *Interpersonal expectations: Theory, research and applications* (pp. 154-178). New York: Cambridge University Press.
Eden, D. (1990). Pygmalion without interpersonal contrast effects: Whole groups gain from raising manager expectations. *Journal of Applied Psychology, 75*, 394-398.
Eden, D., & Shani, A. B. (1982). Pygmalion goes to boot camp: Expectancy, leadership, and trainee performance. *Journal of Applied Psychology, 67*, 194-199.
Elashoff, J. D., & Snow, R. E. (1971). *Pygmalion reconsidered*. Worthington, OH: Charles A. Jones.
Epstein, J. L. (1985). After the bus arrives: Resegregation in desegregated schools. *Journal of Social Issues, 41*, 23-44.
Evertson, C. (1982). Differences in instructional activities in higher and lower achieving junior high English and math classes. *Elementary School Journal, 82*, 329-350.
Farina, A., Allen, J. G., & Saul, B. B. (1968). The role of the stigmatized person in affecting social relationships. *Journal of Personality, 36*, 169-182.
Farina, A., Gliha, D., Boudreau, L. A., Allen, J. G., & Sherman, M. (1971). Mental illness and the impact of believing others know about it. *Journal of Abnormal Psychology, 77*, 1-5.
Feingold, A. (1982). Physical attractiveness and intelligence. *Journal of Social Psychology, 118*, 283-284.
Feingold, A. (1992). Good-looking people are not what we think. *Psychological Bulletin, 111*, 304-341.
Finn, J. (1972). Expectations and the educational environment. *Review of Educational Research, 42*, 387-410.
Fiske, S. T., & Taylor, S. E. (1984). *Social cognition*. Reading, MA: Addison-Wesley.
Fox, R. (1991). Prejudice and the unfinished mind: A new look at an old failing. *Psychological Inquiry, 3*, 137-152.

Frieze, I. H., Olson, J. E., & Russell, J. (1991). Attractiveness and income for men and women in management. *Journal of Applied Social Psychology, 21,* 1039-1057.

Funder, D. C. (1987). Errors and mistakes: Evaluating the accuracy of social judgment. *Psychological Bulletin, 101,* 75-90.

Gersten, R., Walker, H. M., & Darch, C. (1988). Relationship between teachers' effectiveness and their tolerance for handicapped students: An exploratory study. *Exceptional Children, 54,* 433-438.

Gilbert, D. T. (1995). Attribution and interpersonal perception. In A. Tesser (Ed.), *Advanced social psychology* (pp. 99-147). New York: McGraw-Hill.

Gilbert, D. T., & Osborne, R. E. (1989). Thinking backward: Some curable and incurable consequences of cognitive busyness. *Journal of Personality and Social Psychology, 57,* 940-949.

Good, T. L., & Brophy, J. E. (1997). *Looking in classrooms* (7th ed.). New York: Longman.

Gordon, M. T. (1976). A different view of the IQ-Achievement gap. *Sociology of Education, 49,* 4-11.

Gregory, M. K. (1977). Sex bias in school referrals. *Journal of School Psychology, 15,* 5-8.

Hallinan, M. K., & Williams, R. A. (1990). Students' characteristics and the peer influence process. *Sociology of Education, 63,* 122-132.

Hamilton, D. L., Sherman, S. J., & Ruvolo, C. M. (1990). Stereotype-based expectancies: Effects on information processing and social behavior. *Journal of Social Issues, 46,* 35-60.

Harris, M. J. (1989). Personality moderators of expectancy effects: Replication of Harris and Rosenthal (1986). *Journal of Research in Personality, 23,* 381-387.

Harris, M. J., Milich, R., Corbitt, E. M., Hoover, D. W., & Brady, M. (1992). Self-fulfilling effects of stigmatizing information on children's social interactions. *Journal of Personality and Social Psychology, 63,* 41-50.

Harris, M. J., & Rosenthal, R. (1985). Mediation of interpersonal expectancy effects: 31 Meta-analyses. *Psychological Bulletin, 97,* 363-386.

Harvey, V. S. (1991). Characteristics of children referred to school psychologists: A discriminant analysis. *Psychology in the Schools, 28,* 209-218.

Helms, J. E. (1992). Why is there no study of cultural equivalence in standardized cognitive ability testing? *American Psychologist, 47,* 1083-1101.

Hilton, J., & Darley, J. (1985). Constructing other persons: A limit on the effect. *Journal of Experimental Social Psychology, 21,* 1-18.

Hilton, J. L., & Darley, J. M. (1991). The effects of interaction goals on person perception. In M. P. Zanna (Ed.), *Advances in experimental social psychology* (Vol. 24, pp. 235-267). New York: Academic Press.

Hofer, M. A. (1994, December 26). Behind the curve. *The New York Times,* pp. A39.

Itskowitz, R., Abend, T., & Dmitrovsky, L. (1988). The relationship between teachers' self-concept and their tendency to refer students for psychological help. *School Psychology International, 7,* 116-122.

Jackson, P. W. (1968). *Life in classrooms.* New York: Holt, Rinehart & Winston.

Jensen, A. R. (1969). How much can we boost I.Q. and scholastic achievement? *Harvard Educational Review, 39,* 1-123.

Jones, E. E. (1986). Interpreting interpersonal behavior: The effects of expectancies. *Science, 234,* 41-46.

Jones, E. E. (1990). *Interpersonal perception.* New York: W.H. Freeman and Company.

Judd, C. M., & Park, B. (1993). Definition and assessment of accuracy in social stereotypes. *Psychological Review, 100,* 109-128.

Jussim, L. (1986). Self-fulfilling prophecies: A theoretical and integrative review. *Psychological Review, 93,* 429-445.

Jussim, L. (1987). *Interpersonal expectations in social interaction: Self-fulfilling prophecies, confirmatory biases, and accuracy.* Doctoral dissertation, University of Michigan. *Dissertation Abstracts International, 48,* 1845B.

Jussim, L. (1989). Teacher expectations: Self-fulfilling prophecies, perceptual biases, and accuracy. *Journal of Personality and Social Psychology, 57,* 469-480.

Jussim, L. (1990). Social reality and social problems: The role of expectancies. *Journal of Social Issues, 46*, 9-34.
Jussim, L. (1991). Social perception and social reality: A reflection-construction model. *Psychological Review, 98*, 54-73.
Jussim, L. (1993). Accuracy in interpersonal expectations: A reflection-construction analysis of current and classic research. *Journal of Personality, 61*, 637-668.
Jussim, L., & Eccles, J. (1992). Teacher expectations II: Construction and reflection of student achievement. *Journal of Personality and Social Psychology, 63*, 947-961.
Jussim, L., & Eccles, J. (1995). Naturalistic studies of interpersonal expectancies. *Review of Personality and Social Psychology, 15*, 74-108.
Jussim, L., Eccles, J., & Madon, S. (1996). Social perception, social stereotypes, and teacher expectations: Accuracy and the powerful self-fulfilling prophecy. In M. P. Zanna (Ed.), *Advances in experimental social psychology* (Vol. 28, pp. 281-388). San Diego, CA: Academic Press.
Jussim, L., & Fleming, C. (1996). Self-fulfilling prophecies and the maintenance of social stereotypes: The role of dyadic interactions and social forces. In N. McCrae, M. Hewstone, & C. Stangor (Eds.), *The foundations of stereotypes and stereotyping* (pp. 161-192). New York: Guilford Press.
Jussim, L., Madon, S., & Chatman, C. (1994). Teacher expectations and student achievement: Self-fulfilling prophecies, biases, and accuracy. In L. Heath et al. (Eds.), *Applications of heuristics and biases to social issues* (pp. 303-334). New York: Plenum.
Jussim, L., McCauley, C. R., & Lee, Y. T. (1995). Why study stereotype accuracy and inaccuracy? In Y. T. Lee, L. Jussim, & C. R. McCauley (Eds.), *Stereotype accuracy: Toward appreciating group differences*. Washington, DC: American Psychological Association.
Jussim, L., Soffin, S., Brown, R., Ley, J., & Kohlhepp, K. (1992). Understanding reactions to feedback by integrating ideas from symbolic interactionism and cognitive evaluation theory. *Journal of Personality and Social Psychology, 62*, 402-421.
Kenny, D. A. (1994). *Interpersonal perception: A social relations analysis*. New York: Guilford.
Kimball, M. M. (1989). A new perspective on women's math achievement. *Psychological Bulletin, 105*, 198-214.
Krueger, J., & Rothbart, M. (1988). Use of categorical and individuating information in making inferences about personality. *Journal of Personality and Social Psychology, 55*, 87-195.
Kulik, C., & Kulik, J. A. (1982). Effects of ability grouping on secondary school students: A meta-analysis of evaluation findings. *American Educational Research Journal, 19*, 415-428.
Kulik, J. A., & Kulik, C. (1987). Effects of ability grouping on student achievement. *Equity and Excellence, 23*, 22-30.
Kulik, J. A., & Kulik, C. (1992). Meta-analytic findings on grouping programs. *Gifted Children, 36*, 73-77.
Kunda, Z. (1990). The case for motivated reasoning. *Psychological Bulletin, 108*, 480-498.
Kunda, Z., & Thagard, P. (1996). Forming impressions from stereotypes, traits, and behaviors: A parallel-constraint-satisfaction theory. *Psychological Review, 103*, 284-308.
Lerner, M. (1980). *The belief in a just world*. New York: Plenum.
Lareau, A. (1987). Social-class differences in family-school relationships: The importance of cultural capital. *Sociology of Education, 60*, 73-85.
Locke, E. A., & Latham, G. P. (1990). *A theory of goal setting and task performance*. Englewood Cliffs, NJ: Prentice-Hall.
Locksley, A., Borgida, E., Brekke, N., & Hepburn, C. (1980). Sex stereotypes and social judgment. *Journal of Personality and Social Psychology, 39*, 821-831.
Madon, S., Jussim, L., & Eccles, J. (1997). In search of the powerful self-fulfilling prophecy. *Journal of Personality and Social Psychology, 72*, 791-809.
Madon, S., Jussim, L., Keiper, S., Eccles, J., Smith, A., & Palumbo, P. (1997). *The accuracy and power of sex, social class, and ethnic stereotypes: A naturalistic investigation*. Unpublished manuscript.

Marger, M. N. (1991). *Race and ethnic relations* (2nd ed.). Belmont, CA: Wadsworth.
Maruyama, G., & Miller, N. (1981). Physical attractiveness and personality. *Progress in experimental personality research, 10,* 203-280.
McCauley, C. (1995). Are stereotypes exaggerated? Looking for perceptual contrast and cognitive miser in beliefs about group differences. In Y. T. Lee, L. Jussim, & C. R. McCauley (Eds.), *Stereotype accuracy: Toward appreciating group differences* (pp. 215-243). Washington, DC: American Psychological Association.
McCauley, C., & Stitt, C. L. (1978). An individual and quantitative measure of stereotypes. *Journal of Personality and Social Psychology, 36,* 929-940.
McCauley, C., Stitt, C. L., & Segal, M. (1980). Stereotyping: From prejudice to prediction. *Psychological Bulletin, 87,* 195-208.
Meijer, C., & Foster, S. (1988). The effect of teacher self-efficacy on referral chance. *Journal of Special Education, 22,* 378-385.
Merton, R. K. (1948). The self-fulfilling prophecy. *Antioch Review, 8,* 193-210.
Midgley, C. M., Feldlaufer, H., & Eccles, J. S. (1989). Changes in teacher efficacy and student self- and task-related beliefs during the transition to junior high school. *Journal of Educational Psychology, 81,* 247-258.
Miller, D. T., & Turnbull, W. (1986). Expectancies and interpersonal processes. *Annual Review of Psychology, 37,* 233-256.
Minner, S., Prater, G., Bloodworth, H., & Walker, S. (1978). Referral and placement recommendations of teachers toward gifted handicapped children. *Roeper Review, 9,* 247-249.
Myers, D. G. (1987). *Social psychology* (2nd ed.). New York: McGraw-Hill.
Neuberg, S. L. (1989). The goal of forming accurate impressions during social interactions: Attenuating the impact of negative expectancies. *Journal of Personality and Social Psychology, 56,* 374-386.
Neuberg, S. L. (1994). Expectancy-confirmation processes in stereotyped-tinged encounters: The moderating role of social goals. In M. P. Zanna & J. M. Olson (Eds.), *The psychology of prejudice: The Ontario symposium* (Vol. 7, pp. 103-130). Hillsdale, NJ: Erlbaum.
Oakes, J. (1985). *Keeping track: How schools structure inequality.* New Haven: Yale University Press.
Oakes, J. (1987). Curriculum inequality and school reform. *Equity and Excellence, 23,* 8-14.
Oakes, P. J., Haslam, S. A., & Turner, J. C. (1994). *Stereotyping and social reality.* Cambridge, MA: Blackwell.
Palardy, J. M. (1969). What teachers believe—what children achieve. *Elementary School Journal, 69,* 370-374.
Parsons, J. E. (1980). Final Report to the National Institute of Education, Washington, DC. ERIC Document Reproduction Service No. ED 186 477.
Parsons, J. E., Kaczala, C. M., & Meece, J. L. (1982). Socialization of achievement attitudes and beliefs: Classroom influences. *Child Development, 53,* 322-339.
Pettigrew, T. F. (1979). The ultimate attribution error: Extending Allport's cognitive analysis of prejudice. *Personality and Social Psychology Bulletin, 5,* 461-476.
Pyszczynski, T., & Greenberg, J. (1987). Toward an integration of cognitive and motivational perspectives on social inference: A biased hypothesis-testing model. *Advances in experimental social psychology, 20,* 297-340.
Raudenbush, S. W. (1984). Magnitude of teacher expectancy effects on pupil IQ as a function of the credibility of expectancy inductions: A synthesis of findings from 18 experiments. *Journal of Educational Psychology, 76,* 85-97.
Reschly, D., & Lamprecht, M. (1979). Expectancy effects of labels: Fact or fiction? *Exceptional Children, 45,* 55-58.
Rist, R. (1970). Student social class and teacher expectations: The self-fulfilling prophecy in ghetto education. *Harvard Educational Review, 40,* 411-451.
Rist, R., & Harrell, J. E. (1982). Labeling the learning disabled child: The social ecology of educational practice. *American Journal of Orthopsychiatry, 52,* 146-160.

Rokeach, M. (1960). *The open and closed mind.* New York: Basic Books.
Rosenthal, R. (1963). On the social psychology of the psychological experiment: The experimenter's hypothesis as unintended determinant of experimental results. *American Psychologist, 51,* 268-283.
Rosenthal, R. (1974). *On the social psychology of the self-fulfilling prophecy: Further evidence for Pygmalion effects and their mediating mechanisms.* New York: MSS Modular.
Rosenthal, R. (1989, August). Experimenter expectancy, covert communication, & meta-analytic methods. Invited address at the 97th Annual Convention of the American Psychological Association, New Orleans, LA.
Rosenthal, R., & Fode, K. L. (1963). Psychology of the scientist: V. Three experiments in experimenter bias. *Psychological Reports, 12,* 49-51.
Rosenthal, R., & Jacobson, L. (1968). *Pygmalion in the classroom: Teacher expectations and student intellectual development.* New York: Holt, Rinehart, and Winston.
Rosenthal, R., & Lawson, R. (1963). A longitudinal study of the effects of experimenter bias on the operant learning of laboratory rats. *Journal of Psychiatric Research, 2,* 61-72.
Rosenthal R., & Rubin, D. B. (1978). Interpersonal expectancy effects: The first 345 studies. *The Behavioral and Brain Sciences, 3,* 377-386.
Ross, S. I., & Jackson, J. M. (1991). Teachers' expectations for black males' and black females' academic achievement. *Personality and Social Psychology Bulletin, 17,* 78-82.
Roth, B. M. (1995, January 2). We can throw teacher expectations on the IQ scrap heap. *New York Times,* p. A25.
Rowe, D. C. (1995, January 2). Intervention fables. *New York Times,* p. A25.
Rubin, Z. (1973). *Liking and loving.* New York: Holt, Rinehart & Winston.
Rumberger, R. W., Ghatak, R., Poulos, G., Ritter, P. L., & Dornbusch, S. M. (1990). Family influences on dropout behavior in one California high school. *Sociology of Education, 63,* 283-299.
Scarr, S. (1981). *Race, class, and individual differences in I.Q.* Hillsdale, NJ: Lawrence Erlbaum.
Schneider, D. J. (1996). Modern stereotype research: Unfinished business. In C. N. Macrae, C. Stangor, & M. Hewstone (Eds.), *Stereotypes and stereotyping.* New York: The Guilford Press.
Schuman, H., Walsh, E., Olson, C., & Etheridge, B. (1985). Effort and reward: The assumption that college grades are affected by quantity of study. *Social Forces, 63,* 945-966.
Seaver, W. B. (1973). Effects of naturally-induced teacher expectancies. *Journal of Personality and Social Psychology, 28,* 333-342.
Sedikedes, C., & Skowronski, J. J. (1991). The law of cognitive structure activation. *Psychological Inquiry, 2,* 169-184.
Slavin, R. E. (1983). Ability grouping in the middle grades: Achievement effects and alternatives. *Elementary School Journal, 93,* 535-552.
Smith, A., Jussim, L., & Eccles, J. (1997). *Do self-fulfilling prophecies accumulate, dissipate, or remain stable over time?* Unpublished manuscript.
Smith, A., Jussim, L., & Eccles, J., VanNoy, M., Madon, S., & Palumbo, P. (1997). Self-fulfilling prophecies, perceptual biases, and accuracy at the individual and group level. Unpublished manuscript.
Smith, M. L. (1980). Teacher expectations. *Evaluation in Education, 4,* 53-55.
Snyder, M. (1984). When belief creates reality. In L. Berkowitz (Ed.), *Advances in experimental social psychology* (Vol. 18, pp. 247-305). New York: Academic Press.
Snyder, M. (1992). Motivational foundations of behavioral confirmation. In M. P. Zanna (Ed.), *Advances in experimental social psychology* (Vol. 25, pp. 67-114. New York: Academic Press.
Steele, C. M. (April, 1992). Race and the schooling of black Americans. *Atlantic Monthly,* 68-78.
Sutherland, A., & Goldschmid, M. L. (1974). Negative teacher expectation and IQ change in children with superior intellectual potential. *Child Development, 45,* 852-856.
Swann, W. B., & Ely, R. J. (1984). A battle of wills: Self-verification versus behavioral confirmation. *Journal of Personality and Social Psychology, 46,* 1287-1302.

Swim, J. K. (1994). Perceived versus meta-analytic effect sizes: An assessment of the accuracy of gender stereotypes. *Journal of Personality and Social Psychology, 66*, 21-36.

Thorndike, R. L. (1968). Review of *Pygmalion in the classroom*. *American Educational Research Journal, 5*, 708-711.

Thurlow, M. L., Christensen, S., & Ysseldyke, J. E. (1983). Referral research: An integrative summary of findings. University of Minnesota Research Report No. 141. ERIC document 244439.

Tuckman, B. W., & Bierman, M. L. (1971). *Beyond Pygmalion: Galatea in the schools*. Paper presented at the meeting of the American Educational Research Association, New York, NY.

Van der Berghe, P. L. (1997). Rehabilitating stereotypes. *Ethnic and Racial Studies, 20*, 1-16.

Von Baeyer, C. L., Sherk, D. L., & Zanna, M. P. (1981). Impression management in the job interview: When the female applicant meets the male (chauvinist) interviewer. *Personality and Social Psychology Bulletin, 7*, 45-51.

Weber, M. (1930). *The Protestant ethic and the spirit of capitalism*. New York: Scribner.

Weinstein, R. (1985). Student mediation of classroom expectancy effects. In J. Dusek (Ed.), *Teacher expectancy* (pp. 329-350). Hillsdale, NJ: Erlbaum.

Weinstein, R. S., Marshall, H. H., Brattesani, K. A., & Middlestadt, S. E. (1982). Student perceptions of differential teacher treatment in open and traditional classrooms. *Journal of Educational Psychology, 74*, 678-692.

Wentzel, K. R. (1989). Adolescent classroom goals, standards for performance, and academic achievement: An interactionist perspective. *Journal of Educational Psychology, 81*, 131-142.

West, C., & Anderson, T. (1976). The question of preponderant causation in teacher expectancy research. *Review of Educational Research, 46*, 613-630.

Wigfield, A., Eccles, J. S., MacIver, D., Reuman, D., & Midgley, C. (1991). Transitions at early adolescence: Changes in children's domain-specific self-perceptions and general self-esteem across the transition to junior high school. *Developmental Psychology, 27*, 552-565.

Williams, T. (1976). Teacher prophecies and the inheritance of inequality. *Sociology of Education, 49*, 223-236.

Wineburg, S. S. (1987). The self-fulfillment of the self-fulfilling prophecy: A critical appraisal. *Educational Researcher, 16*, 28-40.

Wright, J. D., & Wright, S. R. (1976). Social class and parental values for children. *American Sociological Review, 41*, 527-537.

Ysseldyke, J. E., Algozzine, B., Shinn, M., & McGue, M. (1982). Similarities and differences between low achievers and students classified as learning disabled. *Journal of Special Education, 49*, 223-236.

Ysseldyke, J. E., & Foster, G. G. (1978). Bias in teachers' observations of emotionally disturbed and learning disabled children. *Exceptional Children, 44*, 613-615.

Zanna, M. P., & Pack, S. J. (1975). On the self-fulfilling nature of apparent sex differences in behavior. *Journal of Experimental Social Psychology, 11*, 583-591.

THE ANTECEDENTS AND CONSEQUENCES OF TEACHER EFFICACY

John A. Ross

Expectations are self-fulfilling prophecies. Teachers who believe that students will do well in school have been observed adjusting their behavior to make it so (Good & Brophy, 1990). One type of expectancy that has been consistently linked to teacher attributes, workplace conditions, instructional practice, and student outcomes is teacher efficacy. In this chapter I will identify the antecedent conditions associated with the waxing and waning of teacher efficacy, catalogue its influence on teacher actions and on student achievement, examine the results of interventions to enhance teacher efficacy, outline implications for practice, and suggest directions for further inquiry.

DEFINING TEACHER EFFICACY

Teacher Efficacy is a Type of Self-Efficacy

Teacher efficacy is a self-perception, not an objective measure of teaching effectiveness. It represents teachers' belief that their efforts, individually or collectively, will bring about student learning. Although some researchers equate teacher efficacy with a willingness to take responsibility for student success and failure (e.g., Guskey, 1988), most treat teacher efficacy as a type of self-efficacy. In social learning theory, self-efficacy is a judgment of one's ability to complete a specific action (Bandura, 1986). A teacher faced with a teaching task judges his or her performance capability based on an analysis of the task's requirements, reflections on experience with similar instructional situations in the past, and assessment of resources available. Over time, teachers' predictions for routine situations stabilize as persistent, but not static, performance expectations. Teachers who expect to do well set higher goals and persist through obstacles. Their effectiveness in the classroom increases and their attributions for success to their own effort and ability contribute to higher teacher efficacy. Or conversely, lower expectations lead to lower effort, depressed impact on student learning, and even lower expectations.

Although reflections on the results of one's teaching are the most important contributors to teacher efficacy, other sources of information may be influential. These include observing peers, attempts by colleagues or supervisors to convince teachers they are able to accomplish particular teaching tasks, and physiological responses (e.g., physical symptoms communicating an inability to perform effectively).

Within-Teacher Variation in Teacher Efficacy

Many researchers, but certainly not all, distinguish personal from general teaching efficacy. Personal teaching efficacy is the expectation that one will be able to bring about student learning; general teaching efficacy is the belief that teachers as a group will be able to do so, despite the limitations of children's environments.

Although measurement issues are beyond the scope of this review, it should be noted that teacher efficacy is almost always measured with self-administered surveys that generalize across different teaching tasks. Almost half of the studies conducted to date measured teacher efficacy with items developed by Gibson and Dembo (1984), or adaptations of their scales. These items provide teachers with expressions of teacher competence for recurring situations, such as being able to adjust the difficulty level of an assignment, to which teachers agree or disagree. Items are aggregated into a personal and a general teaching efficacy scale or into a global score. Internal consistencies tend to be moderately high. For example, Ross (1994) reported internal consistencies of .78-.82 for the personal scale and .59-.75 for the general measure. Scores tend to be stable. Ross found test-retest correlations over an eight-month period of .61-.67 and .55-.81 for the two scales.

These instruments may underestimate within-teacher variability. The only studies to decompose the variance in teacher efficacy scores to between- and within-teacher variables found that 21 percent (Ross, Cousins, & Gadalla, 1996) and 41 percent (Raudenbush, Rowen, & Cheong, 1992) of the variance was attributable to within-teacher factors. In these studies, secondary school teachers were more optimistic about their performance when teaching a high track class, with engaged students, in a course they felt prepared to teach and in which they had been successful in the past. The extent to which teacher efficacy was influenced by these teaching assignment characteristics depended upon subject specialization, experience, education, gender, and instructional preferences (Ross, Cousins, & Gadalla, 1996).

Other studies demonstrate that teachers' confidence in their classroom abilities varies across teaching tasks. For example, Ross, Rolheiser, and Hogaboam-Gray (in press) found that teachers identified by their supervisors and peers as exemplary users of cooperative learning techniques had high expectations about their ability to use peer-mediated instruction to meet the learning needs of all children in their classrooms. But negative feelings—guilt, anxiety, and uncertainty—ran through their cognitions about student assessment issues.

ANTECEDENTS OF TEACHER EFFICACY

Teacher efficacy is an outcome of teachers' personal characteristics and the organizations in which they work. But the relationships are all reciprocal (with some obvious exceptions such as gender) because teacher efficacy has generative power. It influences goal setting and persistence and in doing so contributes to mastery experiences (classroom success) that heighten or depress efficacy further. In this section I will review findings linking teacher efficacy to other teacher characteristics and to conditions of teachers' work.

Teacher Characteristics

Gender

Females report higher personal teaching efficacy than males (Anderson, Greene, & Loewen, 1988; Coladarci & Breton, 1991; Lee, Buck, & Midgley, 1992; Raudenbush, Rowen, & Cheong, 1992). One explanation for this finding is that females are overrepresented in the elementary panel and (as described below) teacher efficacy is higher in elementary than secondary schools. Another explanation is that teaching is viewed as a female occupation and female teachers tend to be more in tune with the dominant ideology of schools (Kalaian & Freeman, 1994). The gender-efficacy correlations are small (typically in the .20s). Males have higher teacher efficacy scores in subjects with higher male profiles, such as science (Riggs, 1991) and economics (Enochs, Schug, & Cross, 1996). Gender differences have also been reported in sources of efficacy information. Imants and de Brabander (1996) found that female teachers based their efficacy beliefs on their ability to complete learner-oriented tasks involving teacher-student interaction, such as giving extra help to pupils with learning problems. Males based their feelings of effectiveness on learner-oriented tasks as well but males also brought into their self-efficacy estimates school-oriented tasks that involved interaction with other adults, such as discussing the school development plan at staff meetings.

Experience

General teaching efficacy, belief in the capability of the teacher population to overcome environmental impediments to learning, tends to decline with teaching experience, especially in the early years, including preservice (Bandura 1993; Beady, & Hansell, 1981; Brousseau, Book, & Byers, 1988; Dembo & Gibson, 1985; Hoy & Woolfolk, 1990; 1993; Saklofske, Michayluk, & Randhawa, 1988). Two studies, both using a science-focused measure of general teaching efficacy, provide exceptions to this trend (Ginns & Watters, 1996; Huinker & Madison, 1995).

General teaching efficacy may decline with experience because teachers begin their careers in naive ignorance of the complexities of teaching and the difficulty of reaching all students. As they encounter students with problems that are not readily amenable to instruction, their overconfidence may be replaced with a more realistic view. An alternate explanation is ego-enhancement. Lowering expectations for the teacher population enables individuals to interpret their own teaching results more positively. Personal teaching efficacy rises with experience (Hoy & Woolfolk, 1993; Rubeck & Enochs, 1991), especially following the first practice-teaching episodes in preservice (Cannon, 1992; Housego, 1990; Hoy & Woolfolk, 1990) and in the first few years of teaching (Dembo & Gibson, 1985). Increasing confidence in one's own teaching ability is the result of mastery experiences, classroom events that provide teachers with psychologically credible evidence they are having a positive impact. But in the absence of data on the accuracy of self-appraisals it is difficult to tell how much of the increase is the result of enhanced skill and how much is due to lowered standards. In addition, the relationship between experience and teacher efficacy is reciprocal. Teachers with higher personal teaching efficacy have a stronger commitment to the profession (Coladarci, 1992), are more likely to continue in the career and to seek further training, thereby creating an upward bias in cross-sectional comparisons.

Teacher Certification

Teachers with a graduate degree have higher teacher efficacy than those who do not (Hoover-Dempsey, Bassler, & Brissie, 1987; Hoy & Woolfolk, 1993), most likely because extended training contributes to the acquisition of new teaching skills that increase opportunities for mastery experiences. These cross-sectional findings were confirmed by the longitudinal evidence of Watters and Ginns (1995) that teacher efficacy increased during a graduate course on teaching methods. In apparent conflict with these data, Moore and Esselman (1992) found that teachers with minimal qualifications (i.e., without an undergraduate degree) had higher efficacy, but the finding was likely confounded with grade level, since nondegree teachers are more likely to be found in elementary schools.

Ross, Cousins, and Gadalla (1996) found that the sources of teacher efficacy beliefs were influenced by education level. Teachers with a graduate degree were less influenced (than teachers without graduate training) by feelings of immediate success, perhaps because graduate training provided teachers with a greater appreciation of the uncertainty of teaching outcomes. Success today does not easily predict success tomorrow. Better-educated teachers were more influenced by student engagement levels in the class, perhaps reflecting greater knowledge of the link between motivation and learning.

Other Teacher Cognitions

Teachers with an internal locus of control, a general personality attribute, are more likely to score higher in *total* teacher efficacy (Ashton, Webb, & Doda, 1983; Greenwood, Olejnik, & Parkay, 1990; Haury, 1989; Lucas, Ginns, Tulip, & Watters, 1993; Parkay, Olejnik, & Proller, 1988). When probed more specifically about the causes of student learning, teachers who identify factors under the teachers' control as more important than factors beyond their control are more likely to have higher *personal* teaching efficacy (Brookhart & Loadman, 1993; Czerniak & Schriver-Waldon, 1991; Hall, Burley, Villeme, & Brockmeier, 1992; Hall, Hines, Bacon, & Koulianos, 1992; Kalaian & Freeman, 1990). The explanation for this relationship might be that teachers who believe they are in control exert greater effort. Greater effort typically leads to greater success (mastery experiences) which heightens expectations of future success.[1] This relationship may be mediated by beliefs about the nature of ability. Teachers who believe that student ability can be increased have higher teacher efficacy than teachers who believe that ability is immutable (Fletcher, 1990).

Other studies have found that teacher efficacy correlates with teachers' beliefs about their own learning. Preservice candidates who had higher science achievement in high school or college or who had taken more science courses scored higher on science teacher efficacy measures (Enochs & Riggs, 1990; Enochs, Scharmann, & Riggs, 1995; Haury, 1989). Preservice candidates with greater experience with children in noninstructional settings are more likely to overestimate their teaching abilities (Weinstein, 1988). The influence of prior experience is likely to dissipate as teachers develop expectations based on the outcomes of their teaching.

Summary

Teacher efficacy is associated with teacher characteristics in predictable ways. The findings are reasonably consistent, although the correlations are not large. Teacher efficacy is higher for teachers who are female (except in disciplines with strong male profiles), experienced, well-educated, and take responsibility for the outcomes of their actions. In addition, these between-teacher characteristics influence the sources of information that teachers use to develop expectations.

Workplace Antecedents

Grade

Elementary teachers have consistently reported higher efficacy than high school teachers (Greenwood, Olejnik, & Parkay, 1990; Guskey, 1982; Morrison, Walker, Wakefield, & Solberg, 1994; Parkay, Olejnik, & Proller, 1988) and middle school teachers (Fuller & Izu, 1986; Lee, Buck, & Midgley, 1992; Midgley, Feldlaufer, & Eccles, 1988).

There are two competing explanations for the finding. The first concerns conditions of teacher work that follow from the size and organizational structure of schools. Middle and high schools tend to be large, impersonal organizations in which teachers see a great number of students for brief periods on rotary schedules. These factors inhibit teachers from acquiring the knowledge of student needs essential to good teaching, reducing teacher opportunities for mastery experiences that contribute to higher teacher efficacy. But Lee, Dedrick, & Smith (1991) found that teacher efficacy was higher in larger secondary schools. A counter explanation comes from the finding that higher efficacy scores of elementary teachers have been reported as early as the first two weeks of a teacher education program (Evans & Tribble, 1986). Teacher efficacy might be a personal characteristic influencing choice of grade level when entering the profession. It may also be confounded with gender: Females, who tend to have higher teacher efficacy than males, constitute a larger proportion of the workforce in elementary than in middle and high schools.

Only a few studies investigated the influence of grade assignment on teacher efficacy. The results were mixed. Some found that teacher efficacy declined from grades three to six (Anderson, Greene, & Loewen, 1988; Bandura, 1993). In contrast Raudenbush, Rowen, and Cheong (1992) found that teacher efficacy increased with grade in high school. A speculative explanation is that more able teachers (who, as will be shown below, tend to have higher efficacy scores) may gravitate toward upper classes in high schools, reflecting the greater status of teaching in the senior years. Building principals might encourage this trend since these teachers have the potential to "polish" the graduates. In the elementary panel the reverse might occur. Here the more able might assign themselves to the primary years where the potential for socializing children to the school and the visibility of student development (and hence the intrinsic rewards of teaching) appear to be greater.

Student Ability

Teachers with higher-ability classes report higher teacher efficacy than those with classes composed of less able pupils (Ashton, Webb, & Doda, 1983; Lee, Dedrick, &

Smith, 1991; Riehl & Sipple, 1996; Smylie, 1988). When the effects of orderly behavior are entered into the equation, the impact of student ability is reduced but still significant (Newmann, Rutter, & Smith, 1989; Raudenbush, Rowen, & Cheong, 1992), suggesting that these variables have independent effects even though they are intercorrelated. Teachers of able, orderly pupils are more likely to be successful than teachers of students who pose a greater instructional challenge. Teacher efficacy is higher because there are more opportunities for mastery experiences. An alternate explanation is that teachers with higher efficacy are more likely to be assigned to high-track classes (Midgley, Feldlaufer, & Eccles, 1989). But teacher efficacy is also higher in schools with high achieving students (Beady & Hansell, 1981; Smylie, 1988) and in schools perceived to have well-behaved pupils (Fletcher, 1990), suggesting that ability and orderliness each contribute to efficacy beliefs.

Student ability also influences the sources of information on which teachers build their expectations. Ashton, Webb, and Doda (1983) found that some teachers defined classroom success exclusively in terms of cognitive outcomes, whereas others were more concerned with students' social development. Nurturing teachers would be less threatened by low cognitive achievement. Redefining the criteria for success might represent a coping strategy enabling some teachers to maintain high expectations when assigned to a lower ability class or school.

Student Social Class

The evidence is mixed. Bandura (1993) reported findings from an unpublished path analysis in which low socioeconomic status (SES) combined with high student turnover and absenteeism to create a pattern of low student achievement that reduced teachers' feelings of efficacy. But Rose and Medway (1981) found that teachers in low SES schools scored higher on one personal efficacy measure.

Class Size

Results are mixed. Teacher efficacy has been reported as lower in small high school classes (Raudenbush, Rowen, & Cheong, 1992) and higher in small elementary classes (Hoover-Dempsey, Bassler, & Brissie, 1987). The discrepancy might be attributable to sample differences. In the high school study smaller classes tended to be reserved for remedial students. In the elementary school study there were few segregated classes and greater integration of special-needs students into regular classes.

Teacher Workload

Riehl and Sipple (1996) constructed a composite measure of teacher workload, consisting of number of preparations, proportion of classes in area of teaching expertise, ability track, and achievement of students. Teachers with heavier workloads had lower teacher efficacy. Other researchers found that teachers feeling stressed or burned out had lower teacher efficacy (Bliss & Finneran, 1991; Brissie, Hoover-Dempsey, & Bassler, 1988; Greenwood, Olejnik, & Parkay, 1990). Conditions that increase the difficulty of teachers' tasks diminish confidence in their ability to complete those tasks successfully. In addition, physiological symptoms of stress (e.g., sweaty palms) are an important source of information of self-efficacy information (Bandura, 1993).

School Culture

Collaboration among teachers promotes teacher efficacy (Chester & Beaudin, 1996; Louis, 1991; Morrison et al., 1994; Rosenholtz, 1989; Ross, 1992), especially when it leads to instructional coordination within a school (Hoover-Dempsey, Bassler, & Brissie, 1987; Moore & Esselman, 1992; Raudenbush, Rowen, & Cheong, 1992; Rosenholtz, 1989) or district (Rubeck & Enochs, 1991). In a longitudinal study of fluctuations in teacher efficacy during a period of high stress, Ross, McKeiver, and Hogaboam-Gray (1997) found that collaboration stimulated teacher efficacy in several ways. (a) It contributed to teachers' knowledge of their classroom effectiveness by providing jointly produced measures of students' cognitive and affective performance. This process made it easier for teachers to get evidence that they were successful, and such evidence has the most direct influence on efficacy beliefs. (b) By working together teachers learned about the classroom successes of others. Although vicarious participation is a weaker contributor to self-efficacy, it can increase feelings of self-efficacy if teachers believe their skills are comparable to those of the colleagues they observe (Bandura, 1986). (c) Collaboration created a climate that legitimated help seeking, joint problem solving, and instructional experimentation. Teachers acquired more teaching strategies, which enhanced their effectiveness, thereby increasing perceptions of their current success and expectations for the future. (d) By emphasizing the positive outcomes of their work, collaboration contributed to increased teacher efficacy through peer persuasion. (e) Collaboration reduced duplication of effort, freeing teachers to focus on the development of new teaching strategies that contributed to their feelings of success when implemented.

Increased collaboration might reduce the confidence of some teachers if they received negative feedback on their performance from their peers, if the basis for collaboration was social rather than instructional, or if collaboration reduced teachers' ability to experiment (e.g., through the use of a common exam). This might explain why Ashton, Webb, and Doda (1983) found teacher efficacy to be greater in the low than in the high collaboration school they studied.

Teacher efficacy is enhanced when teachers have greater control of their workplace—when they have opportunities to use their skills (Louis, 1991), control classroom decision making (Moore & Esselman, 1994), and participate in school-wide decisions (Berman, McLaughlin, Bass, Pauly, & Zellman, 1977; Fletcher, 1990; Lee, Dedrick, & Smith, 1991; Moore & Esselman, 1992; Raudenbush, Rowen, & Cheong, 1992; Riehl & Sipple, 1996). Louis found that the strongest organizational influence on teacher efficacy was receiving respect from relevant adults (Louis, 1991) and that demographic influences on teacher efficacy (gender, school composition) were dwarfed by the extent to which teachers believed their school was a professional community (Louis, Marks, & Kruse, 1994). These workplace features confirm teachers' competence and communicate information about their effectiveness, thereby increasing teachers' expectations about their impact on students.

Strong principals enhance teacher efficacy (Lee, Dedrick, & Smith, 1991). By coordinating, supervising, and rewarding teachers, principals can influence teachers' appraisals of their performance, heighten the exchange of vicarious experience, and engage in verbal persuasion. Leadership actions contributing to teacher efficacy include emphasizing accomplishment (Lee, Buck, & Midgley, 1992; Rosenholtz, 1989), increasing teachers' certainty about the worth of their practice (Smylie, 1988), modeling professionalism and inspiring group purpose (Hipp, 1996; Hipp & Bredeson, 1995), providing common planning time to enable teachers to work together (Warren & Payne, 1995), being responsive to teacher concerns (Brissie, Hoover-Dempsey, & Bassler, 1988; Hoy & Woolfolk, 1993; Newmann, Rutter, & Smith, 1989), giving frequent feedback (Chester & Beaudin, 1996), promoting an academic emphasis in the school (Hoy & Woolfolk, 1993), and providing supervision perceived to be useful by teachers (Brissie, Hoover-Dempsey, & Bassler, 1988; Coladarci & Breton, 1991; Lubbers, 1990; Rees, 1986).

Summary

Although evidence of the effect of some organizational variables (grade, social class, and class size) on teacher efficacy is mixed, other relationships have been demonstrated consistently. Teacher efficacy is higher in elementary schools, in schools and classes with high ability and orderly pupils, and when teacher workloads are moderate, school culture is collaborative and participatory, and principals engage in behaviors that confirm teacher competence.

CONSEQUENCES OF TEACHER EFFICACY

Self-efficacy beliefs have generative power. Teachers who believe they will be effective are more likely to set higher goals and to persist through obstacles. Such persistence leads to greater classroom success and these mastery experiences then contribute to higher expectations.

Teacher Outcomes

Teachers with high efficacy beliefs set more ambitious goals for themselves. For example, they are more likely to select instructional strategies to enhance student development rather than to cover the curriculum (Brookhart & Loadman, 1993;

Czerniak & Schriver-Waldon, 1991). They are more likely to use instructional strategies which are powerful, but difficult to acquire, such as small-group techniques (Tracs & Gibson, 1986), cooperative learning (Dutton, 1990), performance-based assessment (Vitali, 1993), and activity-based methods (Czerniak & Schriver-Waldon, 1991; Riggs & Enochs, 1990). Low-efficacy teachers are more likely to rely on approaches, such as whole-class teaching (Ashton & Webb, 1986; Tracs & Gibson, 1986), which are weaker but easier to adopt.[2]

Teachers with greater confidence in their professional competence are also more likely to experiment in the classroom (Allinder, 1994), implement new instructional programs (Berman et al., 1977; Guskey, 1988; Moore, 1990; Rose & Medway, 1981), take responsibility for students with special learning needs (Jordan, Kircaali-Iftar, & Diamond, 1993; Meijer & Foster, 1988; Podell & Soodak, 1993; Soodak & Podell, 1993a 1993b), participate in action research (Cousins & Walker, 1995), and involve parents in school activities (Hoover-Dempsey, Bassler, & Brissie, 1987, 1992). These findings indicate that having high expectations enables teachers to take risks. They do not fear failure because they believe that they will be able to overcome any challenges that might arise.

Teacher efficacy correlates with global measures of competence. For teachers with high efficacy beliefs, commitment to the profession is higher (Coladarci, 1992), absenteeism is lower (Imants & Van Zoelen, 1995), and supervisor ratings are better (Flowers, 1988; Hoover-Dempsey, Bassler, & Brissie, 1987; Riggs & Enochs, 1990; Saklosfske, Michayluk, & Randhawa, 1988; Trentham, Silvern, & Brogdon, 1985).

Teachers with higher expectations set higher goals for students. Teachers with high personal teaching efficacy promote student autonomy (Midgley, Feldlaufer, & Eccles, 1988), confront management problems rather than responding permissively (Korevaar, 1990), and are more successful at keeping students on task (Ashton, Webb, & Doda, 1983). They are more humanistic in their orientations and are less reliant on custodial methods to control the class (Ashton & Webb, 1986; Woolfolk & Hoy, 1990; Woolfolk, Rosoff, & Hoy, 1990).

Summary

Higher teacher efficacy is consistently associated with the use of teaching techniques that are more challenging and difficult, with teachers' willingness to implement innovative programs, and with classroom management practices that promote student responsibility. High expectations of success enable teachers to set higher goals for themselves and others, take risks in experimenting, and learn new methods that contribute to higher student achievement.

Student Outcomes

Higher teacher efficacy contributes to higher student cognitive achievement. In the following studies investigators controlled entry ability because teachers with low self-efficacy are likely to be assigned lower achieving classes (Midgley, Feldlaufer, & Eccles, 1989). Personal teaching efficacy contributes to student achievement in curriculum domains involving language such as reading, language arts, and social studies (Anderson, Greene, & Loewen, 1988; Ashton & Webb, 1986; Moore & Esselman, 1994; Ross, 1992; Tracs & Gibson, 1986; Watson, 1991). In contrast, general teaching efficacy contributes to student achievement in mathematics (Ashton & Webb, 1986; Moore & Esselman, 1992; Ross & Cousins, 1993; Watson, 1991). One explanation for the interaction of efficacy type with subject achievement might be that many teachers view math as a talent that is given and see language as a set of skills that can be acquired. The extent to which teachers believe that natural endowments can be overcome by education (general teaching efficacy) might therefore play a larger role in math and the belief that individual teachers are able to develop student skills (personal teaching efficacy) might come to the fore in language.

Teacher efficacy is also linked to students' affective growth. Higher teacher efficacy is associated with enhanced student motivation (Ashton & Webb, 1986; Midgley, Feldlaufer, & Eccles, 1989; Roeser, Arbreton, & Anderman, 1993), increased self-esteem (Borton, 1991), improved self-direction (Rose & Medway, 1981), pro-social attitudes (Cheung & Cheng, 1997), and more positive attitudes toward school (Miskel, McDonald, & Bloom, 1983). These relationships are reciprocal because able and orderly student populations also contribute to higher teacher efficacy. The strongest evidence that achievement is a consequence of teacher efficacy comes from a longitudinal study (Midgley, Feldlaufer, & Eccles, 1989) that tracked students as they moved from one grade to another. At the beginning of the study researchers found that students with high self-efficacy were more likely to be found in classrooms of high efficacy teachers. When these students moved to a new grade, student self-efficacy continued to be high for students placed in a classroom taught by a teacher with high teacher efficacy. But students' self-efficacy declined if they were assigned to a teacher with low teacher efficacy. The decline was especially large for lower achieving pupils. Students who went from a high- to a low-efficacy teacher had lower motivation than students who experienced low-efficacy teachers in both years. The latter finding suggests that the motivational impact of the collective teacher efficacy of a school is determined by both the mean and the standard deviation of the efficacy distribution.

Teacher efficacy might impact on student achievement in several ways. First, teachers with higher efficacy are more willing to learn about and implement demanding new teaching techniques. Improved teaching might increase achievement.

Second, higher efficacy teachers use classroom management approaches that stimulate student autonomy and reduce custodial control. Student achievement might be higher because these management strategies are more effective in keeping students on task (Woolfolk, Rosoff, & Hoy, 1990).

Third, higher efficacy teachers may be more successful because they attend more closely to the needs of lower ability students. Ashton, Webb, and Doda (1983) found that low-efficacy teachers concentrated their efforts on the upper ability group, giving less regard to lower ability students who were viewed as potential sources of disruption. In contrast, high-efficacy teachers had positive attitudes toward low achievers, built friendly relationships with them, and set higher academic standards for this group than low-efficacy teachers did. This might contribute to higher achievement because lower ability students may be less certain about their competence and more influenced by teacher expectations.

Fourth, teacher efficacy may lead to specific changes in teacher behavior that create changes in students' perceptions of their academic abilities. As student efficacy becomes stronger, students may become more enthusiastic about school work and more willing to initiate contacts with the teacher, processes that impact directly on achievement (Ashton, Webb, & Doda, 1983; Ashton & Webb, 1986). Evidence that teacher efficacy has a delayed impact on student achievement is congruent with this view. (Midgley, Feldlaufer, & Eccles [1989] found that teacher efficacy correlated with achievement in the spring, but not the fall.)

Finally, teacher efficacy may influence student achievement through teacher persistence. Teachers with high perceived efficacy may view student failure as an incentive to greater teacher effort rather than conclude that the causes of failure are beyond teacher control and cannot be reduced by teacher action.

Summary

Teacher efficacy has a positive impact on students' cognitive and affective achievement. High expectations encourage teachers to engage in practices that contribute to enhanced student learning. This results in greater opportunities for mastery experiences that elevate teachers' efficacy beliefs still higher in a reciprocal cycle.

INTERVENTIONS TO STRENGTHEN TEACHER EFFICACY

Teacher efficacy tends to be highly stable following crystallization during the early years of teaching. Research on the consequences of teacher efficacy has led to interventions intended to strengthen teachers' expectations, usually in concert with efforts to influence teachers' instructional practices and/or student achievement. Three types of intervention have generated data on the malleability of teacher efficacy. These studies attempt to increase teacher efficacy by developing teachers' instructional skills, belief systems, and/or workplace conditions.

Skill Development Interventions

The most frequently reported strategy for enhancing teacher efficacy is to strengthen teachers' instructional skills. Attempts to enhance teacher efficacy by modifying the preservice program have had mixed results. Volkman, Scheffler, and Dana (1992) found that preservice teachers who had post-lesson and biweekly conferences with a graduate student coach had higher gains in teacher efficacy than a control group. The feedback procedures may have contributed to higher teacher efficacy by influencing teachers' interpretations of the outcomes of their teaching. The feedback may also have increased teachers' effectiveness (by strengthening their teaching skills), which could have influenced their feelings of success. Vitale and Romance (1992) found that preservice teachers in a course designed to strengthen teachers' disciplinary knowledge (by providing videodisk materials focused on key science concepts) increased concept understanding (compared to preservice candidates in a science methods course that did not have the videodisk materials). The expected enhancement of teachers' expectations of classroom success did not occur, possibly because there was no classroom teaching component to create opportunities for mastery experiences. Guyton, Fox, and Sisk (1991) found that a summer residency followed by a one-year supervised internship had no greater impact on teacher efficacy than the traditional preservice program. The study may have failed to find an impact because it did not measure teacher efficacy until after practice teaching, the period of greatest growth. Also, the validity of the findings was threatened by uncontrolled differences between the treatment and control groups.

Efforts to influence the expectations of experienced teachers through skill development in-service programs have also yielded mixed findings. Bolinger (1988) found that personal teaching efficacy increased in an in-service program based on effective schools research. Bolinger argued that teachers acquired new teaching skills that increased their performance, thereby contributing to greater confidence in their personal teaching abilities. General teaching efficacy was unaffected because the program made no attempt to influence teachers' beliefs about the effectiveness of their peers. Corbitt (1989) found that a similar in-service treatment based on effective schools research had a positive impact on the efficacy beliefs of some but not all teachers. The impact of the experience was mediated by the fit of the effective schools model with teachers' preferred teaching practices.

Dutton (1990) found that an in-service program on cooperative learning increased personal teaching efficacy. The gains were linked to program variables (the provision of group sharing and problem solving during training) and variations in follow-up at the school (opportunities for discussion with colleagues, principal observation and feedback). Professional development programs based on peer coaching also have had a positive impact on teacher efficacy (Edwards & Newton, 1995; Robardey, Allard, & Brown, 1994). Sharing experiences with peers contributes to teacher efficacy by confirming teachers' beliefs about their own success and through vicarious participation in the successes of others.

Ross (1994) found a slight upward trend in teacher efficacy during a six-month in-service program on cooperative learning. Teachers who participated more extensively (e.g., who were more persistent in their attempts to implement cooperative learning techniques in their classrooms) showed higher increases in general teaching efficacy than those who participated to a lesser degree. Activities in which teachers shared their classroom experiences had the greatest impact. Teachers with prior experience in using cooperative learning techniques dominated these sessions. Their reports may have persuaded less experienced teachers that teachers like themselves could be successful.

The most persuasive evidence for a causal connection between in-service experiences and teacher efficacy comes from a study of the implementation of a specific innovation over three school terms (Stein & Wang, 1988). Changes in teacher practice occurred between terms one and two, preceding changes in teacher efficacy that developed between terms two and three. Teachers who made the greatest change in instructional practice showed the greatest increases in teacher efficacy.

Finally, Guskey (1984) found that teachers who used instructional skills acquired in a Mastery Learning workshop became more willing to take responsibility for student outcomes (a measure of teacher efficacy) and the achievement of their students increased. Teachers in the control group and those who attended the workshop sessions but did not implement the treatment showed no gains in student outcome or in teacher efficacy. Guskey also found that teaching self-concept (a global measure of teaching confidence) declined in the treatment group. Guskey argued that the workshop provided teachers with a teaching strategy that contrasted with their existing techniques. When they tested the new method, they found it was superior. This diminished confidence in their existing techniques, demonstrating that teachers' expectations about their effectiveness can increase in one domain while decreasing overall.

Summary

Although the findings from attempts to increase teacher efficacy by strengthening teachers' instructional skills are mixed, these studies demonstrate that teacher efficacy can be increased. The gains were largest among teachers who implemented professional development prescriptions and discussed their interpretations of results with peers.

Interventions Based on Changing Teacher Beliefs

Skill development approaches to strengthening teacher efficacy that provided opportunities for teachers to reflect on their practices might also provide opportunities to examine the cognitive underpinnings of teacher expectancies. Two studies have attempted to change teacher efficacy by overtly challenging teachers' beliefs.

Ohmart (1992) evaluated an in-service program that combined attention to teachers' theories of intelligence with the development of direct teaching skills. The program had an immediate positive impact on participants' teacher efficacy but the effect disappeared on the delayed posttest. Ohmart's intervention relied on lecture and persuasion to influence teachers' conceptions. It may have had only a temporary effect because verbal persuasion is the weakest source of information influencing self-efficacy: "illusory boosts in efficacy are readily disconfirmed by the results of one's own actions" (Bandura, 1993, p. 204). Subsequent information about student outcomes may have swamped the initial glow of success.

Fritz, Miller-Heyl, Kreutzer, and MacPhee (1995) designed an in-service treatment that combined attention to teachers' beliefs about themselves (self-esteem and locus of control) with attempts to develop instructional skills. Teachers who began the year-long program with high teacher efficacy increased expectations about the ability of themselves and their peers to bring about student learning. The effects lasted through the delayed (nine months) posttest. For teachers who entered with low teacher efficacy scores, there were no further declines. The program served to insulate these participants from the declines in teacher efficacy experienced by control group teachers with equally low initial scores. The researchers also found that teachers who participated fully in the teacher sessions but did not implement the classroom component of the program did not change. This suggests that interventions based on changing teacher beliefs need to be combined with opportunities for follow up in the classroom. It also raises the possibility that overt attention to teachers' beliefs may be unnecessary if there is a strong skills development component.

Interventions that Change Conditions of Teacher Work

Externally imposed change can have a negative effect on teacher efficacy. Rosenholtz (1987) found that a minimum-competency-testing program reduced teachers' feelings of success. The testing program forced teachers to cut topics they viewed as important and to increase the pace of instruction beyond what they deemed appropriate. Teacher efficacy declined, except for a small group of teachers with classes similar to those of the test developers, who increased in teacher efficacy. Restructuring depressed teacher efficacy by modifying teachers' attributions, teachers felt less responsible for the outcomes of their teaching. In addition, by intruding into teachers' autonomy and requiring that they teach in a manner they regarded as less effective, the reform told teachers they were not competent.

A government-imposed detracking plan had a negative effect on the teacher efficacy of exemplary mathematics teachers (Ross et al., 1997). These teachers felt capable of teaching different ability groups in separate classes, but found their skills could not be readily integrated to teach a mixed-ability group. Teachers' expectations of success declined because they could not predict whether the new methods would produce student learning in untracked classes. Restructuring depressed teacher efficacy by weakening outcome expectancies. This study also found that the negative effects of restructuring dissipated over time. There was a resurgence of teacher confidence as teachers developed new ways of working with heterogeneous classes and discovered that achievement, particularly of lower ability performers, exceeded teacher expectations. The renewal of teacher efficacy was associated with personal coping strategies (especially certainty about professional goals and control of emotional states) and social processes (particularly collaboration with same-subject peers). The pattern of decline and resurgence in teacher efficacy during implementation of new teaching strategies has also been reported in other studies (Marx et al., 1994; Meyerson, 1995).

Reform efforts intended to increase teacher leadership have had mixed effects. Rosenholtz (1987) found that a career-ladder plan had a negative impact on teacher efficacy because teachers were excluded from setting evaluation standards and poor teachers were promoted. A sense of injustice prevailed that reduced effort. Further decreases in teacher efficacy occurred when teachers were given little feedback by the evaluators and when teachers concluded that the portfolio method of assessment was unrelated to actual teacher performance. In contrast, Rosenholtz found positive effects, as did Ebmeier and Hart (1992), when teachers were involved in designing a career-ladder scheme. Teacher leadership programs may have a positive impact on the teacher efficacy of those selected for leadership roles if they have an opportunity to demonstrate classroom expertise. But teacher leaders tend to judge themselves and to be judged on their consulting, not their teaching skills. Feelings of success as a leader may not transfer to perceived effectiveness in the classroom. The impact of leadership programs on the teacher efficacy of unselected teachers is likely to be negative if teachers interpret their nonselection as a critique of their teaching.

Summary

These studies indicate that restructuring that impacts on classroom practice has a negative effect on teacher efficacy, at least initially. Whether teachers recover depends on their ability to make adjustments in their teaching methods to accommodate the change and on the impact of the new structures on school culture. If the resulting workplace enhances opportunities for collaboration, professional learning, decision-making participation, and supportive administrative leadership, then teacher efficacy may return to its previous level or exceed it. Louis (1991) found that teacher efficacy was highest within schools that had made the greatest changes.

IMPLICATIONS FOR EDUCATIONAL REFORM

Evidence that teacher efficacy contributes to desirable teacher and student outcomes has been replicated in numerous independent studies that vary in teacher characteristics, learner populations, curriculum contexts, and workplace settings. These studies indicate that teachers' expectations about their ability to bring about student learning have generative capacity. Teachers with high expectations set more challenging goals for themselves, more frequently experiment with new instructional methods, implement more powerful teaching techniques, take greater responsibility for student learning (including the learning of less able students), persist through obstacles, and make a substantially greater contribution to students' cognitive and affective achievement than teachers with low expectations. There is no doubt that teacher efficacy matters.

The research base also demonstrates that there are consistent relationships among teacher efficacy, other teacher attributes, and workplace conditions. Interventions designed around these relationships have demonstrated, not always consistently, that teachers' efficacy beliefs can be strengthened and that doing so influences their practices and the learning of their students. The evidence suggests that teacher efficacy is a relatively stable attribute of teachers that crystallizes fairly early in their careers. But it is a dynamic attribute that can be modified by manipulating its predictors.

The most successful interventions are likely to be multiple treatments that combine instructional skill development with explicit attention to teachers' beliefs about their role in guiding student learning and the creation of strong professional cultures. The most powerful source of efficacy information is the teacher's interpretation of the outcomes of classroom action. The surest path to increasing opportunities for mastery experience is through professional development based on powerful new methods for creating deeper student understanding such as reciprocal teaching. But skill development alone is insufficient. Teachers may become more skillful, yet be no more confident about their future success than they were before, if their theories about the immutability of student intelligence are not challenged or if they maintain criteria for defining classroom success that are unattainable or unchallenging for the learners they serve. In addition, strengthened teaching skills will have a greater impact on teacher efficacy if the positive effects of new methods are recognized by peers and principals and if teachers have sufficient control over their professional lives that they can attribute school outcomes to their own efforts. The key is to build professional communities to support collaborative cultures that are focused on the achievement of complex learning outcomes by all students.

Multiple treatment approaches to the enhancement of teacher efficacy will be more effective if they are focused on specific tasks. By treating efficacy as a unitary trait it has been assumed that change in one constituent element will generalize to others. It is more likely that there is a relatively stable core of efficacy beliefs surrounded by dimensions that fluctuate in response to the tasks of specific teaching assignments. Efforts to improve teacher efficacy might focus on helping teachers to meet the demands of a balance of core tasks, such as adapting instruction to meet the needs of less able pupils, with less central (but important) tasks such as disentangling individual from group performance in cooperative assessment.

Professional development planners might aim for a modest degree of overconfidence among teachers. Teachers who believe themselves to be far more competent than an objective appraisal of their skills would warrant will be swiftly disappointed in the classroom. The resulting self-evaluation is likely to be as destructive of teacher motivation as underconfidence. Developing expectations that accurately mirror, or better still, lead performance might be best accomplished in dialogue among teachers. The positive impact of sharing experiences on professional learning (e.g., Ross & Regan, 1993) might lie, in part, in the development of appropriate expectations through social comparison.

Those attempting to track teacher efficacy during systemic reform might be wary of premature evaluation and be guided by a reciprocal model of the relationship between teacher change and teachers' expectations about their effectiveness. Teacher efficacy might fluctuate through restructuring. (a) High teacher efficacy might contribute to experimentation with new teaching ideas by influencing teachers' goal setting. (b) Teacher efficacy could decline as the new techniques disrupted the smoothness of existing practice. (c) Efficacy beliefs might remain depressed even if there was early success if the perceived superiority of the new techniques persuaded teachers of the inadequacy of their routine practice. (d) Teacher efficacy might begin to increase as teachers integrate the new methods into their repertoire and began to enjoy increased student performance consistently. (e) Enhanced teacher efficacy might motivate the search for new skill development opportunities. This multistage conception of the relationship between teacher efficacy and change in teacher practice (Ross, 1995) might supplant the debate between those who believe that alterations in teaching practice precede changes in teachers' beliefs and those who believe the reverse.

PRIORITIES FOR FUTURE RESEARCH ON TEACHER EFFICACY

Research on teacher efficacy has answered many of the questions posed by Rosenholtz (1989) when she described teacher efficacy as being in its infancy, but there are many high priority issues that still compete for researchers' attention. The first concerns the development of teacher efficacy over time. We know very little about how teacher efficacy relates to phases in teachers' careers, whether there are growth spurts and, if so, when they occur and why. Research to date has largely examined the development issue in cross-sectional rather than longitudinal designs and has defined time in years of teaching experience rather than in terms of career stages. In contrast, Huberman (1989) describes a nonlinear picture of progression and regression, with teachers of similar experience tracing dramatically different trajectories.

A second issue concerns the level of generality of teacher efficacy. Researchers have tended to treat teacher efficacy as a unidimensional trait, a practice that might explain the generally low correlations between efficacy beliefs and other constructs. It would make sense to adjust the level of measurement to the needs of the study. When making predictions about global measures like career continuance, a global measure of teacher efficacy is appropriate. When making predictions about the impact of teacher expectations on specific teaching tasks, such as the ability to use real-time data collection and processing software to teach heat and energy concepts in grade-nine science, a much narrower measure is required. The dilemma for researchers is that shifting to more precise measures might produce stronger predictions while weakening the generalizability of the findings.

The third area concerns within-teacher differences. We need to know more about how teachers' confidence in their abilities fluctuates through the school day as they move from task to task or class to class. We especially need to know more about the sources of information that teachers use to develop their expectations in order to elaborate the four broad sources (mastery experience, vicarious experience, persuasion, and physiological response) derived from self-efficacy theory. This line of research might help resolve the uncertainty in teacher efficacy research in how personal and general teaching efficacy, which correlate only moderately well (e.g., $r = .23$ in Ross, 1992), relate to one another.

Finally we need more studies of interventions, as described above, to refine strategies for enhancing teacher efficacy. Of particular interest is the role of the accuracy of teachers' expectations, a topic that has received little attention. The assumption in research to date is that a modest level of overconfidence is most motivating but no studies have tested the hypothesis.

Research on teacher efficacy has matured. The broad outlines of its antecedents and consequences are becoming familiar, even though a host of unanswered questions remain. Research on teachers' expectations in the late 1990s shares some of the concerns of the Pygmalion studies of the early 1970s. In both eras the focus was the influence of teachers' beliefs about future student performance on student learning. In each era researchers attempted to manipulate teachers' expectations in order to change instructional practices and student outcomes. What distinguishes the eras is the motivation for the research. In the Pygmalion studies teachers' expectations were created by researchers telling teachers that randomly selected students were particularly ready for learning. The goal was to understand teaching and learning processes. In the current era teachers' expectations are being influenced by interventions designed to change teachers' skill levels, their interpretations of success, and their relationships with other professionals. The goal is the improvement of teaching and learning, a purpose that brings researchers and practitioners together in a shared agenda .

ACKNOWLEDGMENT

This review was funded by the Ontario Ministry of Education and Training, and by the Social Science and Humanities Research Council of Canada. The views expressed in the paper do not necessarily represent the views of the Ministry or the Council.

NOTES

1. It could also be argued that the association of causal attributions with teacher efficacy is an artifact of measurement. The Rand items used by several of these studies were developed from Rotter's

(1966) locus of control theory and a case has been made that the Gibson and Dembo (1984) instrument used by other studies in this group measures internal and external control (Guskey & Passaro, 1993). But studies that operationalized teacher efficacy as confidence in performing a sample of teacher functions (Brookhart & Loadman, 1993; Haury, 1989; Kalaian & Freeman, 1990) produced similar findings.

2. It is important to note that similar findings are produced by studies using observations of teacher practice (e.g., Tracs & Gibson, 1986) as well as those relying on self-reported practice (e.g., Vitali, 1993). The risk that teachers may be overreporting the use of strategies believed to be desirable is somewhat reduced by evidence that teachers with higher efficacy beliefs receive higher ratings from supervisors (e.g., Flowers, 1988).

REFERENCES

Allinder, R. (1994). The relationship between efficacy and the instructional practices of special education teachers and consultants. *Teacher Education and Special Education, 17* (2), 86-95.

Anderson, R., Greene, M., & Loewen, P. (1988). Relationships among teachers' and student achievement. *Alberta Journal of Educational Research, 34* (2), 148-165.

Ashton, P., & Webb, R. (1986). *Making a difference: Teachers' sense of efficacy and student achievement.* New York: Longman.

Ashton, P. T., Webb, R. B., & Doda, N. (1983). *A study of teacher's sense of efficacy.* Final report to the National Institute of Education, Executive Summary. Gainesville, FL: Florida University. (ERIC Document Reproduction Service No. ED 231 833).

Bandura, A. (1986). *Social foundations of thought and action: A social cognitive theory.* Englewood Cliffs, NJ: PrenticeHall.

Bandura, A. (1993). Perceived self-efficacy in cognitive development and functioning. *Educational Psychologist, 28* (2), 117-148.

Beady, C., & Hansell, S. (1981). Teacher race and expectations for student achievement. *American Educational Research Journal, 18* (2), 191-206.

Berman, P., McLaughlin, M., Bass, G., Pauly, E., & Zellman, G. (1977). *Federal programs supporting educational change. Vol. 8. Factors affecting implementation and continuation.* Santa Monica, CA: Rand Corporation.

Bliss, J., & Finneran, R. (1991, April). *Effects of school climate and teacher efficacy on teacher strees.* Paper presented at the annual meeting of the American Educational Research Association, Chicago.

Bolinger, R. (1988). *The effects of instruction in the Hunter instructional model on teachers' sense of efficacy.* Unpublished doctoral dissertation, Montana State University, Bozeman, Montana.

Borton, W. (1991). *Empowering teachers and students in a restructuring school: A teacher efficacy interaction model and the effect on reading outcomes.* Paper presented at the annual meeting of the American Educational Research Association, Chicago.

Brissie, J., Hoover-Dempsey, K., & Bassler, O. (1988). Individual, situational contributors to teacher burnout. *Journal of Educational Research, 82* (2), 106-112.

Brookhart, S., & Loadman, W. (1993). *Relations between self-confidence and educational beliefs before and after teacher education.* Paper presented at the annual meeting of the American Educational Research Association, Atlanta.

Brousseau, B., Book, C., & Byers, J. (1988). Teacher beliefs and the cultures of teaching. *Journal of Teacher Education, 39* (6), 33-39.

Cannon, J. (1992). *Influence of cooperative early field experience on preservice elementary teachers' science self-efficacy.* Unpublished doctoral dissertation, Kansas State University, Manhattan, Kansas.

Chester, M., & Beaudin, B. (1996). Efficacy beliefs of newly hired teachers in urban schools. *American Educational Research Journal, 33* (1), 233-257.

Cheung, W. M., & Cheng, Y. C. (1997). *A multi-level analysis of teachers' self-belief and behavior, and students' educational outcomes.* Paper presented at the annual meeting of the American Educational Research Association, Chicago.

Coladarci, T. (1992). Teachers' sense of efficacy and commitment to teaching. *Journal of Experimental Education, 60* (4), 232-337.

Coladarci, T., & Breton, W. (1991). *Teacher efficacy, supervision, and the special education resource-room teacher.* Paper presented at the annual meeting of the American Educational Research Association, Chicago.

Corbitt, E. (1989). *The three R's of staff development: Reality, relevance, and relationships.* Paper presented at the annual convention of the Council for Exceptional Children, San Francisco, CA.

Cousins, J., & Walker, C. (1995, June). *Personal teacher efficacy as a predictor of teachers' attitudes toward applied educational research.* Paper presented at the annual meeting of the Canadian Association for the Study of Educational Administration, Montreal.

Czerniak, C., & Schriver-Waldon, M. (1991). *A study of science teaching efficacy using qualitative and quantitative research methods.* Paper presented at the annual meeting of the National Association of Research in Science Teaching, Lake Geneva (WI).

Dembo, M., & Gibson, S. (1985). Teachers' sense of efficacy: An important factor in school achievement. *The Elementary School Journal, 86* (2), 173-184.

Dutton, M. (1990). *An investigation of the relationship between training in cooperative learning and teacher job satisfaction.* Unpublished doctoral dissertation, Portland State University, Portland, OR.

Ebmeier, H., & Hart, A. (1992). The effects of a career-ladder program on school organizational process. *Educational Evaluation and Policy Analysis, 14,* 261-281.

Edwards, J., & Newton, R. (1995, April). *The effects of cognitive coaching on teacher efficacy and empowerment.* Paper presented at the annual meeting of the American Educational Research Association, San Francisco.

Enochs, L. G., & Riggs, I. M. (1990). Further development of an elementary science teaching efficacy belief instrument: A preservice elementary scale. *School Science and Mathematics, 90* (8), 694-706.

Enochs, L. G., Scharmann, L., & Riggs, I. M. (1995). The relationship of pupil control to preservice elementary science teacher self-efficacy and outcome expectancy. *Science Education, 79* (1), 63-75.

Enochs, L. G., Schug, M., & Cross, B. (1996, April). *The development and validation of the economics teacher beliefs instrument (ETEBI).* Paper presented at the annual meeting of the American Educational Research Association, New York.

Evans, E., & Tribble, M. (1986). Perceived teaching problems, self-efficacy, and commitment to teaching among preservice teachers. *Journal of Educational Research, 80* (2), 81-85.

Fletcher, S. (1990). *The relation of the school environment to teacher efficacy.* Paper presented at the annual meeting of the American Psychological Association, Boston, MA (ERIC Document Reproduction Service No. ED 329 551).

Flowers, M. (1988). *The role of preservice personal teaching efficacy in predicting microteaching success (initiating a discussion).* Unpublished doctoral dissertation, University of Akron, Akron, OH. UMI 8817642.

Fritz, J., Miller-Heyl, J., Kreutzer, J., & MacPhee, D. (1995). Fostering personal teaching efficacy through staff development and classroom activities. *Journal of Educational Research, 88* (4), 200-208.

Fuller, B., & Izu, J. (1986). Explaining school cohesion: What shapes the organizational beliefs of teachers. *American Journal of Education, 94,* 501-535.

Gibson, S., & Dembo, M. (1984). Teacher efficacy: A construct validation. *Journal of Educational Psychology, 76* (4), 569-582.

Ginns, I., & Watters, J. (1996, April). *Science teaching self-efficacy of novice elementary school teachers*. Paper presented at the annual meeting of the American Educational Research Association, New York.

Good, T., & Brophy, J. (1990). *Looking in Classrooms* (5th ed.). New York: Harper Collins.

Greenwood, G., Olejnik, S., & Parkay, F. (1990). Relationships between four teacher efficacy belief patterns and selected teacher characteristics. *Journal of Research and Development in Education, 23* (2), 102-107.

Guskey, T. (1982). Differences in teachers' perceptions of personal control of positive versus negative student learning outcomes. *Contemporary Educational Psychology, 7*, 70-80.

Guskey, T. (1984). The influence of change in instructional effectiveness upon the affective characteristics of teachers. *American Educational Research Journal, 21* (2), 245-249.

Guskey, T. (1988). Teacher efficacy, self-concept, and attitudes toward the implementation of instructional innovation. *Teaching and Teacher Education, 4* (1), 63-69.

Guskey, T., & Passaro, P. (1993). *Teacher efficacy: A study of construct dimensions*. Paper presented at the annual meeting of the American Educational Research Association, Atlanta.

Guyton, E., Fox, M., & Sisk, K. (1991). Comparison of teaching attitudes, teacher efficacy, and teacher performance of first year teachers prepared by alternate and traditional teacher education programs. *Action in Teacher Education, 13* (2), 1-7.

Hall, B., Burley, W., Villeme, M., & Brockmeier, L. (1992). *An attempt to explicate teacher efficacy beliefs among first year teachers*. Paper presented at the annual meeting of the American Educational Research Association, San Francisco.

Hall, B., Hines, C., Bacon, T., & Koulianos, G. (1992). *Attributions that teachers hold to account for student success and failure and their relationship to teaching level and teacher efficacy beliefs.* Paper presented at the annual meeting of the American Educational Research Association, San Francisco.

Haury, D. (1989). The contribution of science locus of control orientation to expressions of attitude toward science teaching. *Journal of Research in Science Teaching, 26* (6), 503-517.

Hipp, K. (1996, April). *Teacher efficacy: Influence of principal leadership behavior*. Paper presented at the annual meeting of the American Educational Research Association, New York.

Hipp, K., & Bredeson, P. (1995). Exploring connections between teacher efficacy and principals' leadership behaviors. *Journal of School Leadership, 5*, 136-150.

Hoover-Dempsey, K., Bassler, O., Brissie, J. (1987). Parent involvement: Contributions of teacher efficacy, school socioeconomic status, and other school characteristics. *American Educational Research Journal, 24* (3), 417-435.

Hoover-Dempsey, K., Bassler, O., & Brissie, J. (1992). Explorations in parent-school relations. *Journal of Educational Research, 85* (5), 287-294.

Housego, B. (1990). A comparative study of student teachers' feelings of preparedness to teach. *Alberta Journal of Educational Research, 36* (3), 223-240.

Hoy, W., & Woolfolk, A. (1990). Socialization of student teachers. *American Educational Research Journal, 27* (2), 279-300.

Hoy, W., & Woolfolk, A. (1993). Teachers' sense of efficacy and the organizational health of schools. *Elementary School Journal, 93* (4), 355-372.

Huberman, M. (1989). The professional life cycle of teachers.*Teachers College Record, 91* (3), 31-57.

Huinker, D., & Madison, S. (1995, April). *Impact of methods courses on preservice elementary teachers' science and mathematics teaching efficacy*. Paper presented at the annual meeting of the American Educational Research Association, San Francisco.

Imants, J., & de Brabander, C. (1996). Teachers' and principals' sense of efficacy in elementary schools. *Teaching and Teacher Education, 12* (2), 179-195.

Imants, J., & Van Zoelen, A. (1995). Teachers' sickness absence in primary schools, school climate and teachers' sense of efficacy.*School Organization, 15* (1), 77-86.

Jordan, A., Kircaalilftar, G., & Diamond, P. (1993). Who has a problem, the student or the teacher? Differences in teachers' beliefs about their work with atrisk and integrated exceptional students .*International Journal of Disability, Development and Education, 40* (1), 45-62.

Kalaian, H., & Freeman, D. (1990). Relations between teacher candidates' self-confidence and educational beliefs. *College Student Journal, 23,* 296-303.

Kalaian, H., & Freeman, D. (1994). Gender differences in self-confidence and educational beliefs among secondary teacher candidates. *Teaching and Teacher Education, 10* (6), 647-658.

Korevaar, G. (1990). *Secondary school teachers' courses of action in relation to experience and sense of self-efficacy.* Paper presented at the annual meeting of the American Educational Research Association, Boston.

Lee, M., Buck, R., & Midgley, C. (1992). *The organizational context of personal teaching efficacy.* Paper presented at the annual meeting of the American Educational Research Association, San Francisco.

Lee, V., Dedrick, R., & Smith, J. (1991). The effect of the social organization of schools on teachers' efficacy and satisfaction. *Sociology of Education, 64,* 190-208.

Louis, K. S. (1991). *The effects of teacher quality of work life in secondary schools on commitment and sense of efficacy.* Paper presented at the annual meeting of the American Educational Research Association, Minneapolis.

Louis, K. S., Marks, H., & Kruse, S. (1994, April). *Teachers' professional community in restructuring schools.* Paper presented at the annual meeting of the American Educational Research Association, New Orleans.

Lubbers, J. (1990). *An investigation to determine if principal behaviors can impact teacher efficacy.* Unpublished doctoral dissertation, Michigan State University, UMI 9028670.

Lucas, K., Ginns, I., Tulip, D., & Watters, J. (1993). *Science teacher efficacy, locus of control and self-concept of Australian perservice elementary school teachers.* Paper presented at the annual meeting of the National Association for Research in Science Teaching, Atlanta.

Marx, R., Blumenfeld, P., Krajcik, J., Blunk, M., Crawford, B., Kelly, B., & Meyer, K. (1994). Enacting project-based science: Experiences of four middle grade teachers.*Elementary School Journal, 94* (5), 517-538.

Meijer, C., & Foster, S. (1988). The effect of teacher self-efficacy on referral chance.*Journal of Special Education, 22* (3), 378-385.

Meyerson, M. (1995, April). *Naturalistic assessment: Teachers' concerns and confidence.* Paper presented at the annual meeting of the American Educational Research Association, San Francisco.

Midgley, C., Feldlaufer, H., & Eccles, J. (1988). The transition to junior high schools: Beliefs of pre- and post-transition teachers. *Journal of Youth and Adolescence, 17,* 543-562.

Midgley, C., Feldlaufer, H., & Eccles, J. (1989). Change in teacher efficacy and student self- and task-related beliefs in mathematics during the transition to junior high school. *Journal of Educational Psychology, 81* (2), 247-258.

Miskel, C., McDonald, D., & Bloom, S. (1983). Structural and expectancy linkages within schools and organizational effectiveness. *Educational Administration Quarterly, 19* (1), 49-82.

Moore, P. (1990). *The effect of science inservice programs on the self-efficacy belief of elementary school teachers.* Unpublished doctoral dissertation, University of San Diego. University Microfilms International 90-23069.

Moore, W., & Esselman, M. (1992). *Teacher efficacy, empowerment, and a focussed instructional climate: Does student achievement benefit?* Paper presented at the annual meeting of the American Educational Research Association, San Francisco.

Moore, W. P., & Esselman, M. E. (1994, April). *Exploring the context of teacher efficacy: The role of achievement and climate.* Paper presented at the annual meeting of the American Educational Research Association, New Orleans.

Morrison, G., Walker, D., Wakefield, P., & Solberg, S. (1994). Teacher preferences for collaborative relationships: Relationship to efficacy for teaching in preventionrelated domains. *Psychology in the Schools, 31,* 221-231.

Newmann, F., Rutter, R., & Smith, M. (1989). Organizational factors that affect school sense of efficacy, community, and expectations. *Sociology of Education, 62,* 221-238.

Ohmart, H. (1992). *The effects of an efficacy intervention on teachers' efficacy feelings.* Unpublished doctoral dissertation, University of Kansas, Lawrence, Kansas. UMI 9313150.

Parkay, F., Olejnik, S., & Proller, N. (1988). A study of relationships among teacher efficacy, locus of control, and stress. *Journal of Research and Development in Education, 21* (4), 13-22.

Podell, D., & Soodak, L. (1993). Teacher efficacy and bias in special education referrals.*Journal of Educational Research, 86* (4), 247-253.

Raudenbush, S., Rowen, B., & Cheong, Y. (1992). Contextual effects on the self-perceived efficacy of high school teachers. *Sociology of Education, 65,* 150-167.

Rees, J. (1986). *A study of the relationship between specific job characteristics and teacher efficacy.* Unpublished doctoral dissertation, University of Cincinnati, Cincinnati, OH.

Riehl, C., & Sipple, J. (1996). Making the most of time and talent: Secondary school organizational climates, teaching task environments, and teacher commitment. *American Educational Research Journal, 33* (4), 873-901.

Riggs, I. (1991). *Gender differences in elementary science teacher self-efficacy.* Paper presented at the annual meeting of the American Educational Research Association, Chicago, IL (ERIC Document Reproduction Service No. ED 340 705).

Riggs, I., & Enochs, L. (1990). Toward the development of an elementary teacher's science teaching efficacy belief instrument. *Science Education, 74* (6), 625-638.

Robardey, C., Allard, D., & Brown, D. (1994). An assessment of the effectiveness of full option science system training for third- through sixth-grade teachers. *Journal of Elementary Science Education, 6* (1), 17-29.

Roeser, R., Arbreton, A., & Anderman, E. (1993). *Teacher characteristics and their effects on student motivation across the school year.* Paper presented at the annual meeting of the American Educational Research Association, Atlanta.

Rose, J., & Medway, F. (1981). Measurement of teachers' belief in their control over student outcomes. *Journal of Educational Research, 74,* 185-190.

Rosenholtz, S. (1987). Education reform strategies: Will they increase teacher commitment? *American Journal of Education, 95* (4), 534-562.

Rosenholtz, S. (1989). *Teachers' workplace: The social organization of schools.* New York: Longman.

Ross, J. (1992). Teacher efficacy and the effect of coaching on student achievement. *Canadian Journal of Education, 17* (1), 51-65.

Ross, J. (1994). The impact of an in-service to promote cooperative learning on the stability of teacher efficacy. *Teaching and Teacher Education, 10* (4), 381-394.

Ross, J. (1995). Strategies for enhancing teachers' beliefs in their effectiveness: Research on a school improvement hypothesis. *Teachers College Record, 97* (2), 227-251.

Ross, J., & Cousins, J. B. (1993). Enhancing secondary school students' acquisition of correlational reasoning skills. *Research in Science & Technological Education, 11* (3), 191-206.

Ross, J., Cousins, J. B. & Gadalla, T. (1996). Within-teacher predictors of teacher efficacy. *Teaching and Teacher Education, 12*(4), 385-400.

Ross, J., McKeiver, S., & Hogaboam-Gray, A. (1997). Fluctuations in teacher efficacy during the implementation of destreaming. *Canadian Journal of Education., 22* (3), 283-296.

Ross, J., & Regan, E. (1993). Sharing professional experience: Its impact on professional development.*Teaching and Teacher Education, 9* (1), 91-106.

Ross, J., Rolheiser, C., & Hoagaboam-Gray, A. (in press). Student evaluation in cooperative learning: Teacher cognitions. *Teachers and Teaching.*

Rotter, J. (1966). Generalized expectancies for internal versus external control of reinforcement. *Psychological Monographs, 80* (1), 1-28.

Rubeck, M., & Enochs, L. (1991). *A path analytic model of variables that influence science and chemistry teaching self-efficacy and outcome expectancy in middle school science teachers.* Paper presented at the annual meeting of the National Association for Research in Science Teaching, Lake Geneva, WI.

Saklofske, D., Michayluk, J., & Randhawa, B. (1988). Teachers' efficacy and teaching behaviors. *Psychological Reports, 63*, 407-414.

Smylie, M. (1988). The enhancement function of staff development: Organizational and psychological antecedents to individual teacher change. *American Educational Research Journal, 25* (1), 1-30.

Soodak, L., & Podell, D. (1993a). Teacher efficacy and student problem as factors in special education referral. *Journal of Special Education, 27*, 66-81.

Soodak, L., & Podell, D. (1993b). *Teacher efficacy and bias as factors in special education referral.* Paper presented at the annual meeting of the American Educational Research Association, Atlanta.

Stein, M., & Wang, M. (1988). Teacher development and school improvement: The process of teacher change. *Teaching and Teacher Education, 4* (2), 171-187.

Tracs, S., & Gibson, S. (1986). *Effects of efficacy on academic achievement.* Paper presented at the annual meeting of the California Research Association, Marina del Rey.

Trentham, L., Silvern, S., & Brogdon, R. (1985). Teacher efficacy and teacher competency ratings. *Psychology in the Schools, 22*, 343-352.

Vitale, M., & Romance, N. (1992). Using videodisc instruction in an elementary science methods course: Remediating science knowledge deficiencies and facilitating science teaching attitudes. *Journal of Research in Science Teaching, 29* (9), 915-928.

Vitali, G. (1993). *Factors influencing teachers' assessment and instructional practices in an assessment-driven educational reform.* Unpublished doctoral dissertation, University of Kentucky, Lexington, Kentucky.

Volkman, B., Scheffler, A., & Dana, M. (1992). *Enhancing preservice teachers' self-efficacy through a field-based program of reflective practice.* Paper presented at the annual meeting of the Mid-South Educational Research Association, Knoxville, Tennessee.

Warren, L., & Payne, B. (1995, April). *The impact of middle grades schools' organizational patterns on teachers' sense of efficacy and perceptions of the working environment.* Paper presented at the annual meeting of the American Educational Research Association, San Francisco.

Watson, S. (1991). Cooperative learning and group educational modules: Effects on cognitive achievement of high school biology students. *Journal of Research in Science Teaching, 28* (2), 141-146.

Watters, J., & Ginns, I. (1995, April). *Origins of, and changes in preservice teachers' science teaching self-efficacy.* Paper presented at the annual meeting of the National Association for Research in Science Teaching, San Francisco.

Weinstein, C. (1988). Preservice teachers' expectations about their first year of teaching. *Teaching and Teacher Education, 4* (1), 31-40.

Woolfolk, A., & Hoy, W. (1990). Prospective teachers' sense of efficacy and beliefs about control. *Journal of Educational Psychology, 82* (1), 81-91.

Woolfolk, A., Rosoff, B., & Hoy, W. (1990). Teachers' sense of efficacy and their beliefs about managing students. *Teaching and Teacher Education, 6* (2), 137-148.

TEACHER EFFICACY AND THE VULNERABILITY OF THE DIFFICULT-TO-TEACH STUDENT

Leslie C. Soodak and David M. Podell

Individuals select the teaching profession for a variety of reasons, but one of the reasons most often given is a desire to nurture children and help them learn. However, what happens when teachers are frustrated or even stymied in their efforts to help students? Even the most effective teachers encounter frustration and sometimes failure, and even the most optimistic and confident teachers must face threats to their expectations of student success. The greatest challenge to teachers' expectations of themselves comes from difficult-to-teach students, that is, students who demonstrate significant learning or behavior problems or, in some cases, a combination of both. Because of their difficulties, these students are at the greatest risk for failing in school, being referred for special education services, or being placed outside the regular education environment.

How do teachers respond to situations in which difficult-to-teach students defy their normative expectations? Routinely, teachers make decisions based on their expectations of the likelihood that one choice will be more effective than another.

These decisions may occur in moment-by-moment decisions, such as deciding which student behavior to address and which to ignore, planning decisions, such as grouping and goal setting, and potentially life-changing decisions, such as making referrals of students for evaluation for possible special education placement.

Underlying their decision making are teachers' beliefs about teaching. These beliefs include the conviction that teaching indeed has a genuine impact on students and that teachers themselves can create change in students. Researchers have explored the role of these beliefs, which collectively have been labeled *teacher efficacy*, on teachers' decision making; in this chapter we examine this body of research to determine how teachers' beliefs about the impact of teaching influence the decisions they make about students and, in particular, difficult-to-teach students.

Much of what is known about the effects of teachers' beliefs about themselves and their students suggests that difficult-to-teach students challenge teachers' views of their own efficacy. Bandura (1977) has suggested that individuals' beliefs about their ability to reach a goal influence the likelihood of initiating and persisting in the behaviors directed toward achieving the goal, particularly when the outcomes are difficult to reach. The hypothesized relation among efficacy beliefs and teacher behavior has been supported by research on teachers' responses to students of differing abilities. In a six-year study, Ashton and Webb (1986) found that teachers who had a high sense of their own efficacy tended to have a sense of personal responsibility for students' learning and take pride in student outcomes. Teachers with a low sense of their own efficacy were more likely to attribute outcomes to students' ability and home environments and were less likely to take responsibility for student performance. In addition, teachers with low efficacy spent less time interacting with low-achieving students. Gibson and Dembo (1984) found that, as compared to teachers with high efficacy, teachers with low efficacy were less persistent with low achievers.

Thus, low-achieving students may be at risk when paired with teachers who do not feel efficacious. These teachers may deprive their low-achieving students of much-needed time and attention. Because teachers with high efficacy are more likely to attribute student success and failure to their own efforts, these teachers may view difficult-to-teach students as a challenge rather than as a threat (Hall, Hines, Bacon, & Koulianos, 1992).

To understand the relation between teachers' efficacy beliefs and their decisions about difficult-to-teach students, we begin with an examination of the teacher efficacy construct, which comes largely from the theory of self-efficacy proposed by Bandura (1977) from a social learning paradigm. We describe Bandura's ideas regarding self-efficacy and we then explore how these ideas have been applied to teachers. Specifically, we examine the dimensions of teacher efficacy and its development along the continuum of professional experience. We then synthesize the research conducted by ourselves and others on the relation of teacher efficacy and the decisions teachers make about difficult-to-teach students. Finally, we end by identifying implications of this body of research and directions for future studies.

BANDURA'S THEORY OF SELF-EFFICACY

Bandura theorizes that an individual's behavior is the product, not purely of immediate outcomes (i.e., reinforcements), but of aggregate consequences; the expectation that particular circumstances are necessary for certain outcomes to be reached mediates the power of a reinforcer. Expectations, of which Bandura differentiates two types, drive the behavior of individuals. The first type of expectation is an outcome expectation, that is, the conviction that a particular behavior leads to a specific outcome. For teachers, this might be the belief that ignoring undesirable acts (a behavior) leads to their extinction (an outcome). Second, an efficacy expectation is the belief that one is able to demonstrate a behavior necessary to produce the desired outcome. For teachers, this might be the belief that one is capable of successfully ignoring undesirable acts. Bandura notes that this distinction is important because individuals may recognize the link between a behavior and an outcome, but may not believe that they themselves are capable of performing the necessary behavior.

These beliefs, collectively referred to as an individual's *self-efficacy*, influence three important outcomes: the tendency to initiate behavior, the degree of effort that will be exerted in executing a behavior, and the extent to which the behavior will be sustained when obstacles are encountered. The application of these outcomes to teaching difficult-to-teach students will be explored later in the chapter.

Of the two expectations comprising self-efficacy, Bandura (1977) examines most closely the characteristics and sources of efficacy expectations. He notes that efficacy expectations vary in their magnitude (the level of task difficulty that they encompass), their generality (the extent to which they are applied to specific situations or to a larger set of situations), and their strength (the extent to which they can withstand disconfirming experiences). Efficacy expectations have four sources. First, individuals are particularly influenced by their own performance accomplishments. When they have had repeated successes, an occasional failure is less likely to have a serious negative impact on their expectations. Bandura (1977) notes that "indeed, occasional failures that are later overcome by determined effort can strengthen self-motivated persistence if one finds through experience that even the most difficult obstacles can be mastered by sustained effort" (p. 195). Second, individuals develop efficacy expectations vicariously. When others succeed at a task, individuals draw the inference that they too are able to act in such a way as to reach the desired outcome. Because this source relies on second-hand learning, Bandura notes that its impact is less than that of the first source, personal accomplishment. A third source of efficacy expectations is verbal persuasion: through suggestion, individuals are encouraged in the belief that they can successfully perform actions. Like vicarious learning, verbal persuasion, or exhortation, is a weak influence on efficacy beliefs because individuals tend to draw more heavily on actual personal experience. Fourth, individuals' efficacy expectations are influenced by their emotional and physiological arousal. When individuals experience the physical manifestations of anxiety, they are judging their own coping ability and applying these judgments to their future capacity. Bandura notes that, from a social learning perspective, the experiencing of arousal is both informative and motivating for an individual; arousal may, in fact, energize an individual to engage in a particular action.

The Role of Efficacy Beliefs in the Self System

While Bandura initially applied his theory of self-efficacy to different modes of psychotherapeutic treatment, he more recently has explored its implications for cognitive functioning (Bandura, 1993). Specifically, he contends that self processes mediate human selection and construction of the world; in other words, individuals give meaning and weight to events in their environment through the filter of their beliefs about themselves. In his more recent work, Bandura applies his theory of self-efficacy to four processes: cognitive, motivational, affective, and selection.

Goals, he argues, influences cognitive processes and individuals' self-appraisal of their ability to achieve the goals influences their selection of goals. Those with high self-efficacy set higher goals and commit themselves more resolutely to achieving them. They stay focused on the task before them despite the demands of difficult situations and the possible negative social judgments associated with failure. Influencing one's efficacy beliefs are other belief systems, such as one's beliefs concerning the sources and malleability of one's ability. If individuals believe ability is innate or unmalleable, or if they believe ability declines over time, they will be less likely to harness their abilities to meet challenges that they face. In addition, their beliefs about their performance relative to a reference group influence their self processes. When one perceives the performance of a reference group as exceeding one's own performance, one's sense of efficacy declines as does one's performance. In addition to the effect of other belief systems on self-efficacy, the form of feedback that one receives also has bearing on one's efficacy beliefs. Specifically, feedback emphasizing progress enhances efficacy beliefs and feedback focusing on shortcomings diminishes efficacy beliefs.

In addition to their role in cognitive processes, efficacy beliefs also influence motivation. Beliefs about one's own ability influence goals, the expenditure of effort, one's perseverance when faced with difficulties, and the extent to which one is affected by failure. In particular, Bandura notes that failure does not debilitate people with a strong sense of self-efficacy but, rather, they use failure to spur themselves on to meet difficult challenges. Later in this chapter we will examine how teachers with a strong sense of their own competency react to academic failure among their students.

Individuals' self-efficacy additionally influences their affective processes. Bandura (1988) contends that individuals with high self-efficacy believe that they can deal with threats and are thus unlikely to feel anxiety and stress. Those who doubt their ability to meet threats effectively experience stress and anxiety, both subjectively and physiologically. Interestingly, Bandura believes that efficacy beliefs influence one's tendency to have disturbing thoughts. An aspect of having a positive sense of one's own capability, he claims, is that one believes that thought processes can be controlled and the frequency of having disturbing thoughts can be diminished. Further, lacking a sense of efficacy causes students to feel anxious in academic situations; failure, which reduces their sense of efficacy, causes future failure through the mediating factor of anxiety. Depression is another possible outcome of low self-efficacy through three possible routes: unfulfilled aspirations, failure to seek out social relationships that help an individual deal with stress, and failure to diminish the occurrence of negative, self-defeating thoughts.

Lastly, self-efficacy influences the individual's selection processes, specifically one's selection of activities and social environments. Bandura contends that individuals' sense of their own effectiveness helps to determine their choice of activities, such that those with a low sense of self-efficacy tend to avoid choices that they think will lead to failure. He contends that individuals with higher self-efficacy have greater career options, display greater interest in their work, prepare themselves better for their careers, and persevere and succeed in their work to a greater degree.

SELF-EFFICACY AMONG TEACHERS

The application of Bandura's theory of self-efficacy to teachers began in the late 1970s and has continued to the present; this body of research has examined the development of teacher efficacy as a theoretical construct, the measurement of the construct, exploration of the dimensions of the construct, and the association of teacher efficacy with specific instructional beliefs, practices, and outcomes.

Development of the Teacher Efficacy Construct

An early model of teacher efficacy was developed by Denham and Michael (1981), who hypothesized that teachers' efficacy beliefs mediate their preexisting convictions and measurable consequences, namely teacher behaviors and student outcomes. They identified two aspects of teachers' efficacy beliefs, the belief that teachers in general can bring about change in students and the belief that one can bring about change in one's own students.

This model was elaborated upon and tested by Ashton and Webb (1982, 1986), whose research supported the two-dimensional structure of teacher efficacy and found specific correlates of the two dimensions. For example, they found that teachers with strong beliefs in the efficacy of teaching were more likely to give individual attention to students and show nonverbal signs of acceptance. In addition, teachers with strong beliefs in the effectiveness of their own teaching were more likely to create a more accepting classroom environment and support student initiative.

A third major development in establishing the teacher efficacy construct was the research of Gibson and Dembo (1984) who conducted a three-phase study. The purpose of the first phase was to determine the dimensions of teacher efficacy and the relation of these dimensions to Bandura's notion of self-efficacy. In the second phase they sought to validate the construct of teacher efficacy using a multitrait-multimethod analysis and, in the third phase, they examined differences among teachers with high and low levels of efficacy.

In their first phase Gibson and Dembo used factor analysis to establish the dimensions of teacher efficacy. Two hundred and eight teachers completed a 30-item teacher efficacy survey using a Likert scale. Using a high criterion for factor loadings to determine inclusion of items in their factors (loadings greater than .45), Gibson and Dembo reduced their scale to the 16 items that loaded on either of the two factors that they felt provided the best solution. They found that the two factors were only slightly correlated ($r = -.19$), suggesting that they are discrete entities. The first factor, which accounted for 18.2 percent of the variance, contained items that appeared to relate to teachers' beliefs about their own ability to influence student outcomes. The highest loading item (.61) was "If a student masters a new math concept quickly, this might be because I knew the necessary steps in teaching that concept." Gibson and Dembo consequently labeled this factor *personal teaching efficacy*.

The second factor, which accounted for 10.6 percent of the variance, included items that appeared to relate to the influence of teaching in general on student outcomes. The highest loading item on this factor (.65) was "A teacher is very limited in what he/she can achieve because a student's home environment is a large influence on his/her achievement." This factor was labeled *teaching efficacy*. Gibson and Dembo concluded that these two factors support the two-factor model of teacher efficacy of Ashton and Webb (1982). Overall, they concluded that these two factors were applications to teachers of the outcome expectations and efficacy expectations theorized by Bandura (1977).

In their second phase Gibson and Dembo used a multitrait-multimethod analysis to determine whether evidence regarding teacher efficacy from other sources related to the scale, and whether other constructs pertaining to teachers could be differentiated from teacher efficacy. They employed an alternate measure of teacher efficacy (a "more open-ended" measure in which the teachers selected from a list of variables those that pertained most to student success), two measures of verbal facility, and two measures of flexibility, in each case, one open-ended and one closed-ended. Fifty-five teachers who were also graduate education students were participants in the study. Gibson and Dembo found significant correlations between differing measures of teacher efficacy, as well as discriminability between teacher efficacy and the other two constructs (verbal facility and flexibility) which had low intercorrelations, regardless of method of measurement.

In their third phase Gibson and Dembo explored the relation between teacher efficacy and patterns of teacher behavior in actual classrooms. Using a sample of eight teachers, they examined teachers' use of time in academic tasks (e.g., "whole class," "small group") and nonacademic tasks (e.g., "daily rituals," "transition"). Interestingly, they found a significant difference between high-efficacy and low-efficacy teachers' use of small-group instruction, with low-efficacy teachers using small-group instruction almost twice as much as high-efficacy teachers. They found another significant difference in the use of intellectual games (an activity that Gibson and Dembo classified as nonacademic). Low-efficacy teachers spent two percent of their time in such games, while high-efficacy teachers did not use this method at all. Gibson and Dembo additionally examined teachers' responses to students' answers to questions. They found that, following incorrect responses, low-efficacy teachers sometimes followed up with criticism, a response that none of the high-efficacy teachers demonstrated. In addition, low-efficacy teachers more often followed incorrect responses by moving on to other students or providing the correct answer themselves, while high-efficacy teachers more often provided students with another opportunity to find the correct answer. Although these responses were based on a very small sample of teachers, they support the conclusion that teacher efficacy and teacher behavior are related.

The two-dimensional model supported by Ashton and Webb (1982) and Gibson and Dembo (1984) became the primary way in which teacher efficacy was examined in subsequent studies. Research differed, however, in the way in which the two dimensions were measured.

Measurement of Teacher Efficacy

Early studies on teacher efficacy used a two-item scale believed to measure the two dimensions of teacher efficacy (Armor, Conroy-Osequera, Cox, King, McDonnell, Pascal, Pauly, & Zellman, 1976; Berman & McLaughlin, 1977). These two items, generally referred to as the Rand items, are as follows:

1. When it comes right down to it, a teacher really can't do much because most of a student's motivation and performance depends on his or her home environment.
2. If I really try hard, I can get through to even the most difficult or unmotivated students (Berman, McLaughlin, Bass, Pauly, & Zellman, 1977, pp. 159-160).

Respondents indicate their degree of agreement to these items on a Likert scale. However, as noted by Ashton and Webb (1986), the Rand items are psychometrically limited; they tend to elicit little variability in teachers' responses, reducing the likelihood that researchers will detect relations between teacher efficacy and other variables. Further, the reliability estimates are low, ranging from .33 to .51.

Gibson and Dembo's (1984) study established both the validity of the construct and of their 16-item scale, which had considerably higher reliability (Cronbach's alphas of .78 for the personal efficacy factor, .75 for the teaching efficacy factor, and

.79 for the total scale). Many subsequent studies used the Gibson and Dembo scale, although others, such as Greenwood, Olejnik, and Parkey (1990), continued to use the two-item Rand scale. A small body of research, however, has suggested that the construct might be more complex than first thought and that the interpretation of the factors requires further discussion.

Exploration of the Dimensions of the Teacher Efficacy Construct

The notions that teacher efficacy exists as a two-dimensional construct and that these two dimensions are personal efficacy and teaching efficacy have been challenged by recent research. Guskey (1988), questioning the interpretation of these two dimensions, correlated the efficacy beliefs of 120 teachers, as measured by the two Rand items, with their responses to his own Responsibility for Student Achievement Scale (Guskey, 1981). Two subscales measuring teachers' own expressed responsibility for student success and student failure comprise the scale. Guskey found moderate but significant correlations between teacher efficacy and both of his subscales, suggesting that the personal efficacy dimension might be more accurately divided into two separate dimensions: teachers' beliefs regarding their own ability as it relates to positive and negative student outcomes.

Guskey and Passaro (1994) examined this possibility in a follow-up study of 283 practicing teachers and 59 preservice teachers who completed an adaptation of the Gibson and Dembo (1984) scale. They altered some of the items to ensure that the scale included items that were Personal-External and also negative ("I cannot") and General Teaching-Internal and also positive ("Teachers can"). Guskey and Passaro's factor analysis differentiated negative and internal items from those which were positive and internal. They concluded that Gibson and Dembo confounded the personal versus teaching efficacy dimension with the internal-external dimension, which they likened to the locus-of-control variable used in attribution theory.

A few significant concerns emerge, however, from their interpretation of the data. First, most theorists perceive the locus-of-control variable as a single dimension with internal and external being two ends of a continuum. A second issue is that Guskey and Passaro themselves confounded the internal-external and positive-negative dimensions. Third, Guskey and Passaro imposed a two-factor solution on their data, preventing the possible detection of a model of teacher efficacy that differed from that of Gibson and Dembo. One wonders what Guskey and Passaro might have found had they tested for a three-factor or even a four-factor solution.

Woolfolk and Hoy (1990) were the first to find evidence of a third factor of teacher efficacy which differentiated two types of personal efficacy: teachers' beliefs regarding their responsibility for *positive* student outcomes and for *negative* student outcomes. However, for the purpose of their investigation of correlates of teacher efficacy among prospective teachers, they used the traditional two-factor (personal efficacy/teaching efficacy) distinction.

Interestingly, Woolfolk and Hoy noted a possible discrepancy between how theorists have viewed *teacher efficacy* and how Bandura viewed *self-efficacy*. They observed that theorists and researchers of teacher efficacy have concluded that personal efficacy relates to Bandura's notion of efficacy expectations, while teaching efficacy relates to outcome expectations. However, they pointed out that teaching efficacy is not in fact an outcome expectation because it concerns the ability of teachers to overcome outside influences, not their judgments about the outcomes of their behaviors. They concluded that teaching efficacy, like personal efficacy, is actually an efficacy expectation.

However, the two dimensions of personal efficacy that Woolfolk and Hoy found may in fact reflect the efficacy expectation and outcome expectation components of Bandura's original model. The factor that Woolfolk and Hoy characterized as personal efficacy for positive outcomes includes items such as "When a student gets a better grade than he/she usually gets, it is usually because I found better ways of teaching that student." We suspected that these items could more appropriately be considered as being congruent with Bandura's notion of outcome expectation.

We investigated this possibility in a study of 310 practicing teachers (Soodak & Podell, 1996). We added items to the Gibson and Dembo scale to ensure that positive and negative outcomes were equally represented. We also added items to explore whether teachers' beliefs about their efficacy for learning outcomes were distinct from their efficacy beliefs related to student behavior problems. Finally, to investigate whether the teaching efficacy factor related only to the home (which had been the subject of the original Gibson and Dembo items), we added items relating to other external influences, including heredity, diet, and television violence.

When we imposed a two-factor solution in a factor analysis, our results were similar to those of Gibson and Dembo; however, we found that a three-factor solution yielded the best fit for our data. Table 1 lists the items contributing significantly to the three factors (factor loadings greater than .35). Further, the correlations among the three factors were low or zero.

The first and third factors included items that typically loaded on the traditional personal efficacy factor. Analysis of the items loading on the first factor revealed a distinct pattern. Without exception, these items pertained to teachers' beliefs about their ability to perform the actions needed to promote learning or manage student behavior successfully; we therefore labeled this factor *personal efficacy*.

The four items loading heavily on our third factor were rather different; they pertained to specific student outcomes and how a teacher's behavior can influence them; we consequently labeled this factor *outcome efficacy*.

By dividing the traditional personal efficacy factor into two separate factors, we detected two group differences in our sample that otherwise would have been obscured. First, experienced teachers had greater personal efficacy, using the traditional construction of the factor, than inexperienced teachers. However, when personal efficacy and outcome efficacy were separated, experienced teachers had greater personal efficacy than their inexperienced peers, but the two groups did not

Table 1. Items and Factor Loadings on the Three-Factor Solution

Item	Factor loading
Personal Efficacy	
If my students are difficult to control on a particular day, I am able to gain control of the class.	.68
I can effectively manage my students' behavior.	.65
If a student in my class becomes disruptive and noisy, I feel assured I know some techniques to redirect him quickly.	.64
I can usually help students who show hyperactive behavior.	.61
When a student is verbally abusive to another student, I can usually prevent the conflict from escalating.	.60
If a student is withdrawn and isolated, I know ways of helping the student open up.	.58
If a student did not remember information I gave in a previous lesson, I would know how to increase his/her retention in the next lesson.	.54
When I really try, I can get through to most difficult students.	.48
When a student is having difficulty with an assignment, I am usually able to adjust it to his/her level.	.46
When a student is angry, there is little I can do.	-.41

Item	Factor Loading
Teaching Efficacy	
How much a student learns is influenced more by his or her genetic make-up than by teaching.	.66
Hereditary factors are more important than teaching in determining how well a student does in school.	.64
A teacher is very limited in what he/she can achieve because a student's home environment is a large influence on his/her achievements.	.63
The amount that a student can learn is primarily related to family background.	.61
If students aren't disciplined at home, they aren't likely to accept any discipline.	.48
The hours in my class have little influence on students compared to the influence of their home environment.	.48
Some students are simply uncontrollable.	.47
When students watch too much violence on television, teachers are unable to manage their behavior in the classroom.	.47
I am ill-prepared to help students with emotional problems.	.41
When a student does well in school, it is most likely due to the student's innate ability.	.41

(continued)

Table 1. (Continued)

Item	Factor loading
Teaching Efficacy	
Even a teacher with good teaching ability may not reach many students.	.39
If parents would do more with their children, I could do more.	.39
When a student is angry, there is little I can do.	.36

Item	Factor loading
Outcome Efficacy	
When a student gets a better grade than he usually gets, it is usually because I found better ways of teaching that student.	.76
When the grades of my students improve, it is usually because I found more effective teaching approaches.	.74
When a student does better than usual, many times it is because I exerted a little extra effort.	.56
If a student masters a new concept quickly, this might be because I knew the necessary steps in teaching that concept.	.54

differ in their outcome efficacy. A second difference was that preschool and elementary-level teachers had greater outcome efficacy than junior high school teachers, a difference that would not have been detected had the two factors been collapsed.

We concluded that personal efficacy pertains to teachers' beliefs that they possess particular teaching skills, while outcome efficacy refers to the belief that implementing these skills leads to desirable student outcomes. We further contended that these two aspects of teacher efficacy reflect Bandura's differentiation between efficacy expectations and outcome expectations.

Our examination of the teaching efficacy factor (our second factor) indicated that it was not limited to the influence of the home, but pertained to other external influences as well. Specifically, three items pertaining to the influence of heredity on students and one pertaining to television violence loaded significantly. However, two other items pertaining to the effects of television and two items pertaining to diet did not load on the factor. Interestingly, two items pertaining to extreme emotional or behavioral problems of students contributed to this factor. The presence of these items on the teaching efficacy factor suggests that teachers associate extreme emotional or behavioral problems with factors outside the sphere of teaching, implying that they may link these problems with the influence of the home.

We concluded that these findings suggest the need for continued exploration of the dimensions of the teacher efficacy construct. While the group differences that we detected support the notion that outcome efficacy is a meaningful dimension of teacher efficacy, further demonstration of the relation of outcome efficacy to teacher behavior would be necessary to validate the construct.

The issues of how many dimensions of teacher efficacy exist and their precise nature remain unresolved. However, increasing evidence has been emerging to demonstrate that teacher efficacy plays a critical role in the decision making and behavior of practicing teachers.

THE CRITICAL ROLE OF TEACHER EFFICACY

In their seminal study Gibson and Dembo (1984) found that teacher efficacy relates to teacher behaviors regarding grouping of students and selection of activities. Further research has demonstrated the relationship between teacher efficacy and a variety of other teacher variables. Ashton and Webb (1986), for example, found that teacher efficacy relates to teachers' instructional style. High-efficacy teachers tended to assume more responsibility for their students' learning; further, they were less likely to devote time to student seatwork or to turn control of activities over to students and they tended to have a warmer classroom environment. Those who did not have confidence in the effectiveness of teaching tended to use harsher methods to control students' behavior. In light of their findings, Ashton and Webb (1986) called for future research to focus on the social context of teachers' behavior to promote improvement of teaching and student outcomes.

Despite recognition of the need to examine teacher efficacy within a contextual framework, the vast majority of studies on teacher efficacy have focused on identifying its correlates. For example, studies have found that teachers who are high in personal efficacy also tend to have more humanistic beliefs about student control (Woolfolk & Hoy, 1990), have more positive self-concepts and are more amenable to instructional innovation (Guskey, 1988), and tend to use more effective methods of questioning, instruction, and classroom management (Saklofske, Michayluk, & Randhawa, 1988).

These studies examine teacher efficacy and its correlates at a specific point in time; however, Bandura (1977) noted the critical role of experience in the development of teacher efficacy. A small set of studies had examined the development of teacher efficacy as teachers join the profession. Hoy and Woolfolk (1990, 1993) examined the efficacy beliefs of preservice teachers before and after their practice teaching experience and found that personal efficacy increased, while teaching efficacy decreased. Dembo and Gibson (1985), on the other hand, found a decrease in personal efficacy after the final semester of student teaching, although they also found that teachers' personal efficacy increased with experience. They suggested that the decrease in personal efficacy among student teachers completing student teaching may reflect a diminution of their confidence as they face the prospect of responsibility for their own classroom. Finally, Glickman and Tamashiro (1982) compared first- and fifth-year teachers and found no differences in either personal efficacy or teaching efficacy.

Given these seemingly conflicting findings, we sought to explore the development of teacher efficacy across the career span, from preservice teachers who had not yet student taught to practicing teachers with many years of experience (Soodak & Podell, 1997). Using a cross-sectional design, we surveyed over 600 prospective and practicing teachers. Because their work experience is qualitatively different, we divided our subjects between those in elementary education and those in secondary education. We further divided our sample into seven groups: those taking initial teacher preparation courses with a fieldwork component, those doing student teaching, and practicing teachers with one-two, three-five, six-eight, nine-16, and more than 16 years of teaching experience. The Gibson and Dembo (1984) scale was used to measure the teachers' efficacy beliefs.

In both the elementary and secondary groups, beliefs about the power of teaching to overcome outside influences (teaching efficacy) were unrelated to experience. Beliefs about their own effectiveness as teachers did differ significantly, however, as illustrated in Figure 1. Among the elementary group, personal efficacy beliefs during fieldwork and student teaching were quite high, but these fell dramatically in the first years of teaching. With more years of teaching experience, personal efficacy scores improved, although they never reached the level seen among the preservice teachers. Secondary-level teachers demonstrated a similar (but not statistically significant) pattern, although their initial personal efficacy beliefs were not as high as those of elementary-level teachers, they dropped less dramatically, and they recovered sooner. Although the data are cross-sectional (rather than longi-

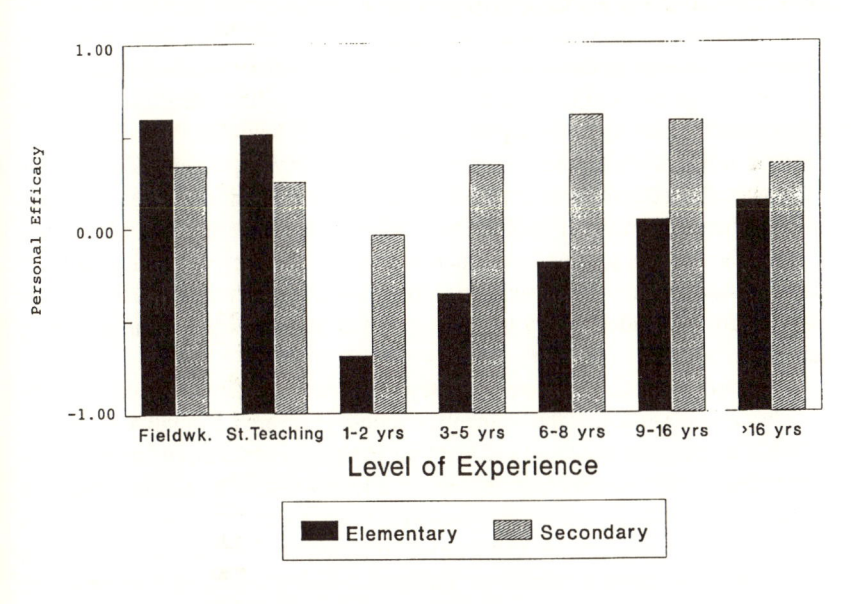

Figure 1. Personal Efficacy of Elementary and Secondary Teachers by Levels of Teaching Experience

tudinal, which would have been preferable although impractical) and the population was self-selected (some having dropped out of the profession altogether), the study reveals what is likely to be an important progression of efficacy beliefs over the career span and demonstrates the pivotal role of teaching experience.

The findings of the study have significant implications for the preparation and support of teachers. Preservice elementary teachers have very high and perhaps unrealistic expectations of their own impact. From such heights, they quickly lose confidence when faced with the demands of their own classrooms. These findings suggest that teacher preparation programs should place student teachers in realistic rather than ideal contexts to help them learn how to cope in classrooms that contain the myriad of problems that they will confront when they become teachers.

Further, these findings suggest that school districts should provide more direct support for new elementary teachers to help them make the transition from student teaching to classroom teaching. Mentoring programs and support groups are two mechanisms that might ease the disillusionment that appears to be associated with the first years of teaching. Given the relationship of personal efficacy to various teacher behaviors, the vulnerability of the new teacher challenges teacher educators and school administrators to provide beginning teachers with experiences that will allow them to confront their beliefs without losing their confidence.

Interestingly, the secondary-level teachers did not show the extreme highs and lows in efficacy beliefs that the elementary-level teachers did. In addition, the secondary group was more homogeneous than the elementary group in their efficacy beliefs. These findings are intriguing; perhaps elementary and secondary teachers possess different motivations, with secondary teachers choosing their profession because of an interest in their content area, while elementary teachers choose their profession out of a desire to nurture young children. An additional finding about teaching efficacy among the two groups supports this possibility. We found that, as compared to secondary teachers, elementary teachers have a greater tendency to believe that teaching can overcome the effects of outside influences such as the home. However, there are a number of plausible explanations for these group differences. For example, the two groups work under dramatically different conditions (e.g., number of students served, opportunities for collegiality with other faculty in their departments). Also, elementary teachers may feel less respected than secondary teachers.

The precipitous drop in personal efficacy beliefs among elementary-level teachers is alarming inasmuch as the elementary school years are a formative period for children's attitudes about school and learning, their self-concepts, and their own efficacy beliefs. The research described earlier indicates that, when teachers perceive themselves as ineffective, they are more likely to be controlling and inflexible. Of particular concern, however, is the potentially dangerous situation that occurs when a teacher with low personal efficacy encounters a student who is difficult-to-teach due to learning problems, behavior problems, or both. Put differently, what is the outcome of a mismatch that brings together a teacher who believes that he or she is ineffective and a student whose qualities differ sufficiently from the norm so as to create a substantial teaching challenge?

TEACHER EFFICACY AND LOW-ACHIEVING STUDENTS

According to Bandura (1986), efficacy beliefs influence the choices people make and the effort they expend in realizing their goals. In the following section we explore the ways in which teachers' sense of their own effectiveness relates to the decisions they make in regard to the most challenging students.

Teacher Efficacy and Referral Tendency

One of the most important decisions teachers make with regard to difficult-to-teach students is whether or not to refer them to special education. Based on practices developed in accordance with federal mandates, students referred to special education are evaluated by a multidisciplinary team and, if an educational disability is diagnosed, the student is declared eligible for special education services. Two critical statistics highlight the magnitude of the decision to refer a student to special education. First, almost all of those referred are ultimately classified and placed in special classes (Algozzine, Christenson, & Ysseldyke, 1982). Second, students who are classified are rarely, if ever, returned to the mainstream (Dillon, 1994). Thus, the teacher's referral of a student to special education virtually ensures that the difficult-to-teach student will be permanently labeled and placed in special education classes for all or part of the school day. In other words, the referred student is almost certain to be educated outside the mainstream and the referring teacher is almost guaranteed of being relieved of the responsibility for teaching the student.

In recent years, many have expressed concern over the dramatic increase in the number of students referred to and placed in special education (e.g., Lipsky & Gartner, 1996; Wang & Walberg, 1988). Teachers often may make referrals to special education based on the assumption that such services will benefit otherwise failing or misbehaving students, but the vast number of students presently served in special education suggests that an "overidentification phenomenon" (Ysseldyke & Algozzine, 1982) may be occurring. There are several factors that contribute to concern over rising referrals, including the expense of educating students in special programs (Lipsky & Gartner, 1996), the negative effects of labeling (Gartner & Lipsky, 1987), and the overrepresentation of students from minority backgrounds in special education classes (Cummins, 1984; Harry, 1992). Perhaps most significant is the lack of evidence demonstrating the efficacy of special education, coupled with evidence of its potential deleterious effects (Carlberg & Kavale, 1980; Gartner, 1986). Clearly, there is overwhelming evidence to suggest that inappropriate placement in special education should be prevented and that, because teachers initiate the vast majority of referrals (Gottlieb, Gottlieb, & Trongone, 1991), efforts to prevent placement must focus on teachers' referral decisions.

Some teachers are more likely than others to refer students to special education and research has begun to shed light on characteristics which underlie teachers' referral decisions. Results of several studies conducted in the early 1980s suggested that teachers' beliefs about themselves and their students may account for underlying differences in their decisions to refer students to special education (Christenson, Ysseldyke, & Algozzine, 1982; Riffle, 1985; Smart, Wilton, & Keeling, 1980). As evidence emerged linking teachers' efficacy beliefs to their behavior toward low-achieving students, researchers began to explore the role of efficacy in teachers' referral decisions.

One of the first studies to explore the relation between teacher efficacy and referral decisions was conducted in the Netherlands. Meijer and Foster (1988) found that teachers who had greater confidence in their own teaching ability were less likely to refer students than were teachers who had low efficacy beliefs. Further, they found that the student's problem related to referral chance: teachers were more likely to refer students with combined learning and behavior problems than students with either a learning or a behavior problem alone. While this study supports the notion that confidence in one's teaching relates to referral decisions, it had several inherent limitations. First, the correlations obtained were small (e.g., the correlation between efficacy and referral chance was -.14). Second, differences in educational practices limit the generalizability of the findings, particularly given that the Netherlands provides special education services in segregated schools. Last, only one dimension of teacher efficacy was explored, precluding investigation of the role of one's beliefs about the efficacy of teachers in general.

An unpublished study by Miller (1987, described in Miller 1991) offers additional support for the relation between teacher efficacy and teachers' referral decisions. Based on a survey of first-, second-, and third-grade teachers, Miller found that high-efficacy teachers referred fewer students for special education services than did low-efficacy teachers. Interviews with a small number of teachers with high and low efficacy scores revealed several interesting differences between the groups; specifically, high-efficacy teachers reported a greater sense of responsibility for the performance of low-achieving students and a greater number and range of teaching strategies used with this student population. While this study supports the hypothesis that efficacy relates to differences in referral decisions and instructional practices, lack of information regarding the sample (e.g., the number of teachers involved) and procedures (e.g., whether teachers differed in one or more dimensions of efficacy) limits the interpretation of the reported findings.

Despite limitations in the methods employed in these studies, the investigations conducted by Meijer and Foster and by Miller strongly suggest a relation between teacher efficacy and teachers' decisions to refer students to special education. To clarify this relationship we conducted a series of four studies (Soodak & Podell, 1993; Podell & Soodak, 1993; Soodak & Podell, 1994; Soodak, Podell, & Lehman,

1998). In the first two studies we examined teachers' referral decisions. The first study sought to confirm earlier evidence of the importance of efficacy in teachers' referral decisions, while the second study explored how teachers' efficacy beliefs relate to the biases they may hold. Because teachers may not rely exclusively on their own judgment in determining which students to refer, but may also be influenced by systemic and administrative factors in making this decision (Christenson, Ysseldyke, & Algozzine, 1982), we asked teachers to judge the appropriateness of a student's placement in general education, as well as whether they would refer a student to special education.

Study 1: Teacher efficacy, student problem type, and referral decisions (Soodak & Podell, 1993). The purpose of this study was to determine whether teacher efficacy relates to teachers' decisions regarding the placement and referral of students with different types of problems. In addition, we compared the responses of regular and special educators to determine whether professional training and experience relate to judgments of students. Consistent with the earlier studies that explored the relation between efficacy and referral (Meijer & Foster, 1988; Miller, 1991), we predicted that general education teachers with greater efficacy would be more likely to retain students with problems in the regular classroom. Further, consistent with the findings of Meijer and Foster (1988), we expected that students with both learning and behavior problems would be judged inappropriate for regular education and referred to special education more often than students having only one type of problem.

This study involved 192 teachers (96 general educators and 96 special educators). All had been teaching for a minimum of one year. Each teacher read a case study that described one of three hypothetical second-grade students. In each of the case studies we held constant the student's gender, grade, and background, while we varied the student's problem such that the student had a learning problem, a behavior problem, or both. Teachers indicated on a Likert scale the degree to which they agreed with the student's current placement in general education and the decision to refer the student to special education. In addition, teachers completed the Gibson and Dembo (1984) scale.

We performed multivariate analyses separately on each of the two dependent variables, that is, judgments regarding general education placement and special education referral. In addition to the categorical variables of professional group and student problem type, we entered the continuous variables of personal efficacy and teaching efficacy into a regression equation. Personal efficacy and teaching efficacy scores consisted of unweighted sums of responses to the items included on the two factors emerging from the principal axis factor analysis performed on participants' responses to the Teacher Efficacy Scale.

We detected a significant interaction between teachers' professional group and personal efficacy in the analysis of teachers' judgments of regular class placement. As predicted, regular educators with a greater sense of efficacy were more likely to perceive regular education placement as being appropriate for students having difficulties. This finding is consistent with previous research suggesting that teachers who believe in the value of their own efforts take responsibility for student outcomes (e.g., Ashton & Webb, 1986). In contrast, special educators' judgments of the appropriateness of a regular class placement were not related to their sense of efficacy.

Results further indicated that, as expected, teachers were more likely to find students with either a learning or a behavior problem to be more appropriately placed in regular education than were students with combined problems. Similarly, students with combined learning and behavior problems were more likely to be referred to special education than were students with behavior problems only. As noted in previous research (Meijer & Foster, 1988), students exhibiting multiple problems are at greatest risk of being referred to special education.

This initial study supports the assertion that personal efficacy is a critical belief underlying teachers' decisions pertaining to difficult-to-teach students. In addition, we found that personal efficacy and teaching efficacy interact in their relation with teachers' judgments such that teachers with both high personal efficacy and high teaching efficacy were most likely to regard general education as the appropriate placement for this student. While this study does suggest the importance of teacher efficacy in their thinking about difficult-to-teach students, the need to consider additional factors that might mediate teachers' decisions (i.e., their beliefs about their students) led to a second study.

Study 2: Teacher efficacy and bias in referral decisions (Podell & Soodak, 1993). Several important points influenced the design of the second study. First, models of teachers' thinking posit that teachers' intuitive beliefs about their students contribute to their perceptions and actions (Clark & Peterson, 1986). These preconceptions both expedite decision making and heighten opportunity for bias, that is, the likelihood of responses based on interpretation of only some of the available information (Hamilton, 1981). Second, ambiguous information is most susceptible to bias (Darley & Fazio, 1980). Last, biases in judgments about students are likely to contribute to the overrepresentation of students from minority backgrounds in special education (Huebner & Cummings, 1986). Thus, in the second study we explored the hypothesis that referral decisions relate to teachers' beliefs about their own effectiveness and their preconceptions based on specific student characteristics. In particular, we examined the characteristics of socioeconomic status (SES) of a student's family and the etiology of a student's learning problem. We expected to find that teachers low in efficacy are most likely to be biased by information about the student and the lack of information regarding etiology.

Two hundred and forty regular education teachers responded to questions pertaining to one of six case studies and completed the Teacher Efficacy Scale (Gibson & Dembo, 1984). The case studies described a third-grade male student who, although well behaved, was having difficulty in reading and was unable to concentrate. The case studies were identical except for changes in the suggested etiology of the student's academic difficulties (medical, environmental, or unspecified) and the SES of his family (high or low). In the medical condition teachers read that there were complications during the boy's birth. In the environmental condition teachers read that there were conflicts in his family. In the unspecified condition teachers read that neither the boy's medical background nor his family history appeared to be remarkable. In the low-SES condition the boy's father and mother were described as being a security guard and a waitress, respectively. In the high-SES condition the boy's father and mother were described as being an executive in a large financial institution and a sales representative, respectively. As in Study 1, participants were asked the degree to which they agreed with the student's current placement in a regular class and the decision to refer the student to special education.

We conducted regression analyses on each of the dependent variables with etiology, students' SES, and teachers' personal efficacy and teaching efficacy scores as independent variables. Evidence of an interaction between the students' SES and teachers' personal efficacy, illustrated in Figure 2, supports the contention that per-

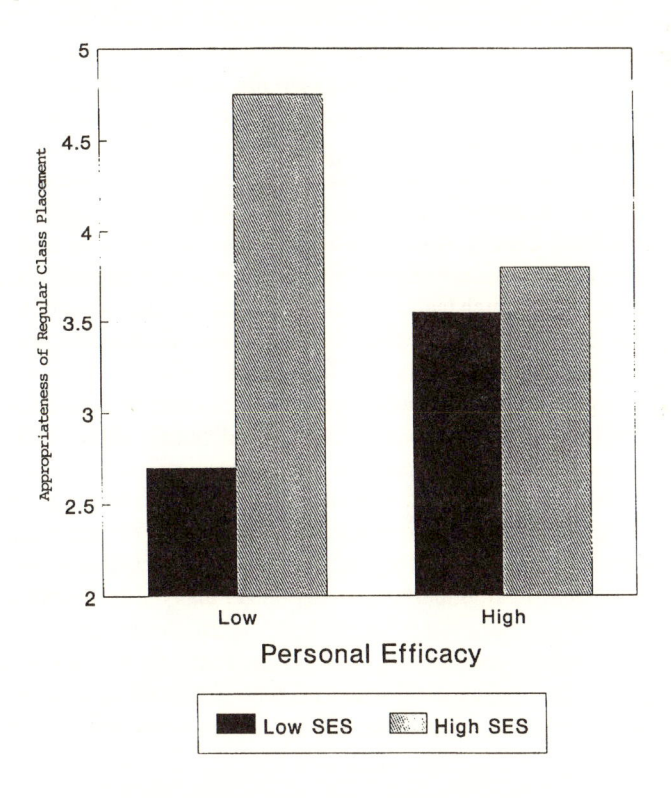

Figure 2. Teachers' Personal Efficacy, Students' SES, and Teachers' Judgments about the Appropriateness of Regular Education Placement

sonal efficacy mediates bias in teachers' decisions about referring students. For low-SES students, teachers with a greater sense of personal efficacy tend to perceive the regular education placement to be more appropriate for a student with academic difficulties. Teachers with a low sense of efficacy, on the other hand, tend to perceive regular education to be a less appropriate placement for low-SES students with learning problems. Personal efficacy did not relate to teachers' placement judgments about students from high-SES backgrounds. In other words, teachers' decisions about poor children are susceptible to bias when teachers perceive themselves as ineffectual. Thus, evidence of the mediational role of students' SES qualified the relationship between efficacy and referral detected in Study 1. Although personal efficacy, and not teaching efficacy, interacted with SES, presence of a significant correlation between teaching efficacy and teachers' referral decisions suggests that both dimensions play a role in teachers' referral judgments.

A second source of bias investigated in this study concerned teachers' assumptions about the causes of a student's learning problem. As expected, teachers were more likely to refer students whose problems were unexplained than those whose problems appeared to be medically based or environmentally induced. Given that referral decisions are most often made in the absence of complete or accurate information regarding etiology, this finding suggests that many referred students may be at risk of bias. The hypothesized interaction between etiology and teacher efficacy was not supported.

Collectively, these two studies provide converging evidence of the importance of teacher efficacy in teachers' decisions pertaining to students with learning and behavior problems. Clearly, teachers who perceive themselves as highly efficacious are more likely to accept responsibility for low-achieving students; the optimal situation is one in which teachers feel confidence in their own abilities as well as confidence in the teaching profession. Most vulnerable, it seems, are poor students who are assigned to teachers with low personal efficacy.

Thus, findings pertaining to teachers' referral decisions are consistent with Bandura's hypothesis regarding the relation between individuals' sense of efficacy and their willingness to persevere. Research on teachers' referral decisions suggests that teachers' confidence in their own abilities and their confidence in their profession underlie their willingness to maintain responsibility for teaching students whom they perceive to be difficult to teach. However, the decision to refer a student to special education, albeit important, is only one of many instructionally relevant decisions that teachers make about their students. Therefore, in a third study we employed an open-ended response format to investigate the interventions that teachers spontaneously suggest when faced with a student with learning problems. In this way, teachers were free to suggest any strategy and were not limited to placement options.

Study 3: Teacher efficacy and instructional interventions (Soodak & Podell, 1994). In the third study we sought to determine whether teachers' assumptions about themselves (i.e., their sense of personal and teaching efficacy) or their students (i.e., their beliefs about the cause of students' problems) relate to their suggestions for addressing the needs of difficult-to-teach students. Participants in the study included 110 elementary education teachers who had been teaching for one to 21 years, with a mean of 13. Each teacher read a case study describing a third-grade male student who was experiencing difficulty in reading and occasional problems with self-control. The case study stated that the student had moved to an apartment with his mother and brother during the previous year, following the divorce of his parents. It also indicated that his teacher used a basal reader and ability grouping for reading instruction. Further, his reading and behavior problems had recently begun to disrupt the class. After reading the case study, teachers responded to three questions: (a) List all the ways you can think of to address the situation, (b) Which of these do you think would be most effective?, and (c) What do you think is the cause of this situation? Teachers were also asked to complete the Gibson and Dembo Teacher Efficacy Scale.

Overall, each teacher made an average of six suggestions for meeting the needs of the student. We categorized teachers' suggestions as either pertaining to changes in teaching strategy (teacher-based interventions) or suggestions involving resources outside the classroom (non-teacher-based interventions). The frequencies of each suggestion are shown in Table 2.

Analysis of the responses overall indicated that teachers suggested more non-teacher-based interventions than teacher-based interventions. One hundred and four of the 110 teachers suggested non-teacher-based interventions. More than three-quarters of the teachers suggested some form of parent participation as a means of addressing the student's problem, making this the most prevalent category of suggestions. Almost two-thirds of the teachers suggested some form of assessment, and more than half specifically mentioned the need for an evaluation by a special education team. The third most frequently cited non-teacher-based intervention was outside services, such as counseling or remedial help. Teacher-based interventions were more often directed at the student's learning problems than at his emotional and behavioral needs. More than three-quarters of the teachers suggested instructional modifications including peer tutoring, the use of high-interest materials, and the use of whole-language or literature-based reading materials. Slightly less than half of the teachers suggested the use of behavior modification.

Although teachers readily suggested ways to address the student's problems, they were reluctant to claim that their suggestions would be effective. Comparison of teachers' suggestions and their beliefs regarding effectiveness indicated that they tended not to have confidence in the effectiveness of those strategies. The number of teachers who believed that a particular suggestion was effective relative to the number who spontaneously made that suggestion exceeded 50 percent in only one category, namely, outside services. Clearly, these data indicate that teachers have a tendency to seek solutions outside the classroom when working with difficult-to-teach students and, further, that they tend to lack confidence in the effectiveness of many of the strategies they suggest.

Table 2. Teachers' Suggested Interventions (in descending order of frequency)

Intervention	N	%
Teacher-based		
Behavior modification	42	38.2
Peer-tutoring	38	34.5
Supporting students' self-esteem	30	27.3
Tutoring by the teacher	25	22.7
Using high-interest materials	19	17.3
Using whole language/literature approaches	18	16.4
Giving the student additional responsibilities	18	16.4
Changing assignments	16	14.5
Using an auditory approach	14	12.7
Focusing on reading in the content areas	11	10.0
Changing classroom groups	10	9.1
Using cooperative learning	8	7.3
Introducing journals/writing	8	7.3
Using a phonics approach	8	7.3
Using manipulatives	5	4.5
Non-teacher-based		
Interdisciplinary team assessment	57	51.8
Giving parents suggestions	45	40.9
Counseling for the student	38	34.5
Try to increase parent involvement	35	31.8
Investigate home situation	28	25.5
Remedial reading instruction	20	18.2
Consult with psychologist or counselor	20	18.2
Resource room services	14	12.7
Solicit suggestions from the student	14	12.7
Increase family's involvement	12	10.9
Assessment by the teacher	11	10.0
Assess visual ability/perception	10	9.1
Consult with other teachers	10	9.1
Consult with principal	10	9.1
Assessment by reading specialist	8	7.3
Retain in grade	7	6.4
Prevention program	5	4.5

Note: Only interventions suggested by five or more teachers are listed.

This study explored the relation between teachers' suggestions and their sense of personal and teaching efficacy by classifying teachers according to the proportion of teacher-based suggestions to non-teacher-based suggestions. We formed three groups: (a) teachers who made more teacher-based suggestions, (b) those who made more non-teacher-based suggestions, and (c) those who made an equal number of each type of suggestion. Results indicated that teachers who made more teacher-based suggestions had greater personal efficacy than did teachers who made more non-teacher-based suggestions. No difference was found between the three groups in teaching efficacy. These findings suggest that personal efficacy plays a role in teachers' tendency to take personal responsibility for finding solutions to the problems of the difficult-to-teach student. It may be that teachers must not only believe that the intervention they are suggesting can be effective, but they must also have confidence in their ability to implement the intervention effectively.

We further explored teachers' acceptance of responsibility for their students through the analysis of teachers' perceptions of the causes of students' problems. When asked about their beliefs regarding the cause of the difficulties of the student described in the vignette, teachers most often cited problems in the home, such as the parents' divorce or the absence of the father from the home. A smaller majority believed that the problems were intrinsic to the student, mentioning factors such as learning difficulties and low self-esteem. The teachers' tendency to attribute the student's learning problems to his family situation or to his emotional state, rather than to school factors, suggests that teachers believe that students' problems are likely caused by factors other than teaching and learning. Thus, teachers are no more willing to accept responsibility for contributing to students' difficulties than they are for seeking solutions that involve their direct efforts.

Results of the present study also indicate that teachers' beliefs about the causes of students' problems relate to the types of suggestions they offer to address the student's needs. Those seeking parent involvement were likely to see the home as the cause of the student's problems and those opting for teacher-based interventions were likely to attribute problems to school factors. The importance of this relation lies in the teachers' tendency to attribute problems to home factors that may cause teachers to relinquish responsibility for remediating students' difficulties.

Results of this investigation confirm earlier findings of the importance of teachers' beliefs about themselves and their students in their decisions pertaining to difficult-to-teach students. In this study the relation between teacher efficacy and teachers' decisions was confirmed and extended, first, by exploring teachers' instructional decisions, which included but were not limited to placement and referral decisions and, second, by employing a more experienced sample consisting of elementary teachers only. Results of this study indicate that teachers are more likely to seek solutions to the problems of difficult-to-teach students outside the classroom than to pursue interventions that they themselves can implement. Perhaps most important, teachers' choices of interventions may be influenced by their beliefs about their ability to affect change in their students and their beliefs about the causes of a student's difficulties.

The studies described thus far demonstrate the salient role of teacher efficacy in teachers' decisions pertaining to students with learning problems who are already placed in their classrooms. Further, these studies provide evidence of the complex nature of teachers' thinking. Apparently, teachers simultaneously consider multiple sources of information, including their efficacy beliefs, in determining their plans and actions. We conducted an additional study to explore more thoroughly the interactive nature of teachers' thinking about difficult-to-teach students. In this study we explored teachers' willingness to work with students who have already been classified. Further, this study explored teacher, student, and school factors that may influence teachers' acceptance of students with disabilities.

Study 4: Teacher efficacy and inclusion (Soodak, Podell, & Lehman, 1998). In addition to efforts aimed at reducing the number of students referred to special education, many seeking to reform current practice advocate the inclusion of all students with disabilities in general education. The inclusion movement has grown in direct response to the ever-increasing number of students placed in special education classes, the seemingly permanent nature of these placements, and their questionable efficacy (Lipsky & Gartner, 1996; Stainback & Stainback, 1992). Supporters of inclusion maintain that students with disabilities have the right to be educated with their nondisabled peers and that general educators must share in the responsibility for educating all students. Most concur that, to be successful, inclusive education requires that general educators receive the supports needed to teach students with disabilities in their classrooms.

Research has indicated that teachers' willingness to work with students with disabilities is a critical factor in the successful implementation of inclusion (Hasazi, Johnston, Liggett, & Schattman, 1994). Clearly, the degree to which teachers' theories about teaching and learning are reflected in a given innovation will influence the effort they expend implementing the innovation (Clark & Peterson, 1986; Sarason, 1982). Further, research has demonstrated that teachers' perceptions of their own effectiveness relate to their receptivity toward instructional innovation. Guskey (1988) surveyed 120 practicing teachers following a staff development program and found that highly efficacious teachers who enjoyed teaching and who were confident in their abilities were most willing to implement the particular instructional innovation for which they had been trained. Therefore, in the fourth study we investigated teachers' willingness to include students with disabilities in their classes and sought to identify factors that might influence their acceptance of these students.

The factors considered in this study do not reflect all possible variables relating to teachers' thinking about inclusion, but rather are factors that have been shown to be important in prior research and which may be amenable to change. The first factor, teacher efficacy, had been linked to teachers' referral and instructional decisions in our earlier studies. The second factor, teachers' use of differentiated instructional practices, is one of the most consistently cited conditions necessary for successful implementation of inclusion (Lipsky & Gartner, 1996; Scruggs & Mastropieri, 1994). Specifically, successful inclusion has been associated with instructional practices such as flexible grouping, cooperative learning, and peer support. Student disability was the third factor considered in this study. Results of prior research had suggested that teachers' attitudes toward inclusion vary as a function of their perceptions of the specific disability labels as well as their beliefs about the demands that the students' instructional and management needs will place on them (Center & Ward, 1987; Houck & Rogers, 1994). Last, school climate was investigated as a possible influence on teachers' willingness to include students because prior research has identified variables such as administrative support and collaboration as powerful indicators of teachers' attitudes toward inclusion (Cross & Villa, 1992; Villa, Thousand, Meyers, & Nevin, 1996).

We surveyed 188 teachers who had been teaching for one to 29 years with a mean number of 9.3 years. Each teacher read a hypothetical scenario in which their principal told them that their school was planning to include a student with a disability in their class. The case study described the student as having one of five disabilities: a hearing impairment, a learning disability, mental retardation, a behavior disorder, or a physical handicap. We investigated teachers' responses to inclusion through a semantic differential task in which they responded to a set of 17 pairs of adjectives along a four-point continuum. Teachers also completed (a) the Gibson and Dembo Teacher Efficacy Scale, (b) a survey of questions pertaining to the frequency with which they engaged in specific teaching practices, and (c) a school climate scale, which required responses to statements about their perceptions of characteristics of their school.

Factor analysis of the semantic differential scale yielded two independent dimensions of teachers' responses to inclusion. Based on an examination of the items loading on each factor, we labeled the factors hostility/receptivity and anxiety/calmness. We then performed regression analyses using the two factors as the dependent variables.

Both teaching efficacy and personal efficacy related to the first dependent variable, teachers' hostility/receptivity toward inclusion. However, teachers' use of differentiated instruction and their perceptions of opportunities for collaboration mediated the relation between the two dimensions of teacher efficacy and hostility/receptivity to inclusion. An analysis of the interaction between teaching efficacy and instructional practices on hostility/receptivity, illustrated in Figure 3, indicates that teachers who use differentiated instructional practices and have a high sense of teaching efficacy are most likely to accept a student with disabilities into their classrooms. On the other hand, teachers with a low sense of teaching efficacy remain hostile to inclusion regardless of their use of differentiated instructional practices. Thus, teachers' instructional practices seem to relate to their willingness to include students with disabilities only if teachers believe that teaching in general has a positive effect on students.

Results also indicated that teachers' sense of personal efficacy interacted with their perceptions of opportunity for collaboration with other teachers within their school, as illustrated in Figure 4. Only teachers who do not perceive themselves as

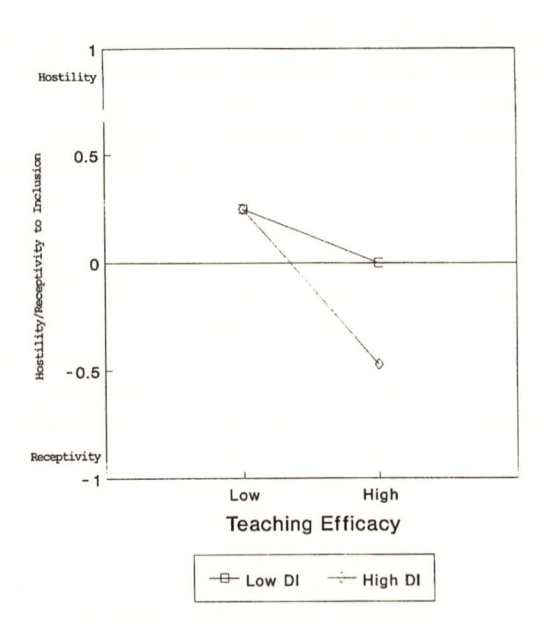

Figure 3. Teachers' Teaching Efficacy, Use of Differentiated Instruction (DI), and Hostility/Receptivity Toward Inclusion

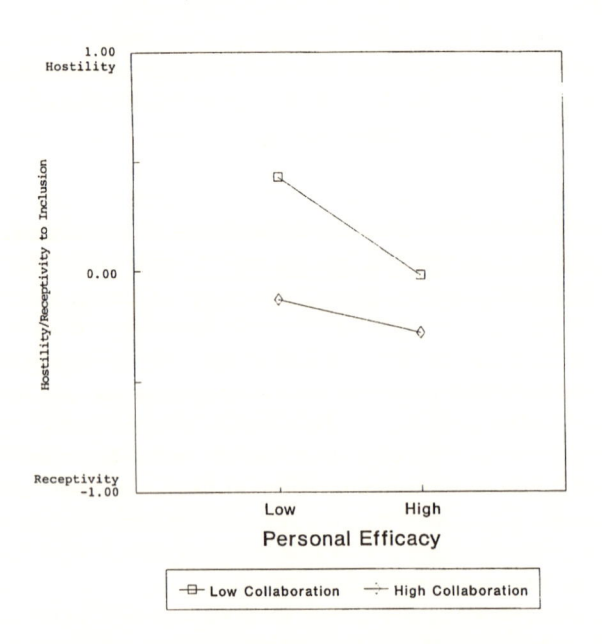

Figure 4. Teachers' Personal Efficacy, Perceptions of Opportunities for Collaboration, and Hostility/Receptivity Toward Inclusion

having opportunities to collaborate and who also have low personal efficacy are more hostile toward inclusion. Teachers who have opportunities to collaborate are more receptive to inclusion, regardless of their beliefs about their own effectiveness. This finding suggests that, at least in the situation of including a student with disabilities, providing teachers with the support of other professionals may compensate for their low personal efficacy.

This study also revealed that personal efficacy related to the second dependent variable, teachers' anxiety toward inclusion. The findings indicate that when teachers' sense of personal efficacy is high, they are less anxious about inclusion. Perhaps personal efficacy (but not teaching efficacy) related to anxiety because personal efficacy reflects a more personal belief rather than a general belief about one's profession.

Additional teacher, student, and school factors related to teachers' willingness to include students with disabilities in this study. Results indicated that such willingness varies depending on the nature of the students' disability. Teachers are unreceptive to the inclusion of students with mental retardation, behavior disorders, and, with greater experience, learning disabilities. But of these disabilities, they are only anxious about the inclusion of students with mental retardation. Other than the opportunity to collaborate, the only school factor that related to teachers' affective responses to inclusion was class size. Not surprisingly, the greater the number of students in the class, the more anxious teachers were about including a student with a disability.

As in our prior studies, teachers' thinking emerged as a complex process wherein several factors are simultaneously considered in decisions about students. In Study 4 receptivity toward inclusion was associated with higher teacher efficacy, inclusion of students with physical rather than cognitive or behavior disorders, years of teaching experience, use of differentiated teaching practices, and collaboration. Collectively, these factors accounted for a considerable proportion (43.6%) of the variance in teachers' hostility/receptivity toward inclusion. Lower anxiety about inclusion was associated with high teacher efficacy, inclusion of students with learning or behavior disorders, and small class size. These variables accounted for 19.8 percent of the variance in teachers' anxiety/calmness, suggesting that teachers' fear of inclusion may be influenced by factors other than those considered in this study.

This series of four studies demonstrates the importance of teachers' sense of efficacy in their decision making about difficult-to-teach students, regardless of whether the decisions involve the referral of students to special education, the appropriateness of placement in regular education, the strategies that may be used to ameliorate problems, or the inclusion of classified students into general education. These studies suggest that students with learning problems are particularly vulnerable when paired with teachers who possess a low sense of efficacy. The following summary of findings highlights the significance of teacher efficacy in decisions pertaining to students with learning problems:

1. Teachers with high personal efficacy are more likely to agree with regular class placement than are those with low personal efficacy.
2. Teachers with both high personal efficacy and high teaching efficacy are most likely to retain students with difficulties in their classes.
3. Teachers with low personal efficacy do not consider regular education to be appropriate for underachieving students from low-SES families.
4. Teachers with low personal efficacy offer more non-teacher-based suggestions for working with students with learning problems, whereas teachers with high personal efficacy suggest more teacher-based interventions.
5. Teachers with a low sense of teaching efficacy are more hostile toward inclusion, regardless of whether they engage in differentiated teaching practices. Teachers with high teaching efficacy who engage in differentiated teaching are most receptive to including a student with disabilities in their classrooms.
6. Teachers with low personal efficacy and few opportunities to collaborate with their colleagues tend to be more hostile toward inclusion.
7. Teachers with a low sense of personal efficacy are more anxious about including students with disabilities in their classes.

IMPLICATIONS AND FUTURE DIRECTIONS

Implications of the research on teacher efficacy, as they apply to the theoretical construct and to the practice of teaching, are discussed in the following section.

Elaboration of the Construct

Efficacy beliefs, and more specifically, teachers' efficacy beliefs, are complex phenomena. Consistent with the notion initially put forth by Bandura (1977), teacher efficacy has been conceptualized as consisting of at least two dimensions: teachers' beliefs about their own effectiveness (personal efficacy) and their beliefs about the effectiveness of the teaching profession (teaching efficacy). Although the link between these dimensions and Bandura's notion of efficacy expectations and outcome expectations has been questioned by us, among others (Guskey & Passaro, 1994; Woolfolk & Hoy, 1990), there is consensus that both personal and teaching efficacy contribute to teachers' thoughts and actions. It may be important to note that our research suggests that personal efficacy may play a greater role in teachers' thinking about difficult-to-teach students than does teaching efficacy. Given that research suggests that personal efficacy may be more accurately conceptualized as encompassing two separate dimensions (i.e., either personal and outcome efficacy, as we suggest, or beliefs regarding positive and negative student outcomes, as Guskey and Passaro [1984] suggest), future research should explore the role of each dimension in teachers' thinking. Specifically, additional research should determine the relations among the different aspects of teacher efficacy and actual teacher behavior.

While discrete dimensions of efficacy beliefs may comprise personal efficacy and these dimensions may differentially affect teacher behavior, a somewhat different uncertainty persists about the more general sense of teaching efficacy. Findings of our research on teachers' decisions pertaining to difficult-to-teach students suggest that teachers' beliefs about the teaching profession serve as a base level of confidence. High teaching efficacy may be a necessary, but not sufficient, condition for accepting responsibility for difficult students. Teaching efficacy related to very few of the dependent variables investigated in our research. Specifically, teaching efficacy interacted with other factors in its relation with teachers' decisions pertaining to placement and inclusion, whereas personal efficacy was both independently and interactively associated with decisions pertaining to placement, referral, and inclusion. Further, we found a developmental pattern in personal efficacy and not in teaching efficacy, suggesting that teachers' views about the impact of their profession on students are unaffected by their personal successes and failures. We suspect that teaching efficacy may reflect a general sense of confidence in the profession that has little to do with self-efficacy. Further research is needed to explore the validity of this dimension of teacher efficacy in order to determine whether teachers' beliefs about their professions are relevant to their efficacy beliefs. A second possibility that should be explored is whether efficacy beliefs of teachers are domain specific, as Pajares (1996) contends. This possibility might explain the weak relation between teaching efficacy and teachers' decisions regarding difficult-to-teach students.

The research presented in this chapter supports the developmental nature of teacher efficacy, or more specifically, the personal efficacy dimension. As hypothesized by Bandura, experience plays a critical role in determining teachers' personal efficacy beliefs. Findings that the sense of efficacy of preservice elementary teachers is extraordinarily high and then plummets as they enter the profession are particularly troubling and suggest that teacher training programs are not adequately preparing students for the realities of teaching. The precipitous drop in teachers' personal efficacy may have devastating effects on students with educational difficulties given that teachers with low efficacy are more likely to avoid responsibility for difficult-to-teach students and to be biased by factors unrelated to student ability in their instructional decisions. Thus, teachers may be especially susceptible to bias and likely to make poor decisions early in their careers. It is also likely, but yet unsupported, that teachers' experiences with difficult-to-teach students influence the recovery of their efficacy beliefs. Future research needs to determine whether successful experiences with students with learning or behavior problems early in one's career promotes improvements in personal efficacy. Conversely, it is possible that novice teachers who experience negative outcomes in their interactions with difficult-to-teach students are among those who leave teaching within their first few years of their careers. Further research should examine whether the possible attrition of low-efficacy teachers is preventible or whether it represents a strengthening of the teaching profession.

Results of our research support those of prior investigations (Greenwood, Olejnik, & Parkay, 1990; Guskey, 1982; Parkay, Olejnik, & Proller, 1988) that suggest a somewhat different pattern of development in efficacy beliefs of elementary and secondary educators. Our research indicates that secondary educators do not emerge from teacher preparation with an inflated sense of their own effectiveness, they do not have their efficacy beliefs challenged once they enter the profession, and, as a group, they are more homogeneous in their efficacy beliefs than are their counterparts on the elementary level. Differences between the efficacy beliefs of elementary and secondary educators may reflect underlying motivational differences and differences in their teaching experiences. The finding indicating differences among the two groups raises two as-yet unexplored questions. First, are group differences attributable to level of teaching or are other variables, such as gender, involved? Second, is it advantageous, in terms of teacher retention and student achievement, to have less volatile shifts in efficacy beliefs, as observed among secondary educators, and, if so, can developmental patterns be modulated?

Implications for Practice

We and others have shown that teacher efficacy interacts with student, teacher, and school factors in its relation to teachers' thinking about difficult-to-teach students. High-efficacy teachers, particularly those who differentiate instruction and collaborate with their colleagues, take greater responsibility for teaching all students. On the other hand, teachers who believe themselves to be ineffectual are likely to relinquish responsibility for low achievers to others and are likely to be biased in their decisions about students. These findings have several important implications. First, assuming that teachers' experiences with difficult-to-teach students determine their sense of efficacy, teachers need to gain more positive experiences with such students. More specifically, drawing on the four sources of efficacy beliefs identified by Bandura, teachers may need to experience, reflect on, and be praised for their own performance accomplishments, and they may need to be made aware of the accomplishments that other teachers have had in working with difficult-to-teach students. Improved teacher effectiveness will likely have a positive effect on teachers' perceptions of efficacy; however, this relation needs to be verified in future studies. It is possible that the relation between teacher effectiveness, teacher efficacy, and teachers' thoughts and actions is complex and multidirectional. In other words, whereas it may be appropriate to assume that increased teacher efficacy will promote positive changes in teacher effectiveness, this assumption is not yet supported by research. It is critical that further research employ models of teachers' thinking that reflect the reciprocal effects among variables so as to illuminate more accurately the complexity of teachers' decision making.

Second, research on the role of efficacy in teachers' thinking about difficult-to-teach students suggests that present efforts aimed at reducing the number of students in special education may be somewhat misguided. Traditionally, pre-referral efforts have focused on the student, primarily through the practice of providing remedial services to ameliorate academic difficulties (Fuchs et al., 1990; Graden, Casey, & Bonstrom, 1985). Although some programs attempt to prepare teachers to meet diverse students' needs, there has been a lack of attention to teachers' underlying beliefs and biases that may influence their willingness to work with difficult students. Our research indicates that teachers' beliefs about themselves and their students may account for a substantial portion of the variance in their referral decisions. Thus, future efforts aimed at reducing the number of students in special education may be made more effective by addressing teachers' underlying beliefs.

The research discussed in this chapter provides preliminary evidence of the importance of teacher efficacy in the success of inclusive education. Teachers' beliefs about their own ability to bring about desired outcomes in their students seem to underlie their willingness to accept responsibility for difficult-to-teach students in their general education classes before and after the students are labeled. The role of teachers' sense of efficacy in their accommodation of students with disabilities in general education parallels the multiple roles of efficacy in the self system as discussed by Bandura (1993). First, teachers' sense of efficacy influences, and is influenced by, their cognitions. This is seen in teachers' willingness to accept responsibility for students whose problems have a known, or perceived, etiology. It is further seen in the relations among teacher efficacy, professional collaboration, and teachers' receptivity toward inclusion, which suggest that feedback and support are catalysts for positive attitudes toward inclusion. Second, the receptivity to inclusion of high-efficacy teachers suggests that these teachers are motivated to persevere when faced with difficult students. Not surprisingly, teachers already equipped with the necessary skills, for example, the ability to engage in differentiated teaching practices, are among those who are most willing to meet a difficult challenge. Third, evidence indicating that anxiety toward inclusion is related to low efficacy is consistent with Bandura's contention that individuals' self-efficacy influences their affective processes. And last, teachers' sense of efficacy seems to influence the choices and opportunities that they allow themselves. Teachers with low efficacy were not willing to maintain students with disabilities in their classes and were not receptive to the idea of including labeled students who had previously been segregated from the mainstream.

Brownwell and Pajares (1997) conducted a recent study that confirms and in some ways extends our findings regarding the importance of teachers' efficacy beliefs in determining their willingness and ability to include students with disabilities in general education. This study employed path analysis to determine the effects of teachers' efficacy beliefs on their perceived success in working with classified students in mainstream classes. Consistent with Bandura's assertion that efficacy beliefs are contextual judgments of capability, Brownwell and Pajares developed a set of items specifically designed to measure teachers' judgments of their capability to teach and manage students with learning and/or behavior problems in mainstream settings. In addition to teachers' sense of efficacy, this study involved many of the variables that we considered in our research, such as support from the building principal, collegiality among teachers, and students' SES. In addition, the authors considered the potential effects of preservice and in-service preparation on teachers' perceptions of success in teaching students with learning and behavior problems. Results indicated that teachers' efficacy beliefs had a strong direct effect on their perceptions of success in teaching in inclusive settings. Further, teacher efficacy mediated the effects of perceived collegiality and preservice preparation on teachers' perceptions of success. Based on these findings, the researchers concluded that fostering teachers' efficacy beliefs may facilitate the successful implementation of inclusion and that preservice preparation appears to play an important role in promoting teachers' sense of efficacy in dealing with students with learning and behavior problems. The findings of this study, coupled with those of our research, demonstrate the critical role of teachers' efficacy beliefs in the success of inclusive education.

Because there have been few studies on the relation of teacher efficacy and the inclusion of students with disabilities, there are numerous questions that require further study. For example, if the relation between efficacy and behavior is reciprocal, as Pajares (1996) suggests, does experience with inclusion change teachers' efficacy beliefs and, ultimately, their behavior toward students with disabilities? Or is there a minimal degree of efficacy necessary for teachers to benefit from their initial experiences with inclusion? In practical terms, does inclusive education depend on volunteerism? Do teachers' cognitive and affective responses to inclusion, which are each uniquely linked to teachers' efficacy beliefs, differentially relate to their behavior toward students in inclusive classrooms? Or, more simply, *how* do efficacy beliefs relate to teachers' ability to implement inclusive education?

In addition to greater understanding of the relation between efficacy and inclusion, the question of the degree to which teacher efficacy is malleable remains unanswered. It is presently unclear as to whether teachers' sense of efficacy is at all modifiable. We conducted a pilot study on the stability of efficacy by inducing teachers' memories of success and failure with students with learning difficulties and found no effect. These memories, whether of positive or negative experiences, were unrelated to teachers' efficacy beliefs among our small sample, suggesting that teacher efficacy is stable and perhaps not easily modifiable. On the other hand, preliminary research suggests that personal efficacy can be enhanced through staff development activities (Fritz, Miller-Heyl, Kruetzer, & MacPhee, 1995). Given the salient role of teacher efficacy in teachers' thinking about including students with disabilities, future research should determine whether efficacy beliefs are modifiable and whether modifying these beliefs affects outcomes related to the acceptance of difficult-to-teach students.

REFERENCES

Algozzine, B., Christenson, S., & Ysseldyke, J. E. (1982). Probabilities associated with the referral to placement process. *Teacher Education and Special Education, 5,* 19-23.

Armor, D., Conroy-Osequera, P., Cox, M., King, N., McDonnell, L., Pascal, A., Pauly, E., & Zellman, G. (1976). *Analysis of the school preferred reading programs in selected Los Angeles minority schools* (Report No. R-2007-LAUSD). Santa Monica, CA: Rand Corporation.

Ashton, P., & Webb, R. (1982, March). *Teachers' sense of efficacy: Toward an ecological model.* Paper presented at the annual meeting of the American Educational Research Association, New York.

Ashton, P., & Webb, R. (1986). *Making a difference: Teachers' sense of efficacy and student achievement.* New York: Longman.

Bandura, A. (1977). Self-efficacy: Toward a unifying theory of behavioral change. *Psychological Review, 84,* 191-215.

Bandura, A. (1988). Self-efficacy conceptions of anxiety. *Anxiety Research, 1,* 77-98.

Bandura, A. (1993). Perceived self-efficacy in cognitive development and functioning. *Educational Psychologist, 28,* 117-148.

Berman, P., & McLaughlin, M. (1977). *Federal programs supporting educational change: Vol. 2, Factors affecting implementation and continuation.* Santa Monica: Rand Corporation.

Berman, P., McLaughlin, M. W., Bass, G., Pauly, E., & Zellman, G. (1977). *Federal programs supporting educational change: Vol. 7, Factors affecting implementation and continuations.* Santa Monica: Rand Corporation.

Brownwell, M. T., & Pajares, F. (1997, March). *Teacher efficacy and perceived success in mainstreaming students with learning and behavior problems.* Paper presented at the annual meeting of the American Educational Research Association, Chicago.

Carlberg, C., & Kavale, K. (1980). The efficacy of special class versus regular class placement for exceptional children: A meta-analysis. *Journal of Special Education, 14,* 295-309.

Center, Y., & Ward, J. (1987). Teachers' attitudes towards the integration of disabled children in regular schools. *The Exceptional Child, 31,* 41-56.

Christenson, S., Ysseldyke, J., & Algozzine, B. (1982). Institutional constraints and external pressures influencing referral decision. *Psychology in the Schools, 19,* 342-345.

Clark, C. M., & Peterson, P. S. (1986). Teachers' thought processes. In M. C. Wittrock (Ed.), *Handbook of research on teaching* (pp. 255-296). New York: Macmillan.

Cross, G., & Villa, R. (1992). The Winooski school system: An evolutionary perspective of a school restructuring for diversity. In R. Villa, J. Thousand, W. Stainback, & S. Stainback (Eds.), *Restructuring for caring and effective education: An administrative guide to creating heterogenous schools* (pp. 219-237). Baltimore: Paul H. Brookes.

Cummins, J. (1984). *Bilingualism and special education: Issues in assessment and pedagogy.* Cleveland, England: Multilingual Matters, Ltd.

Darley, J., & Fazio, R. H. (1980). Expectancy confirmation processes arising in the social interaction sequence. *American Psychologist, 35,* 867-881.

Dembo, M., & Gibson, S. (1985). Teachers' sense of efficacy: An important factor in school achievement. *The Elementary School Journal, 86,* 173-184.

Denham, C. H., & Michael, J. J. (1981). Teacher sense of efficacy: A definition of the construct and a model for further research. *Educational Research Quarterly, 5,* 39-63.

Dillon, S. (1994, April 7). Special education soaks up New York's school resources. *The New York Times,* p. A16.

Fritz, J. J., Miller-Heyl, J., Kruetzer, J. C., & MacPhee, D. (1995). Fostering personal efficacy through staff development and classroom activities. *Journal of Educational Research, 88,* 200-208.

Fuchs, D., Fuchs, L. S., Bahr, M. W., Fernstrom, P., & Stecker, P. M. (1990). Prereferral intervention: A prescriptive approach. *Exceptional Children, 56,* 503-513.

Gartner, A. (1986). Disabling help: Special education at the crossroads. *Exceptional Children, 53*, 72-76.
Gartner, A., & Lipsky, D. K. (1987). Beyond separate education: Toward a quality system for all students. *Harvard Educational Review, 57*, 367-395.
Gibson, S., & Dembo, M. H. (1984). Teacher efficacy: A construct validation. *Journal of Educational Psychology, 76*, 569-582.
Glickman, D., & Tamashiro, R. (1982). A comparison of first-year, fifth-year, and former teachers on efficacy, ego development, and problem solving. *Psychology in the Schools, 19*, 558-562.
Gottlieb, J., Gottlieb, B. W., & Trongone, S. (1991). Parent and teacher referral for a psychoeducational evaluation. *Journal of Special Education, 25*, 155-167.
Graden, J. L., Casey, A., & Bonstrom, O. (1985). Implementing a prereferral intervention system: Part II. The data. *Exceptional Children, 51*, 487-496.
Greenwood, G., Olejnik, S., & Parkay, F. (1990). Relationships between four teacher efficacy belief patterns and selected teacher characteristics. *Journal of Research and Development in Education, 23*, 102-107.
Guskey, T. R. (1981). Measurement of the responsibility teachers assume for academic successes and failures in the classroom. *Journal of Teacher Education, 32*, 44-51.
Guskey, T. R. (1982). Differences in teachers' perceptions of personal control of positive versus negative student learning outcomes. *Contemporary Educational Psychology, 7*, 70-80.
Guskey, T. R. (1988). Teacher efficacy, self-concept, and attitudes toward the implementation of instructional innovation. *Teaching and Teacher Education, 4*, 63-69.
Guskey, T. R., & Passaro, P. D. (1994). Teacher efficacy: A study of construct dimensions. *American Educational Research Journal, 31*, 627-643.
Hall, B. W., Hines, C. V., Bacon, T. P., & Koulianos, G. M. (1992, April). *Attributions that teachers hold to account for student success and failure and their relationship to teaching level and teacher efficacy beliefs.* Paper presented at the annual meeting of the American Educational Research Association, San Francisco.
Hamilton, D. L. (1981). Cognitive representation of persons. In E. T. Higgins, C. P. Herman, & M. P. Zanna (Eds.), *Social cognition: The Ontario Symposium* (Vol. 1, pp. 135-160). Hillsdale, NJ: Lawrence Erlbaum.
Harry, B. (1992). *Cultural diversity, families, and the special education system: Communication and empowerment.* New York: Teachers College Press.
Hasazi, S. B., Johnston, A. P., Liggett, A. M., & Schattman, R. A. (1994). A qualitative policy study of the least restrictive environment provision of the Individuals with Disabilities Education Act. *Exceptional Children, 60*, 491-507.
Hoy, W., & Woolfolk, A. (1990). Socialization of student teachers. *American Educational Research Journal, 27*, 279-300.
Hoy, W., & Woolfolk, A. (1993). Teachers' sense of efficacy and the organizational health of schools. *Elementary School Journal, 93*, 355-372.
Huebner, E. S., & Cummings, J. A. (1986). Influences of race and test data ambiguity upon school psychologists' decisions. *School Psychology Review, 15*, 410-417.
Houck, C. K., & Rogers, C. J. (1994). The special/general education integration initiative for students with specific learning disabilities: A "snapshot" of program change. *Journal of Learning Disabilities, 27*, 58-62.
Lipsky, D. K., & Gartner, A. (1996). Inclusion, school restructuring, and the remaking of American society. *Harvard Educational Review, 66*, 762-796.
Meijer, C. J. W., & Foster, S. F. (1988). The effect of teacher self-efficacy on referral chance. *Journal of Special Education, 22*, 378-385.
Miller, P. (1987). *The relationship between teacher efficacy and referral of students for special education services.* Unpublished doctoral dissertation. Virginia Tech, Blacksburg, VA.
Miller, P. S. (1991). Increasing teacher efficacy with at risk students: The sine qua non of school restructuring. *Equity and Excellence, 25*, 30-35.

Pajares, F. (1996). Self-efficacy beliefs in academic settings. *Review of Educational Research, 66*, 543-578.
Parkay, F., Olejnik, S., & Proller, N. (1988). A study of relationships among efficacy, locus of control, and stress. *Journal of Research and Development in Education, 21*, 13-22.
Podell, D. M., & Soodak, L. C. (1993). Teacher efficacy and bias in special education referrals. *Journal of Educational Research, 86*, 247-253.
Riffle, C. A. (1985). Factors influencing regular classroom teachers' referral practices. *Teacher Education and Special Education, 8*, 66-74.
Saklofske, D. H., Michayluk, J. O., & Randhawa, B. S. (1988). Teachers' efficacy and teaching behaviors. *Psychological Reports, 63*, 407-414.
Sarason, S. B. (1982). *The culture of the school and the problem of change.* Boston: Allyn and Bacon.
Scruggs, T., & Mastropieri, M. A. (1994). Successful mainstreaming in elementary science classes: A qualitative study of three reputational cases. *American Education Research Journal, 31*, 785-811.
Smart, R., Wilton, K., & Keeling, B. (1980). Teacher factors and special class placement. *Journal of Special Education, 14*, 217-229.
Soodak, L. C., & Podell, D. M. (1993). Teacher efficacy and student problems as factors in special education referral. *Journal of Special Education, 27*, 66-81.
Soodak, L. C., & Podell, D. M. (1994). Teachers' thinking about difficult-to-teach students. *Journal of Educational Research, 88*, 44-51.
Soodak, L. C., & Podell, D. M. (1996). Teacher efficacy: Toward the understanding of a multi-faceted construct. *Teaching and Teacher Education, 12*, 401-411.
Soodak, L. C., & Podell, D. M. (1997). Efficacy and experience: Perceptions of efficacy among preservice and practicing teachers. *Journal of Research and Development in Education.*
Soodak, L. C., Podell, D. M., & Lehman, L. (1998). Teacher, student, and school factors as predictors of teachers' responses to inclusion. *Journal of Special Education, 31*, 480-497.
Stainback, S., & Stainback, W. (1992). *Curriculum consideration in inclusive classrooms: Facilitating learning for all students.* Baltimore: Brooks.
Villa, R. A., Thousand, J. S., Meyers, H., & Nevin, A. (1996). Teacher and administrator perceptions of heterogeneous education. *Exceptional Children, 63*, 29-45.
Wang, M. C., & Walberg, H. J. (1988). Four fallacies of segregrationism. *Exceptional Children, 55*, 128-137.
Woolfolk, A. E., & Hoy, W. K. (1990). Prospective teachers' sense of efficacy and beliefs about control. *Journal of Educational Psychology, 82*, 81-91.
Ysseldyke, J. E., & Algozzine, B. (1982). *Critical issues in special and remedial education.* Boston: Houghton Mifflin.

EXPECTATIONS AND BEYOND:
THE DEVELOPMENT OF MOTIVATION AND LEARNING IN A CLASSROOM CONTEXT

Pekka Salonen, Erno Lehtinen, and Erkki Olkinuora

INTRODUCTION: BROADENING PERSPECTIVES

Theoretical views of differences in academic performance have recently been subject to several reappraisals and a continuous broadening of perspectives. Cognitive psychology challenged the simplistic IQ-based explanations of high achievement as well as organic and psychological deficit explanations of learning disabilities (Cole & Bruner, 1974; Cicourel et al., 1974; Ericsson & Charness, 1994). A prominent cognitive view of differences in learning problems depicts high achievers as active, autonomous, and strategic learners; in contrast, learning disabled children are characterized as passive learners, whose problems in academic learning are based on poorly developed strategic skills, lack of metacognitive awareness, and insufficient self-regulation (Baker & Brown, 1984; Torgesen, 1977; Torgesen & Licht, 1983; see Wong, 1985, 1991 for reviews). In particular, researchers amalgamizing cognitive views with a Vygotskian-influenced sociocultural approach interpreted such "passivity" in terms of incomplete sociocultural mediation or internalization of symbolic tools (see Feuerstein, Rand, Hoffman, & Miller, 1980; Brown & Ferrara, 1985). This view has inspired the development of several cognitive training programs for regular classroom and special educational settings (e.g., Palincsar & Brown, 1984; Paris & Jacobs, 1984; Palincsar, 1986; Stone & Wertsch, 1984; Brown, Palincsar, & Purcell, 1986). Even though instructional mediation models such as "reciprocal teaching" and "scaffolding" have yielded very promising results with many disadvantaged student groups (e.g., Palincsar & Brown, 1984; Paris & Jacobs, 1984), a considerable number of subnormally performing students persistently resist the most advanced efforts to mediate higher level cognitive strategies and metacognitive skills (e.g., Brown, Campione, & Barclay, 1979; Pressley, Levin, & Bryant, 1983).

Motivation research has opened interesting complementary perspectives to individual differences. Among the most influential have been models and assumptions connected to helplessness and attribution theories (Seligman, 1975; Abramson, Seligman, & Teasdale, 1978; Licht, 1983). The general motivational model of the development of subnormal performance combines aspects of helplessness, attribution, and self-esteem theories (see Licht, 1983). As a result of frequent failures experienced by certain children from their earliest school years, they come to perceive themselves as helpless, that is, lacking in control of the causes of their difficulties. These children's beliefs about their abilities and about the stability and uncontrollability of the causes of their difficulties are expected to lead them to lower their expectations and decrease their achievement efforts. Thus, the probability of failure is increased and their doubts about their ability to overcome the difficulties are strengthened. Such motivational beliefs will foster dysfunctional adaptive efforts such as avoidance of academic tasks (Adelman & Taylor, 1983), an external locus of control, and lowered self-esteem (Black, 1974; Durrant, 1993).

The importance of Rosenthal and Jacobson's (1968) classic study of the Pygmalion in the classroom has become apparent during the last decades. This research not only induced a vast number of studies with fruitfully controversial results and ongoing debate (Brophy, 1983; Cooper & Good, 1983; Elashoff & Snow, 1971; Rosenthal, 1985; West & Anderson, 1976), but it also anticipated a new approach or paradigm to research on classroom learning. The Rosenthal and Jacobson study anticipated the following expansions of the focus: (1) from purely objective antecedents of academic performance to its subjective codeterminants, (2) from simplistic linear causal explanations to more complex interactionist models underscoring reciprocal causation and cyclical processes, (3) from purely cognitive interpretation of academic performance to motivational and emotional codeterminations, (4) from intra-individual antecedents of cognitive performance to its social (or sociopsychological) underpinnings.

There is increasing evidence that learning and academic performance cannot be explained in purely objective, individual, or cognitive terms. Subjective beliefs, expectations, goals, emotional processes, as well as socio-cognitive transactions should be recognized as essential constituents of academic achievement. In this chapter we present an integrative theoretical model which can be used in analyzing the situational dynamics of student-task-teacher interaction (i.e., the microgenesis) as related to students' long-term development and to the institutional-cultural (macro level) structures of the school (e.g., teacher and student roles, identity management, participation structures, assessment structures). More specifically, we construct theoretical links between students' and teachers' situational adaptations, students' progressive and regressive learning careers, institutional-cultural frame factors, and social regulation mechanisms. This endeavor is based on our own theoretical and empirical work on the cumulative long-term formation of progressive and regressive learning careers (Lehtinen et al., 1995).

DIVERGING LEARNING CAREERS

The polarization hypothesis, originally and mainly tested within strictly tracked (segregated/ability grouped) school settings, claims that assigning students to different curricular programs based on their abilities and interests divides students into pro- and anti-school attitudes and behaviors. These behaviors and attitudes tend to become more divergent over time (Berends, 1995). There is ample evidence of educational polarization within tracked curriculum settings (Hargreaves, 1967; Lacey, 1970; Ball, 1981; Abraham, 1989). Researchers have also stressed the negative consequences of tracking not only on the development of school attitudes and attitude-related behaviors but also on cognitive skills, as well as motivation- and self- related beliefs (Berends, 1995; Oakes, 1985, 1992; Oakes & Lipton, 1992; Urdan, Midgley, & Wood, 1995).

Yet, teachers who work in "inclusion" settings commonly report that in desegregated, heterogeneous classrooms, there is a wide and hard-to-reduce gap between high- and low-achieving students, both in terms of cognitive skills, motivation, and school attitudes (see Fuchs & Fuchs, 1994; Gerber & Semmel, 1984; Kauffman, 1989; Kauffman, Gerber, & Semmel, 1988; Semmel, Gerber, & MacMillan, 1994; Minke, Bear, Deemer, & Griffin, 1995). It has been demonstrated that inclusive programs and desegregated classrooms are far from optimal for a considerable number of exceptional children (Pugach, 1995; Schofield, 1993; Hepler, 1994). Many numbers of learning and reading disabled students persistently fail to make gains in inclusive programs (Zigmond, Jenkins, Fuchs, Deno, & Fuchs, 1995; for reviews see also Sindelar & Deno, 1979; Madden & Slavin, 1983; Carlberg & Kavale, 1980). For example, research on reading and text comprehension skills in regular classrooms (Stanovich, 1986; Juel, 1988; Vauras, Kinnunen, & Kuusela, 1994) provides evidence for cumulatively diverging developmental courses, that is, a phenomenon called the "Matthew effect" (Stanovich, 1986). It seems that, as stated, "['] generously will be given to anybody who already possess a lot, whereas even the little the disadvantaged has got is likely to be taken away."

Our colleagues Vauras, Kinnunen, and Kuusela (1994) investigated the development of students' text processing skills from third to fifth grade. Third graders in Finnish comprehensive schools (which comprise grades one-nine, no ability grouping) were assigned on the basis of teacher interviews into three groups: high, average, and low achievers. The children's comprehension of expository texts was examined at micro (sentence), local (paragraph), and global (whole text) level processing skills. The results showed a clear progression in average and high achievers' local- and global-level construction of coherent meaning units. This development was most striking in the extreme group of top achievers whose text-processing skills were in the fifth grade highly comparable to those of expert, adult learners. In contrast, low achievers showed practically no progression of the higher (local and global) level text-processing skills during the two years. Rather, in relation to increasing demands, their learning careers could be characterized as regressive. The same developmental patterns were found in students' grade-point averages. Since text-comprehension skills are among the most fundamental prerequisites for all subsequent classroom learning, it is likely that the gap between low achievers and others continues to widen during the following school years. Particularly alarming in this finding is that there is a group of low achievers, representing about 20 percent of the population, that seems to derive no benefit from the general instructional settings (Vauras, Kinnunen, & Kuusela, 1994). Similar results have been reported in writing skills. Bereiter and Scardamalia (1993) concluded that the poorer writers they studied approach the tasks in ways that minimize opportunities for growth, whereas the better writers maximize them. The result is a multiplier effect, where the more expert keep gaining in expertise while the less expert make little progress.

The Vauras, Kinnunen, and Kuusela (1994) study affords a deepening perspective to the results of a recent international comparative study (Elley, 1992). The latter study, using text-comprehension tasks largely comparable to the tasks of the Vauras, Kinnunen, and Kuusela (1994) study, showed that at third and seventh grades Finnish comprehensive school students displayed the most advanced text-processing skills. Such average-oriented results may be used as evidence for the overall efficacy of the school system, but what usually remains hidden is that desegregated school students may be latently partitioned into vulnerable and resilient subgroups (Pugach, 1995; Schofield, 1993). The comprehensive school may provide children with "equality of opportunity," but the differing developmental tracks indicate that the polarizing selection mechanism begins its harsh job very early even in an inclusive setting. Based on socially conditioned cognitive prerequisites and motivational predispositions, it labels and divides the students into potential "loser" and "winner" groups, and allocates them to differential instructional transactions and educational pathways (Kivinen & Rinne, 1996). Our empirical results indicate that diverging learning careers are closely related to the formation of students' motivation (Salonen, Lepola, & Niemi, in press) and to the emerging patterns of classroom interaction (Hamalainen & Lehtinen, 1989).

THE CLASSROOM AS AN ENVIRONMENT FOR STUDENT ADAPTATION AND LEARNING

Learning and cognitive development are not isolated functions or ends as such. Both are embedded in a total process of adaptation that increases an individual's capacity to act successfully under varying environmental demands, and to control the complex physical, mental, and social conditions of life.

According to modern biological, psychological, and systems conceptualizations (Buckley, 1967; von Bertalanffy, 1969; Bronfenbrenner, 1979; Charlesworth, 1976; Piaget 1961; Schoggen, 1989; Sameroff, 1983; Tinbergen & Tinbergen, 1985), adaptation is not simply a passive adjustment to environmental changes, aimed at the restoration of internal homeostasis or fit between organism and environment. Rather, adaptation comprises active efforts to assimilate environmental objects into preexisting structures and the parallel process of active modification (accommodation) and reorganization of such structures (Piaget, 1971).

A central characteristic of progressive development is the growing capacity of self-governed activity and self-regulation. A parallel developmental process in the motivational domain leads to the increasing of sense of control or feeling of self-efficacy (Bandura, 1989). Maladaptive, or regressive, development diminishes the functionality of an individual's activity with regard either to immediate situational demands or to future change (Crittenden & DiLalla, 1988).

Classroom adaptation relates not only to the above demands and structures but also to complex inter- and intrapersonal regulatory mechanisms, used in striving for personal adaptive goals. Both teachers and students attempt to fulfill their own needs and to adapt to each others' needs, expectations, and emotions through various socio-cognitive and emotional strategies: (a) they make reciprocal typifications and categorizations (e.g., labeling processes) concerning the Other and the Self, (b) they present their social identities to each other (Goffman, 1972), (c) they try to maintain and restore their personal sense of situation control, and (d) they try to manage (enhance, defend) their own identities. In sum, classroom activities are not purely cognitive, that is learning- or task-focused, but they often serve to protect the participants' self-worth (Covington & Beery, 1976; Thompson, 1994), to manage threats to their identities (Barga, 1996; Rueda & Mehan, 1986; Sheer & Weigold, 1995), to maintain or restore the stability and order of interaction/discourse (Edwards, 1994; Knapp, Shields, & Turnbull, 1995; Macbeth, 1990; Mehan, 1979; Schultz, 1994).

Institutional-cultural frame factors, interpersonal regulatory mechanisms, and personal adaptive strivings contribute to the expectations and causal interpretations of the participants (students and teachers). For instance, a teacher who works under constraining time allocation and participation structures and experiences strong pressure for professional success, may concentrate more on her or his own identity management than, for example, on task-focused attempts to individualize the instruction. When a low-achieving student tries to cope with a threatening learning situation through avoidance behavior, the teacher is likely to increase her or his own sense of well-being by defensive externalizing of the cause of failure. Consequently the teacher is likely to accuse the student once again of being disabled.

Situational Demands and Student Adaptation

There are specific structural frame factors and regulatory mechanisms in the classroom: time allocation structures, participation and turn-allocation structures, social control structures, and assessment structures. These structures involve social norms and rules that are assumed to be shared. From the point of view of student adaptation, classroom learning and performance settings comprise of a complex array of simultaneous situational demands (see Blumenfield, Pintrich, & Hamilton, 1987; Nicholls & Thorkildsen, 1988, 1989). Learning tasks as such have certain objective requirements to be fulfilled, or certain levels of difficulty to be overcome. Teachers have their implicit expectations, preferences, and rules to be met. Teachers also provide rewards to be striven for and criticism to be avoided. In addition, students need to maintain their internal sense of well-being and self-worth in public academic performance situations that comprise continuous interpersonal comparison and evaluation. Students thus must cope with a complex demand structure that comprises often inconsistent situational demands. For instance, the child may feel that at the same time he or she should try hard to understand the principles of a task, respond quickly, perform decently, and defend against being laughed at.

Task-intrinsic demands arise purely from the nature of the task, from its cognitive-logical requirements, or from the matters of intellectual substance (see Nicholls & Thorkildsen, 1988, 1989). Such demands require and induce direct task- or problem-focused activities from the student. To respond adequately to the questions of fact, comprehension, or application tasks, the student has to construct the objective task requirements (e.g., correspondence with reality, coherence of meaning, the "functionality" of the application) and to formulate a solution that fulfills those requirements. As students conceive this, they understand that the accuracy of these solutions can be checked out by anyone at any time, and it is not merely a matter of opinion or social agreement. When the teacher and the student co-orient to a task- intrinsic demand and the teacher gives feedback on the student's performance, both teacher and student agree that it is the task-intrinsic principle that can be submitted to intersubjective verification.

There are two main categories of *task-extrinsic* demands. Demands of the first type arise from the experienced social requirements of the situation, whereas demands of the second kind arise from an individual's need to maintain and restore his or her own emotional well-being. Social demands extend from somewhat globally shared norms, rules, and expectations (typical to most schools, classrooms, and teachers) to a specific teacher's personal preferences and situation-specific whims. The subcategory of social conventions (Nicholls & Thorkildsen, 1988) represents social demands of the global type. It consists of shared norms and rules that typically regulate classroom interactions. Intersubjectively shared social conventions are fundamental prerequisites for smoothly flowing interaction. Nevertheless, such conventions are arbitrary, can be changed, and may vary across different societies or subcultures (Nicholls & Thorkildsen, 1988). Examples of social conventions are habits of good conduct (e.g., ways of addressing, sufficient self-control, requesting permission) as well as habits of responding and producing (e.g., responsiveness, seriousness, carefulness, quickness, neatness).

Another classroom-relevant subcategory of social demands could be called moral demands. Moral imperatives are like matters of logic and empirical law in the sense that they are understood to be universally applicable and obligatory (Nicholls & Thorkildsen, 1988). Teachers, for instance, may feel they have universal moral grounds to take offense at certain behavior and accuse students of laziness, dishonesty, or arrogance.

In addition to general social and moral conventions students also adapt to more specific and occasional person- and situation-bound social demands. A student may become sensitized to an individual teacher's person-specific expectations (e.g., personal preferences), or to a teacher's occasional expectations arising in specific situations (e.g., whims).

The Conflicting Roles of the Teacher and Their Impact on Teacher-Pupil Interactions

The role of the teacher is multifaceted, and even internally contradictory. This creates certain adaptive problems for teachers. We can assume within the systemic point of view that the behavioral solutions of teachers to these adaptive pressures may, in many cases, be in conflict with the social and learning needs of pupils. There are many potential ways of classifying the different roles that the teacher must play within the school context. For instance, on some occasions the teacher acts as a facilitator of learning, on other occasions she may act as a socializing agent of society, as a creator and supporter of discipline, as an evaluator, as a therapist, as a classroom manager, and so on.

The way the teacher moves from one role to another between different situations is very significant from the point of view of classroom culture and climate, on the one hand, and personal and interactional relationships, on the other. It may be based on explicit pedagogical principles, but for some teachers it is founded more on intuitive, spontaneous situational judgments, of which he or she may not be even aware. However, from the point of view of its consequences, the very central feature may be the consistency versus inconsistency of this behavioral pattern of the teacher. The degree of consistency depends on many factors, some of which are linked to the professional skills of the teacher due to the impact of teacher education, amount of experience as a teacher, and his or her personal qualities.

We can, however, point to many cultural and institutional, that is sociological, factors which may support role conflicts. These frequently cause unintentional situational behavior in teachers and produce inconsistency in the ways teachers combine their varying roles. Lundgren (1981, pp. 24-38) has analyzed so-called frame factors, which direct the educational processes in the classrooms. Frame factors are determined outside the teaching process. The most central are those linked with curriculum, which govern teaching via goals; those linked with administrative apparatus, which constrain the instructional processes; and those linked with the judicial apparatus, which regulate teaching activities through formal rules. Frame factors impact time allocation within teaching, participation structures, teaching methods, evaluation practices, and so on. In addition to those factors mentioned above, expectations targeted at teachers are influenced by teaching traditions, parental pressures, and the societal functions of education. For instance, the selection function of school education is in conflict with the function of cultural transmission and with equality objectives. Such conflicts cause internal contradictions in teachers' roles. The selection function produces certain demands for assessment of pupils (the teacher as an evaluator and gatekeeper) and may disturb the teacher's role as a facilitator of learning who is responsible of the well-being and optimal development of all pupils.

When we try to concretize how role conflicts and adaptive needs of teachers structure their interactive relations to certain pupils we must pay attention to the subjective interpretations included in teachers' and pupils' orientations in different situations and to their dynamic interplay. Gutierrez and Larson (1995) demonstrated how classroom culture based on strongly discipline-oriented scripts and a monolithic vision of what counts as learning actually prevents certain task-focused activities and easily leads to teachers' misinterpretations and strong controlling reactions, especially with pupils from minority cultures. McNeil (1988) has argued that when schools exercise bureaucratic controls, tension develops between the contradictory goals of educating students and of controlling and processing them. From the point of view of the quality of classroom culture, certain kinds of control systems and methods of assessment may influence which kind of goal structure prevails in classrooms: one that triggers mastery objectives (task-focused goals) versus one that triggers performance objectives (ability-focused goals) (cf. Dweck & Leggett, 1988; Ames, 1990). Edwards (1994) has analyzed how control-oriented discipline is conducive to pupils' alienation from learning because it reduces intrinsic motivation and strengthens instrumental, calculative performance motives or negative attitudes and resentment.

According to Frankel and Snyder (1978), self-worth protective individuals are vulnerable to withdrawal of effort in situations felt by them to be threatening (probable failure). Because teachers feel that one of their most crucial duties is to get pupils to do their best, the "laziness" of some pupils in those situations means that a teacher has failed to fulfill his or her pedagogical intentions. In order to shelter his or her professional identity (self-worth as a teacher), the teacher becomes prone to defensively attribute his or her failure to external causes, that is to the laziness or lack of motivation of the pupil, which also arouses emotions of anger toward that kind of pupil. Thus, the behavior of the pupil and the attributional dynamics behind the interaction episode elicit teacher behavior (for instance, pressing the pupil to try harder) which may increase the apparent threat of failure and self-worth on the part of the pupil and strengthen avoidance behavior.

The teacher's attempts to increase task-difficulty or demand level, linked with his or her role of facilitator of pupils' learning and development, may be felt by some pupils to be an increased threat, thereby eliciting non-task-oriented avoidance behavior. In turn, this will hinder the realization of the teacher's intentions and further complicate the interaction between the teacher and certain pupils.

STUDENTS' MOTIVATIONAL ORIENTATIONS, COPING STRATEGIES, AND LEARNING

Low-achieving children differ from their high-achieving peers not only in regard to cognitive-strategic skills but also in the ways they are motivationally, emotionally, and socially "tuned" to academic performance situations. Differences have been found on a variety of motivational and socio-emotional dimensions. When compared with normally or high-achieving peers, subnormally performing children have been found to show a lower academic self-concept (Eshel & Klein, 1981), a lower sense of control (Stipek & Weisz, 1981) or self-efficacy (Schunk, 1989), and lower expectations of success (Butkowsky & Willows, 1980), as well as more negative self-evaluations and self-deprecatory attributions (Pearl, 1982). Children with learning problems have been characterized as performance, ability, or emotion focused, extrinsically motivated, ego-oriented, or helpless, whereas normally or high performing children have been described as learning focused, intrisically motivated, task oriented, or mastery oriented (Haywood, 1968; Thomas, 1979; Licht, 1983; Kistner, Osborne, & LeVerrier, 1988). Most of these dimensions overlap and accentuate the basic distinction between task focused, or task intrinsic, and non-task focused, or task extrinsic, motivation.

The way in which the social environment responds to the young child's actions contributes to the development of generalized motivational and socio-emotional tendencies (Burgess & Conger, 1978; Crittenden, 1981; Crittenden & DiLalla, 1988). The transactions between a child and parent leading to adaptive or maladaptive developmental courses are reciprocal and cyclical in nature (Sameroff, 1975). Just as parents and other adults influence the course of socialization during childhood, the child participating in a transaction can be viewed as a source of influence over his or her own development (Marcus, 1975). As regards the early antecedents of motivational directedness, it is of particular importance how parents respond to and control the child's adaptive efforts in everyday learning and problem situations.

Self-reinforcing transactional cycles seem to be essential to the early development of motivational and socio-emotional dispositions. Parents tend to reinforce the particular child's dominant behavior at the time and shift their interactional styles accordingly (Marcus, 1975). For instance, dependent behavior in children has been found to elicit greater encouragement of dependence and directiveness from parents, whereas independent conduct elicits greater encouragement of independence and nondirectiveness (Osofsky, 1971; Osofsky & O'Connell, 1972; Marcus, 1975; Yarrow, Waxler, & Scott, 1971).

An Adaptation Model for Motivational Orientations

From the point of view of student adaptation, academic learning and performance situations represent extremely complex environments. Learning tasks as such have certain objective requirements. Teachers demand compliance and acceptable performances, compare students and their learning products, give rewards and criticism, provide and fade support, control the students' conduct, and have their expectations, preferences, and rules. Students must adapt to such multilevel demand structures that consist of many simultaneous and often inconsistent requirements.

We constructed a three-part model for the child's adaptation and motivational orientation to learning and performance situations. The orientations are derived from three adaptive goal structures that are formed and actualized during children's learning and social reward histories. Since the motivational orientations are actualized in a certain set of learning situations, they refer to an individual's more or less "global" (i.e., generalized and stabilized) tendency to act in a similar manner in forthcoming situations with similar cues. Nevertheless, the orientations should not be considered as trait-like in nature but as essentially interactionist constructs, that is, as a dynamic typology (Harter & Jackson, 1992; Magnusson, 1985; Mischel, 1979). A typical set of coping strategies tends to be launched when a certain orientation tendency interacts with proper situational cues. This model (see Figure 1) consists of the most essential elements of the socially guided learning situation: the student, the teacher, and the learning task with a curriculum-based learning goal. The model is extremely condensed, and does not cover the actual complexity of the classroom environment, but it suffices

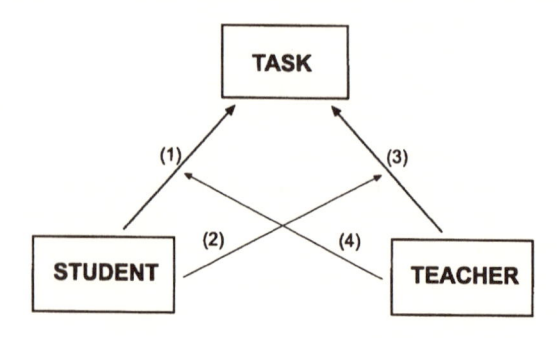

Figure 1. Student's and Teacher's Complementary Task Orientation

to present the basic alternatives for student motivational orientations and the complementary relationships between student's and teacher's reciprocal adaptations. The relationships described by the model refer to the student's (or the teacher's) subjective perspectives: to which aspect of the situation the student (or the teacher) is sensitized, how he or she constructs the situational demands, and how he or she tries to cope with them. The model makes it possible to analyze parallel motivational, affective, and cognitive aspects of task-oriented and non-task-oriented adaptation in instructional settings.

Task Orientation

In task orientation the student's adaptation is dominated by an intrinsically motivated tendency to approach, explore, and master the challenging aspects of the environment. Relationship (1) (Figure 1) represents the student's sensitization to a learning task and his or her tendency to make sense of the task content and to master the learning goal. Intrinsic or mastery motivation elicits not only direct, or "egocentric," task-related curiosity and activity in the student but also interest in the guiding adult's interpretations concerning the task at hand. This is due the fact that most learning tasks and goals are socially prestructured and their meanings cannot be extracted without social perspective taking and social mediation (Lave, 1988; Rogoff, 1990; Brown, Collins, & Duguid, 1989; Mehan, 1979; Cicourel, 1973). Relationship (2) refers to the student's tendency to socially co-construct the task, that is, to complement his or her task interpretation with the perspectives of the teacher (or parent). The child thus coordinates the (interpreted) adult perspective (Rel. 2) to his or her own interpretations of the task (Rel. 1) and, through such interpersonal "decentrations" (Feffer, & Suchotliff, 1966; see also Piaget, 1971), corrects and reorganizes his or her own task assimilations. In Vygotskyan terms, Relation 2 could be described as students' participation in culturally and socially supported activities in their zone of proximal development (Vygotsky, 1978). For the task-oriented student, challenges provided by the learning task (Rel. 1) predominate over all other situational demands. Although the students with task orientation are also involved in constructing the teacher's (or parent's) views of the task (Rel. 2), this is only instrumental to their striving for self-governed learning and independent mastery of learning tasks.

In trying to meet the objective requirements (e.g., the inherent logic or rule system) of the task, the child explores and mentally transforms the task elements, as well as refines his or her task-related plans and schemata. Sustained, intense task-related manipulations and elaborations are characteristic of task orientation. Due to strong intrinsic motivation, the student is fully concentrated on the task and ignores incidental stimuli, so the integrity of action is maintained. Any inconsistencies, obstacles, or teacher prompts and criticism are interpreted as challenges to be responded to with growing persistence and cultivated strategies and not, for instance, with avoidance, inhibition, or immediate help seeking.

The central characteristics of task orientation are highlighted through examples drawn from student interviews. The examples were selected on the basis of typicality criteria, that is, they represent the most common response tendencies that were found in interview data gathered for different studies from over 300 students representing grade levels one to eight (see Lehtinen et al., 1995).

Mastery motivation (Harter, 1975) was indicated throughout most interviews with high performers. These students liked challenging tasks and enjoyed the intellectual "resistance' provided by such tasks. When asked what kind of tasks high performers prefer, they usually said that they like "brain exercise,: "tasks that demand some thinking," or "tasks that are not too easy" (see also Harter, 1974). These verbal responses were in line with the subjects' actual behavior in a level-of-aspiration test (where they were given tasks of different difficulty levels). A high-performing fourth grader expressed his mastery motivation as follows:

Experimenter (E) (after S had chosen a difficult Finnish language task): Why did you choose this kind of task?
Student (S): Because my grade in Finnish was very good...
E: Which kind of school tasks do you like most?
S: Difficult and very difficult.
E: How come?
S: I don't like the easy ones...
E: (showing an easy task): Would you like to try this?
S: I could look at it but I would still like to try more difficult ones.
E: If you were given such (easy) task, how would you feel?
S: Dull
E: Why?
S: Because I like challenges and hate everything too easy

Task-oriented students' mastery motivation is manifested through their preference for new and challenging tasks over old and familiar ones. The following comment of the above fourth grader is very illuminating and typifies task-oriented high performers' beliefs:

E: Which kind of tasks do you like more, those that you don't quite yet master or those that you already master well?
S: The new ones...because you have to clear up something you can't yet do.

Task-oriented high performers' persistence is indicated in their tendencies toward independent struggle despite difficulties and toward a return to interrupted tasks. The tendency toward mastery cannot be explained merely by the motivational fact that high performers tend to relate their momentary level of aspiration to their own (high) level of performance. Many of the best performers (beginning from about 10 years of age) stressed in our interviews the cognitive meaning of difficult tasks. They saw such tasks as tools for self-evaluation and for developing their skills. They emphasized that demanding tasks are the best way to assess their own "real" level of mastery. For instance, in their opinion, "mechanical counting problems" and "memory tasks" do not offer such possibility for checking their own level of mastery or understanding as word problems do. Several high performers referred to the following experience: that only by doing word or "thinking" problems one gets the feeling of certainty; one knows that the solution is right as "the parts finally fit together." The following excerpt from the interview of a high-performing sixth grader illustrates this point:

> E: Which kind of math tasks do you like most?
> S: Word problems.
> E: How come?
> S: Cause one learns to get insight, to understand what it all is about. Sometimes when you can't do it, you can keep the paper and try again.

These interviews illustrate important cognitive and metacognitive components of task orientation. They reflect the students' understanding of the general meaning of problem-solving strategies or processes. The students see that they have learned something important only after having understood how they reached the right solution. The interviewed students seemed to get a sense of satisfaction in mastery and solution: not from "finding the answer" as such, but from the mental transformation and reorganization of the task elements until the parts fit together to form a meaningful, coherent cognitive structure. As the students seemed to understand, such structures of cognition and action are extremely adaptive in school learning. They guide the students' hierarchically organized strategic activity that can be applied to different tasks that have similar logical requirements but may vary greatly with regard to unessential details. This reflects some of the most common cognitive characteristics of task orientation found in high-performing children (from nine-ten years up). They commonly utter that it is nice "to think" or "to use your brains"—to get the full certainty of how to reach the solution.

Characteristic of task orientation is the tendency to assess the level of own mastery and the need for additional effort. This metacognitive goal presupposes task-related goal setting and some kind "objective" attitude toward one's own performance. This, in turn, is not possible without a more or less generalized positive academic self-concept. For instance, the student must have a relatively intact sense of self-efficacy in academic performance situations (Bandura, 1989; Schunk, 1989). For such students, errors or lack of skill are not ego related or social problems to be ashamed of. Instead, they simply reflect the fact that he or she has not yet fulfilled the task requirements.

Task-oriented students typically make a distinction between task-related goals and social goals. For them, the social demands of a learning situation are secondary or submitted to the task requirements. This is reflected in the following interview with a high-performing sixth grader. Although this student sees the teacher's opinions and needs as important, he tends not to comply blindly with the teacher (as socially dependent students are likely to do).

> E: Do you feel that it is most important to learn whatever the teacher stresses?
> S: Yes, because often the things that the teacher emphasizes are important in English language...also in language learning in general...and not only in the teacher's opinion.

Sometimes high-performing students are obliged to make a compromise between social demands and task requirements. In such cases the students are often metacognitively aware of their goal and strategy selection. The following interview, made after a foreign language test comprising many small items, illustrates this point:

> E: Which is more important, to do right or to do quickly?
> S: (sixth grader): To do right.
> E: Why?
> S: If you do quickly, you make careless mistakes; so it doesn't help if you had done it quickly. It's not useful, so it's better to think about the tasks really carefully...though to be fair I didn't do it.
> E: Well, why didn't you do it?
> S: I thought that this test had to be done quickly...I thought there was a time limit...still, I was ready in time.

This student probably thought in the performance situation about the requirements of correctness and speed. He preferred speed, though at the same time was aware that the probability of errors would increase. The student had adopted from school practices the conception that, in certain kinds of tests (e.g., in tests comprising a large number of small items), one has to do as many items as possible, in spite of the fact that at such speed one cannot properly fulfill the task requirements. Task-oriented students may also begin to feel that purely social speed criteria—unrelated to the goal of accurately meeting the task requirements—are essential in typical formative and summative school tests, and thus that these tests are often illogically composed; a lengthy series of small questions.

Although a task-oriented student, when needed, will utilize adults' guidance and feedback to develop his or her own cognitive skills, the student still tends to relate social support to his or her own goals and cognitive structures. Such help-seeking has been called "instrumental," that is, "the help requested is limited to only the amount and type that is needed to allow the child to solve a problem or attain the goal in question for himself or herself" (Nelson-LeGall, 1981, p. 227). A task-oriented student will not engage in "executive" help-seeking, that is, use the teacher or parents as the source of correct answers. The following interviews reflect the need for independent mastery, a disposition that most clearly contrasts task-oriented from non-task-oriented students.

> E: If you cannot do your homework, if it is very difficult, what do you do?
> S: Well, I'm the lucky one whose father happens to be an English teacher. I usually ask him answers...or, I mean, how can the task be put in such way...where he explains the principle of how the task should be done.
> E: Do your parents help you in doing your homework?
> S: No, and I would not like that.
> E: Why?
> S: Somehow...I want to learn myself.

Task orientation is likely to lead to cumulative reinforcing of the generalized positive academic self-concept. Task-oriented students in our interview studies perceived themselves as good learners but considered this in a matter-of-fact, not an ego-enhancing, manner.

Teachers' Expectations and Responses With Regard to Task-Oriented Behavior

Task orientation may be unconditionally "truth-oriented" or more prosocial by nature. In the first case the student rigorously strives for the values of clarity, coherence, and groundedness. Such students are typically critical of inconsistent or incomplete presentations given by teachers or textbooks. Their need for truth, independent understanding, and self-governed task control (Rel. 1 in Figure 1) is overwhelming and it overcomes the need for social consensus and adjustment (Rel. 2 in Figure 1). Some teachers feel that their in-depth questions, critical comments, and arguments disturb the normal course of lessons and consider them as "intellectual troublemakers." Our observations indicate that among the top achievers, such "truth extremism" is not rare. The other subgroup of task-oriented students, the more prosocial or conformist ones, also have a rather strong need for self-governed understanding and mastery (Rel. 1), but they are much more likely to socially accommodate and adjust their interpretations (Rel. 2) and not to risk the social balance too much. The majority of high achievers belong to the latter group.

The way the teacher encounters a rigorous or prosocial task-oriented student depends on the teacher's own orientation. Task-oriented teachers show a strong intrinsic interest and mastery motivation toward the conceptual-logical and instructional aspects of the task. Such teachers typically manifest not only in-depth interest in the subject matter and "the logic of things" (strong Rel. 3; Figure 1), but also a well-developed sense of the student's understanding (Rel. 4; Figure 1). Since task-oriented teachers usually show high self-confidence, they do not interpret task-related uncertainties and controversies as disturbing or threatening. According to their leading idea "let the logic of things decide," such teachers expect and promote student independence, divergent thinking, and cognitive conflicts. They tend to allow and support intellectual dissidents' testing of opposite views against the teacher's and other participants' perspectives. Social balance comprises teacher's and student's symmetrical task directedness with reciprocal joint activity (even "partnership") assumptions and interest in each other's perspectives (see Figure 1). This leads both participants to seek out challenges that often surpass curricular demands. One possible disadvantage of such intellectual endeavors is the lack of consideration of the variety of formal curricular demands, or students' different characteristics, which may result in uneven school success.

There is a more conformist type of teacher orientation that tends to form a balanced system with student prosocial task orientation. These teachers orient themselves according to shared societal expectations (coming from principals, colleagues, and parents) concerning the "normal" practices and efficiency of schooling. Even though they may emphasize cognitive and metacognitive skills, such as summarizing, selecting, transforming, and comparing, these strategies are applied task-immanently, as in constructing coherent text bases from textbook chapters. The instruction rarely transcends subject-matter contents. To teach properly, one must dutifully control the process of going through the "important" instructional contents. Although these teachers could be characterized as task oriented in the sense that they are really involved in instructional contents and methods, their underlying static conception of knowledge is likely to lead to a somewhat "inert" (see Bereiter & Scardamalia, 1993) learning process.

It is obvious that the prosocial variety of task-oriented adaptation corresponds to most teachers' conception of ideal student behavior. Students with such conduct are considered as "bright," "easy to manage," and "nice." These students progress with little teacher effort and without any conflicts. They spontaneously fulfill most teachers' expectations concerning the intellectual and motivational quality of the work (e.g., coherence, independence, persistence), but they are also considered as showing high-grade prosocial skills in their work (e.g., flexibility, empathy, conformity). Teachers, believing in the student's intellectual abilities and skills, tend to allow them more independence and to give them more challenging tasks in the context of conventional academic learning. They also express their beliefs in the student's capacity and give intellectually encouraging feedback with the purpose of integrating the student into the academic community. This, in turn, reinforces in the student's continued school enjoyment, mastery motivation, as well as cognitive strategies and practices embedded in curricular demands and educational tradition. Such a combination typically leads to excellent school achievement but the skills learned may be weakly generalized to out-of-school settings (cf. Resnick, 1987; Brown, Collins, & Duguid, 1989).

Independent elaboration and critical argumentation, as well as possible uneven school success manifested by "rigorously" task-oriented students, however, may cause "didactic uncertainty" or social imbalance since it does not follow conventional expectations. Teachers with lower self-confidence respond to students' activ-

Expectations and Beyond

ity beyond the instructional content as threatening. This elicits principally two kinds of defensive efforts in the teacher. First, the teacher may try to overly control the "too independent" student, for instance by demanding literal following of the instructional form and content or by requiring more drill and practice to guarantee sufficient mastery of task content. Second, the teacher may try to avoid conflict through withdrawing from the dialogue (for instance ignoring) and trying to normalize the situation (for instance turn reallocation, shifting the theme).

Insecure teachers with a weak in-depth interest in the nature of things, as well as self-assured teachers with firm conceptions of content-based truths and "right" methods, are likely to experience intellectual ambiguities and controversies as distracting. According to their epistemic beliefs, all relevant knowledge is "inert" by nature. Consequently they tend to interpret students' elaborations and critical considerations as deviant and fruitless. In cases where the teacher's presentation or guidance appear inconsistent or inconceivable as related to the student's cognitive structure, a "rigorously" task-oriented student tends to seek clarity and consistency, even though this may increase the possibility of social conflict with the teacher. The following observation protocol shows that even very young extremely task-oriented children may argue with the teacher while trying to find the logic underlying the problem or the teacher's utterances. The teacher, in turn, tries to balance the situation through defensive ignoring acts.

Peter, one of the best in his grade, was observed during a mathematics lesson at the end of the first school year (at the age of eight years). Peter had a tendency to interrupt instructional activities while trying to express his large knowledge, reasoning, and opinions—which all were a couple of years ahead of his grade level—through long and eager presentations. Peter put up his hand during the teacher's presentation. The teacher, anticipating the usual disruption, tried to continue her presentation as if she did not notice the sign. As Peter almost stood up on his chair, gesticulating eagerly, the teacher finally responded.

> Teacher (T) (a bit irritated): What's the matter this time, Peter?
> Peter (P): (after having found a printing error in the math book) Why there are two number elevens in this book?
> T: (intending to return as soon as possible to the interrupted presentation) Well, there may be anything in books.
> P: But not dragons.
> T: Sure you may find dragons in dragon books.
> P: But not in math books! (collapses and looks depressed for a couple of minutes)

This student's behavior did not indicate any sign of arrogance or self-assertion. He just wanted to get an explanation to the problem he had found. When he remarked that the teacher had tried to pass his question over by giving an arbitrary answer, he derived from the teacher's answer a logical corollary ("reductio ad absurdum") that highlighted the inconsistency of the teacher's argument.

The above example not only illustrates a task-oriented child's intrinsic need for consistency and clarity, but also the great self-reliance that is needed to strive for this goal in a dispute with authority. Maybe due to the teacher's prejudice against "disturbances" caused by Peter, and due to her lack of interest in the in-depth questions that surpass formal curricular demands, she had a strong tendency to normalize the situation by repressing the disturbing impulses. Yet, she could have taken a marked didactic advantage by asking Peter what he had in mind and by prompting the class into discussion about the severity of "apparent clashes due to misprints" and what "genuine clashes" like "1+1 = 1" would do to our arithmetical thinking.

Social Dependence Orientation

In social dependence orientation the student's adaptation to the learning situation is dominated by social motives, such as seeking help, approval, and affiliation from the authority figure (Crombie, Pilon, & Xinaris, 1991; Crutchfield, 1962; Haywood, 1968; Harter, 1974). Relationship 5 (Figure 2) is dominating.

The student has a strong tendency to monitor, and to adjust to, the teacher's (parent's) performance-related wishes and expectations in order to receive social approval. When trying to interpret the teacher's momentary behavioral cues and to respond acceptably to social demands, the student shifts the responsibility for initiating and controlling the academic activities to the teacher. A socially dependent student is not inclined to meet independently the objective task demands or to construct consistent task representations (i.e., weak or nonexistent Rel. 1). Neither is the student interested in the teacher's view of the task or in co-constructing the task content with the teacher (i.e., weak or nonexistent Rel. 2). Social dependence, as described here, refers to an "executive" type of help-seeking activity and "emotional" aspects of dependence, because the child's intention is to have an adult cope with the task on his or her behalf, and to use questions and bids of help primarily as a means of attaching to the adult (Nelson-LeGall, 1981, pp. 226-227).

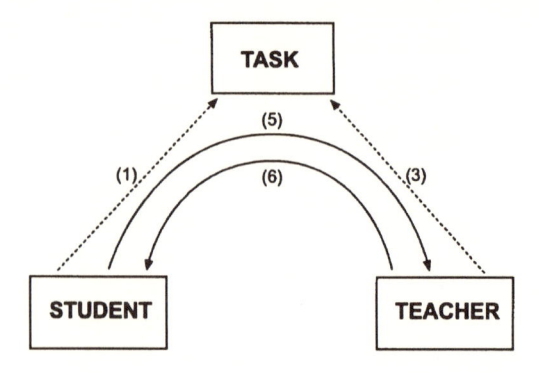

Figure 2. Student's and Teacher's Complementary Social Dependence Orientation

Social dependence can be manifested by two main types of coping strategies for getting support and help: (1) babyish (regressive) appealing behaviors (e.g., helpless gaze, appealing smile, baby talk), and (2) more advanced social tactics for eliciting supportive cuing from the adult (e.g., imitating, "gift of gab," enticement, persuasion).

As a reflection of the social balancing acts typical of social interaction at home and in school, both helplessly smiling babyish children and socially active, "nice" pupils who guess uninhibitedly and give fluent but inconsistent answers tend to be far more over-helped and rewarded than called to account (Holt, 1964). In school, social dependence leads to superficial processing of learning materials; the child will be increasingly bound to form associative links between the elements of learning materials and teacher acceptance cues (Olkinuora & Salonen, 1992).

The following example illustrates the associative and inconsistent nature of knowledge characteristic of a student with a long history of dependence-type coping. Maria, a girl with a very nice outlook and good manners, was followed from seventh to ninth grade (from 14 to 16 years of age). Her knowledge base and skills turned out to be exceedingly fragmentary and fragile in our diagnostic assessments. Besides basic skills in mechanical reading, writing, and arithmetic, very little was built up during the previous school years and regular resource room periods. Characteristic of her performance was the nonexistence of stable, hierarchical knowledge structures. Practically no firm hierarchical organization was found among grammatical, mathematical, geographical, or biological concepts. The answers given to the same question on different occasions seemed to be quite random and often contradictory. Maria seemed to draw her "answer words" from a totally mixed pool of grammatical, mathematical, geographical, or biological terms. The following excerpts from mother language and math diagnostic tests illustrate confusion concerning even the basic grammatical and mathematical categories:

Question (Q): What are the tenses?
Maria (M): Active, passive, singular, plural.
Q: What are the main constituents of the sentence?
M: Verb, adjective, pronoun, particle, numeral, noun.
Q: What are the cases?
M: Particle, perfect.

Q: What kind of numbers are whole numbers?
M: Ten, twenty, thirty.
Q: Are there whole numbers that are not tens?
M: (hesitantly) No.
Q: What kind of number is a whole number?
M: (hesitantly) An odd number.
Q: What kind of number, then, is an even number?
M: I don't know

Due to the lack of any task-oriented aim, this socially dependent student did not have any interest in creating order among task elements and knowledge domains. Her social directedness caused her simply to collect and remember knowledge elements, for instance "grammatical answer words," and to guess blindly which of them might be the "right one" this time.

Teacher Expectations and Responses With Regard to Socially Dependent Behavior

Since one of the major goals of the educational system is prosocial behavior and compliance, students with positive attitudes and good manners contribute positively to the teachers' efforts to establish and maintain social balance. It is not surprising that socially dependent students, who are tuned full time to teachers' wishes, fulfill most teachers' expectations as regards social management. Social balance will be established through the complementary functions of a student's dependent role and the teacher's reciprocal role as "emotional caregiver" (see McLaughlin, 1991). Caring and loving children is a cultural "script" related to mothering (Acker, 1995). This script guarantees the teacher's sensitivity to the students' emotional needs and personal perspectives and is also an important component of the teacher's professional skill. In the case of over-empathy (dominating Relation 6, weak Relation 3; Figure 2), however, students' emotional needs dominate the teacher's interpretation and the teacher is likely to respond to the students' appealing and helpless behaviors (dominating Relation 5, weak Relation 1; Figure 2) with over-helping and overprotective responses. Just like a dependent student, an overempathic teacher has a weak self-governed relationship to the task and subject matter. The balanced social situation is not based on the teacher's and student's coorientation toward the task, as is the case in complementary task orientation, but on the reciprocal attachment needs between the teacher and the student. Even though the latter may form a stable and balanced socioemotional situation, the teacher who devotes her or himself to fulfilling students' needs cannot consistently carry out instructional procedures or serve as a model of in-depth interest in the logic of things.

This example illustrates how Maria continued to give seemingly random "answer words" until the teacher accepted one. If the teacher did not signal acceptance, Maria pursued a "trial and error" technique, no matter how contradictory the answers might be. Such sessions might last over half an hour, and Maria never showed any sign of frustration or emotional tension. The following excerpt from an eighth-grade resource room session illustrates Maria's dependence-type behavior in socially guided instructional situations.

The teacher gives the task $18 - 12 = ?$

M (with an insecure, inquiring look): Sixteen?. No I mean six? (looks inquiringly at the teacher)
T: Which?

M: (looks inquiringly at T) Six?
T: Yes

Teacher gives the task 18 + (-12) = ?

M: (hesitantly) Then one should add these it results minus six? (looks inquiringly)
T: How come it results minus six if you added them?
M: Cause there is that plus (points at the + -sign).
T: Why?
M: (hesitantly) Minus six? (looks inquiringly)
T: (looks at M)
M: (hesitantly): Plus twelve? (looks inquiringly)
T: Try once more.
M (hesitantly): Plus six? (looks inquiringly)
T: Is that right?
M (hesitantly): Yes? (looks inquiringly)

Despite very poor learning results, Maria managed to barely scrape through exams and grades on the grade-per-year basis. Her responses in tests were confused but she remembered enough "answer words" to almost regularly get some "hits" in simple achievement test items. Additionally, Maria's appealing behavior elicited mainly positive responses in the teachers. They responded to her neatness, kindness, and willingness to receive help with sympathy and support. One of the teachers crystallized the situation by explaining that the very lowest grades are not assigned to nice and pleasant students, no matter how weak they are, but to poor students who are negative and ill-mannered.

Maria's extreme dependence was based on a strong attachment to her high-achieving elder sister, who showed task orientation in all school subjects. Over the school years the sister regularly helped Maria and practically prepared the school tasks for her. Maria obviously had shifted the intellectual responsibility to her sister and begun to expect step-by-step guidance and ready-made solutions. The following episode illustrates Maria's dependent coping style in the presence of her sister.

Maria (14 years of age) and her sister (18 years of age) were observed in a joint problem-solving situation. Maria was instructed to calculate the price of wax carpet needed to cover the floor of a certain two-room apartment (a floor plan with length and width measures was given) and prompted to work independently. She could ask for her sister's help if she could not get along.

The sister took the task paper and began to read silently. Maria looked at her sister and occasionally the task paper but seemed not to read systematically. At times both looked at each other and began to giggle as if to reinforce the bond between them. After reading, the sister began to calculate and think aloud. Maria watched on.

S (pointing at the floor plan and looking smilingly at M): Well, then?
M: I can't (both look giggling at each other).

The sister continued calculating and thinking aloud.

    S: (smilingly):You too, Maria, say something.
    M: (giggling): I dunno.

The sister continued to calculate and elaborate the procedure. Maria did not participate in cognitive elaboration. She merely made a couple of content-irrelevant comments concerning the formal aspects of calculation. Several times they both looked at each other and giggled.

Characteristic of the interaction described above is that from the very beginning the older sister took the task-oriented, intellectually active, and leading role, whereas Maria took the dependent, passive role. However, to accommodate the instructions the sister made two weak attempts to prompt Maria's initiation. Maria responded to both trials by signaling a need for help and the sister did not try to shift the responsibility. Maria and her sister exemplify a social system that strives for maintaining social balance through the reciprocal acts of coordinating roles and expectations according to established scripts (see Feffer, 1970; Mehan, 1979). The situation in which Maria is required to take the intellectual responsibility is felt not only by Maria but also by her sister as causing imbalance. Therefore they prefer to resort to the accustomed script. Thus, reciprocal smiling and giggling served not only to reinforce the emotional bond but also to maintain social balance in a situation in which the change of safe conventional roles was required.

Since Maria had practiced needlework, we embedded a practical math problem in a crochet task to see if a personally meaningful problem context might elicit task orientation. In addition, we tried to make dependence-type coping strategies less tempting by reducing the social guidance available. The following excerpt illustrates how powerfully dependence-type coping efforts are maintained in familiar social situations.

Maria and her sister were given a picture of a bedspread composed of little square crochet pieces sewn together. Maria was instructed to work as independently as possible and her sister had been told to avoid helping her. Maria had to compute the total amount of thread required for this crochet work on the basis of the size of the bedspread, of the size of a square piece, and the amount of thread needed for each piece. Maria waited for the sister's initiative for long periods and several times said, smiling, that she could not do the task. Maria then directly asked for the solution to the problem. As instructed, the sister rejected requests for help and prompted Maria to do independent work. After a long period of waiting, Maria finally made some hesitant trial and error attempts, looking inquiringly and smiling at her sister at every step. Maria's activities were quite thoughtless and severely disorganized.

A plausible explanation for Maria's behavior could be that the mere presence of an authority figure (sister, teacher) or some situational cues (a "school-type" setting) cause her to shift the intellectual responsibility to that guiding person.

Our observations suggest that sharpening developmental polarization into task orientation and social dependence among other sibling pairs is not exceptional. It is plausible that due to slight temperament and/or treatment differences one of the siblings may adopt a more active and independent role, whereas the other tends to adopt a more passive role. The parents may contribute to such differentiation by stereotyping and treating the siblings correspondingly (for instance, "our energetic one" versus "our timid one"). As the siblings mirror their roles in each other, the original roles and reciprocals will be gradually reinforced and stabilized.

The above findings illustrate the paramount point we put forward in this paper: Social balancing mechanisms, both in family and classroom systems, may actually increase motivational imbalance between participants, and contribute to the differentiation of learning careers. For instance, the task-oriented participant is likely to derive growing self-confidence and self-governed strategic skills from guiding the dependent one, whereas the dependent participant becomes increasingly addicted to the help and stagnates with regard to the development of cognitive skills.

A typical teacher role that is complementary to a student's social dependence is the "controller" role characteristic of "teacher (and curriculum) programmed" learning environments. Due to widely shared efficiency expectations (felt to come from principals, colleagues, and parents), many teachers tend to think that to obtain the expected instructional results one must direct and control the processes in their classrooms. Under such pressures, many teachers focus on formal and concrete criteria of instructional quality and progress, such as covering the prescribed "important" instructional content, being on time, following the textbook order, or applying close "objective" evaluation procedures. According to the study of Mikkila, Olkinuora, and Laaksonen (1997), students in this kind of learning environment are regarded as being less responsible for their own learning than the students of less textbook-bound teachers. Fluent, progressive instruction is seen as a balanced didactic "normal form" by more "conformist" teachers, who are regulated more by interiorized social expectations than by intrinsic interest in the learning tasks and subject matter. Routinization may be based on the teacher's anticipation of ego-related threats such as loss of classroom control and deviant student behavior (extremely strong Relation 8, weak Relation 3; Figure 3).

A strictly controlling teacher role is "suited" to socially dependent students who are disposed to adjust to adults' expectations and feedback with submission and giving up of their intellectual responsibility. The balance, however, is not as complete and stable as in mutual emotional attachment. Socially dependent students willingly accommodate to clear-cut social rules, but may not receive as much emotional support, help, and detailed feedback as they wish. On the contrary, the teacher often responds to their appealing and feedback-hunting strategies by ignoring or even rejecting. The following case example illustrates such imbalance.

Jani is a chubby, babyish third grader with a lovable wide-eyed appearance and a strong tendency toward social dependence of the "babyish" variety. Over the first two school years, Jani had a motherly, supportive teacher who had responded to his appealing and help-seeking coping efforts with empathy and step-by-step guidance. In third grade Jani got a strict male teacher who expected more "mature," that is, more independent and adult-like, conduct.

We observed Jani and his teacher in an achievement test setting at the end of the third grade. A sentence completion-type text-comprehension task (a simple story with some missing words) was administered to the class. The students were instructed to work independently and ask the teacher for help only "in an emergency."

T (giving instructions): Now, first fill in your name, your school, and your grade on the lines at the top of the paper.
J (looks in a confused way at other participants and experimenters in the room, looks helplessly at the teacher and puts up his hand)
T: (with a bit irritated voice) Jani, what do you have in mind?
J (timidly): Do I put XX School (the school's name) on the top?
T (with an annoyed and loud voice): I just said it! Didn't you listen?! Your name, school, and grade.

Jani does not start working on the task. For long periods of time, he does not look at his test paper, but glances around hesitantly and inquiringly. He looks regularly at the other students working on their tasks and the teacher who walks around in the classroom. Every now and then he fills in a couple of words that turn out to be rather random guesses with little relationship to the semantic context. The teacher, noticing Jani's inattentiveness, is annoyed, steps forward, and points sharply at the task-paper. Jani hunches, startled, nearer the paper.

The teacher walks around toward Jani's desk. Jani follows the teacher with sidelong glances and hunches timidly as the teacher approaches. Just when the teacher passes by, Jani starts hesitantly to put up his hand, obviously with the intention of getting help. However, Jani seems not to have enough courage to request help, and he suddenly transforms the intended help-seeking gesture to the light stroking of ear and hair.

The teacher approaches Jani's desk again and Jani follows with sidelong glances. As the teacher passes, Jani hunches over the paper and freezes as if he was concentrating on the task.

After a while the teacher approaches again. Jani, looking a bit frightened, hunches over the paper. Without raising his gaze, Jani cautiously shifts his hand so that the pencil points to a certain row of the text (as if intending "I know that asking for help is improper but you may help if you want"). The teacher notices this slight cue and gives a general direction, "Think about the meaning of the sentence." Jani, who apparently expected a more concrete hint, remains in a state of confusion.

The above example illustrates how the teacher, expecting independence and concentration on the task, responded to overwhelming help-seeking and continuous off-task behavior with strictness and irritation. Our further observations indicate that the teacher's strict demands for independence and refusals of help contributed to the re-

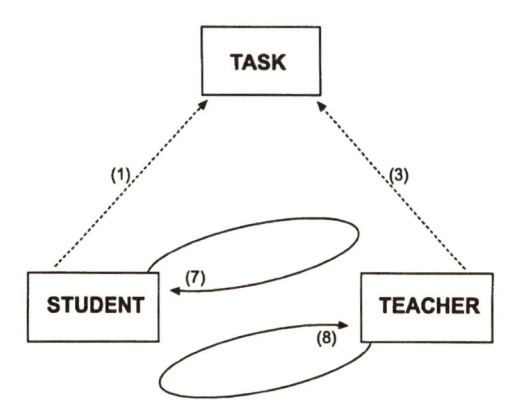

Figure 3. Student's and Teacher's Complementary Ego-Defensive Orientation

duction of Jani's dependence-type coping behavior over the third and fourth grade. Even though Jani's explicit requests for help were gradually inhibited and a more "balanced" social situation was created, this happened at the cost of Jani's growing withdrawal, avoidance, and inhibition. In other words, instead of task-oriented coping behavior, Jani shifted increasingly to ego-defensive coping efforts.

Ego-Defensive Orientation

In an ego-defensive orientation the student's adaptation to the learning and performance situation is dominated by self-defending and self-protecting motives.

The student thinks of him or herself as an object-like entity rather than as an active agent, and is sensitized to situational factors suggesting ego-related threat or risk, such as task-difficulty cues and signs anticipating the teacher's negative responses (Relation 7, Figure 3). Novelties and ambiguities will not be interpreted as challenging starting points for task-related efforts but rather as discomforting obstacles that suggest loss of personal control, or failure, and should be avoided.

Task-approach demands given by the teacher, together with strong inhibitory forces, induce a motivational-emotional conflict that may lead to regressive disorganization of the already established structures of action. Situational stress factors (e.g., social demands, task-difficulty cues, evaluation pressures) inducing tension may have extensive destructive effects on performance, ranging from the disorganization of automatized perceptual processes and motor dysregulation to the deterioration of higher cognitive functions (Olkinuora & Salonen, 1992; Tinbergen & Tinbergen, 1985; Wine, 1979).

An ego-defensively oriented child tends to alleviate the motivational-emotional conflict or postpone the intensification of tension either by using inhibited, passive-avoidant-type behaviors (e.g., withdrawing, daydreaming, freezing), with substitute activities (e.g., selecting easier substitute tasks, playing), or by active-manipulative ("acting out") social behaviors (e.g., whining, defiance, aggression, tantrums).

Characteristic of the ego-defensive orientation is a weak mastery motivation in academic performance situations. The child does not experience a new (or a new type of) task or teacher prompt as challenging and does not enjoy the intellectual resistance provided by such demands. By contrast, the child's low sense of self-efficacy leads him or her to interpret such tasks or teacher demands as threatening. When encountering new tasks, low-achieving children tend to emit expressions like "It's too hard for me," "I can't do it," "I don't like this," or "Let's do something nice." The following interview with a low-achieving fourth grader illustrates these tendencies, which are indicated by the avoidance of new (or assumed to be difficult) tasks and by the preference for familiar (or easy) tasks:

> E: Which kind of tasks do you like more, new tasks that you don't quite yet master or old tasks that you already master well?
> S (fourth grader): The old ones.
> E: How come?
> S: An emergency may happen, if you have chosen a new task, so you cannot do it.

This student's response clearly indicates a poor sense of self-efficacy. New tasks are seen as sources of insecurity. They may cause "emergencies" that threaten the unsteady personal control. Many interviews revealed that the main concern of an ego-defensive child seems not to be the difficulty itself but the anticipated social consequences, that is, expected negative/positive evaluation. Here, again, the extrinsic aspects of an ego-defensive student's motivation are highlighted. Low-achieving students typically believe that doing familiar tasks better secures good grades, so the uncertainty regarding teacher feedback will be reduced.

Certain task-difficulty cues (e.g., "word problem," "long text," "big numbers") activate avoidance tendencies in ego-defensive students:

> E (asking before the performance of a math word problem): How do you like the task?
> S (sixth grader): It's too long, I don't want to begin it.
> E: Why?
> S: Cause it's a word problem!

Low-achieving, ego-defensive students typically prefer low levels of aspiration and report avoidance tendencies when they encounter difficulties during performance. Their low persistence is indicated in their tendency to give up and to avoid returning to interrupted tasks. These tendencies are illustrated in the following interviews:

Expectations and Beyond 137

> E: Imagine that you are doing quite difficult homework and nobody is at home to help you. Which do you like more, to continue or to stop trying?
> S (fourth grader): Stop trying...usually next day I'll ask the teacher.
> E: Why do you want to stop trying?
> S: I don't understand...and I've had enough of thinking.
>
> E: Do you like thinking about tasks you did not complete or you made mistakes in?
> S: No.
> E: Why?
> S: I get tired...I'm dead tired in the morning..tired of thinking...it's no fun at all.

The latter example in particular indicates typical extrinsic motivational characteristics of ego-defensiveness, that is, the lack of (intrinsic) interest in learning tasks and the concentration on product aspects of performance. Such centering on product may be indicated merely by worry connected to giving a wrong answer, as it is in the case of pure ego-defensiveness, or, additionally, by a tendency toward "executive" help-seeking (Nelson-LeGall, 1981), where the child still relies on some kind of rescue in the form of ready-made solutions from the teacher. In the latter case, ego-defensiveness is mixed with a tendency toward social dependence.

As stated earlier, ego-defensive children tend to be involved in development that undermines their sense of self-efficacy and leads to a more or less generalized poor academic self-concept. In our interviews many low-achieving third and fourth graders and older children expressed a rather generalized negative academic self-concept. The following examples are illustrative:

> E: Let's discuss for a while what you think of yourself as a student, as a learner....Think about yourself as a student trying to learn, to understand and to remember things. Tell me something about yourself, as a learner.
> S (fourth grader): Poor.
> E: Tell me what kind is a poor student?
> S: Don't know.
> E: Try to think about it.
> S: A student who gets poor grades and doesn't answer.
> E: Do you think you are such a student?
> S: Yeah.
> E: In every subject and situation?
> S: Not quite in every subject.
> E: In which subject you think you are not poor?
> S: In craft...

Teacher Expectations and Responses With Regard to Ego-Defensive Behavior

Teachers' self-confidence is largely based on their belief in being able to promote students' learning and prosocial activities. Students' relatively permanent avoidance and acting out behaviors stand in conflict with this basic expectation, and are interpreted as threatening. Teachers with a tendency to be sensitized to ego-related threats (dominating Relation 8, Figure 3) are especially likely to feel accused of being unable to control student behavior. They tend to attribute the causes of the permanent problem to stable factors, such as their own weakness or inability. Since these teachers feel that their professional ego is at stake, such a conflict is extremely disturbing. A possible balancing mechanism may be based on the teachers' self-serving attributional bias called "defensive externalization," although the empirical evidence of this mechanism is not unequivocal (see Miller, 1995; Tollefson, Melwin, & Thippavajjala, 1990; Medway, 1979). It is plausible that many teachers are tempted to attribute the cause of a pervasive problem to an external, stable factor, such as student deviance, disability, or deficit in social background, because defensive externalization alleviates the conflict and gives an excuse or justification for the teacher's failure and withdrawal. Parents as well as teachers may hope for a similar "release" from being possibly accused of causing, or not being able to control, their child's problem behavior. As if made to order for social balancing purposes, medical models of special education and clinical psychology provide the teacher and parents with a handy collection of "official releases," diagnostic labels referring to alleged "constitutional" deficits (minimal brain dysfunction, attention deficit hyperactivity disorder, dyslexia, dyscalculia, etc.) (see Miller, 1995; Sarason & Doris, 1979; Stanovich, 1993). Through such labeling/attribution processes, the teacher can show to her or himself, and to others, that reducing demands, letting the student be, or referral is "justified." Correspondingly, the student's avoiding or deviant coping strategies prove successful so that the student, as well as the teacher, feels less threatened. The achieved socio-emotional balance is based on the status quo of reciprocal avoidance and withdrawal (Figure 3). It will, however, continuously distort the student's task-related activity.

The teacher may try to control the avoidant student by, for instance, demanding strictly active efforts and concentration on task. Second, the teacher may try to avoid the conflict through withdrawing from the dialogue (for instance, ignoring) and through trying to normalize the situation (for instance, turn reallocation, shifting the theme).

Such social balancing is illustrated by the case of Heli, a female student, who was followed from third to sixth grade. She had great difficulties with text comprehension, restlessness, and lack of concentration, yet she was a rather fluent reader. At the beginning of the follow-up, Heli's responses to various motivational disposition measures were conflicting. A self-concept of attainment-rating scale (Nicholls, 1978) indicated a very optimistic (but unrealistic) academic self-concept: Heli rated herself in all subjects among the best students in her grade, whereas the teacher rated her as next to last (25th of 26). In contrast, she demonstrated an extremely low sense of control in a questionnaire concerning typical situations of scholastic success and failure. Also, Heli's disposition toward fear of failure and her low sense of self-efficacy were systematically indicated in most performance situations. If tasks representing several levels of difficulty were given, she immediately selected the easiest task and said that she did not like more difficult ones. She was extremely sensitive to even slight task-difficulty cues or teacher prompts.

The following cumulative process was observed in dozens of situations. While encountering a new task she immediately groaned "Too long a text!," "I can't do it!," "I don't like this!," or "Gimme an easier one!." If she did not succeed in her initial attempt to avoid the task, she tried a mixture of mainly non-task-oriented coping strategies. Social dependence coping strategies (e.g., appealing, babyish chatting, smooth-tongued questioning aimed at eliciting direct teacher help or "piloting") prevailed at first, but then, if the teacher did not give tangible hints or continued to ask for independent performance, ego-defensive strategies began to predominate. Avoidance behaviors varied from the passive-inhibited modes (e.g., absent-minded staring, averting gaze from the task, turning away, or "silent treatment") to active, manipulative forms of behavior (e.g., attempts to leave the field, engaging in substitute and subsidiary activities, stereotyped rituals and mannerisms, signaling tiredness, fierce demanding, whining, or intentional tantrums). Symptoms of motivational conflict and emotional tension typically increased after only one or two experienced difficulties or task-approach requests. "Short-cut" behaviors with a low level of organization (e.g., responses to arbitrary situational cues) were increasingly observed. The frequency of substitute actions, as well as the intensity of restlessness and social conflict behavior, increased rapidly during such situations. In many instances of extreme tension, regressive disorganization reached even automatized skills, distorting fine motor regulation (e.g., impairing of writing). Very typically, after a recent failure, the depressive emotional state was transferred to the next task, and was indicated by self-blaming thoughts and negative expectations.

The case of Heli provides evidence that the cyclic processes leading to cumulative reinforcing of non-task-oriented coping tendencies are interactively formed during long-term development and maintained by situational conditions. With time, interaction with the remedial teacher resulted, with increasing frequency, in extreme forms of avoidance and conflict behavior. The frustrated teacher sometimes finally gave up efforts to elicit the independent task-related activity required at Heli's grade level. Having just responded with extreme avoidance, conflict behavior, and disorganization of action (tantrums), she shifted almost immediately to social dependence coping when she realized that the teacher had given a series of routine school tasks and was no longer asking difficult questions. Heli, now appealing, smiling, and chatting, completed the series without a sign of fatigue or frustration. It seemed that such episodes increased the teacher's tendency to "balance" the social situation through allowing a very low level of aspiration. After such a tension-reducing change, Heli tried to maintain the established comfortable situation through working cheerfully on easy tasks.

Heli was observed in situations in which her father was guiding her task performance. In these situations she launched similar sequences of situational transactions whenever she was faced with difficulties. The father had a tendency to demand and even force Heli to concentrate on tasks. Heli's avoidance and conflict behaviors escalated rapidly and culminated in open social conflict. The tearful daughter frequently threatened: "I'll tell Mom." The father, then, apparently having a bad conscience, gave up his demands and worked on the task for the daughter. Heli, in turn, almost immediately took the role of a "good girl." In an interview the father showed awareness that Heli "is playing a social game" in which she enters in alliance with her emotionally more sensitive mother against the occasionally stern father. According to the father, Heli manipulates the mother to provide unnecessary help in homework situations in which he would require more independent work. The father admitted that he, too, "has the tendency to relent" in the face of the daughter's extreme frustration. Both parents reported that such episodes had been rather frequent during the first school years.

Heli's coping behavior did not change much from third to sixth grade, except that more extreme forms of conflict and avoidance behavior came to the forefront and babyish appealing behavior decreased. Task-oriented behavior was still very occasional and was restricted to situations in which Heli felt secure.

Heli's self-concept of attainment changed dramatically from the third to the sixth grade. In three years her optimistic academic self-concept collapsed completely. Heli now believed that she was the poorest student in her grade (30th of 30), in all subjects. She said that she did not expect a better position in the future. The interviews and other disposition measures revealed a systematic change toward negative expectations, depressive moods, and self-blaming causal attributions. In contrast to the third grade data, Heli's observed behavior-in-context coincided more unequivocally with her disposition measures.

RECIPROCAL FORMATION OF ORIENTATION TENDENCIES: EMPIRICAL OBSERVATIONS IN CLASSROOMS

In order to gather empirical evidence on the reciprocal development of orientation tendencies in classrooms, we made a study (Hamalainen & Lehtinen 1989; see also Vauras, Lehtinen, Kinnunen, & Salonen, 1992) which was aimed at investigating how the interaction between the teacher and individual students is related to the development of the student's orientation. Methodologically, we presume that the teacher's influence on students' motivation is mediated by the reciprocal reorganization of individual interpretations during repeated micro events. In order to analyze the motivationally relevant micro events of classroom interaction, a qualitative diagnosis system was developed. Our analysis was aimed not at classifying just the

Expectations and Beyond

events of external behavior, but at interpreting the intentions behind the different acts and the subjective meanings of the interaction events for the participants.

Six teachers and their 125 fifth-grade students served as subjects. Detailed interactional analysis concerned 17 learning-disabled and 20 top-achieving students. The analysis of students' orientation tendencies was based on assessments by teachers and extensive observations during the eight-month diagnostic and therapeutic period (approximately 16 hours of observations per student, in individual, small group, and classroom situations). The intensity of maladaptive coping tendencies (for ego-defensive and/or socially dependent behavior) was rated on the basis of all motivational assessments.

In order to analyze the quality of the teacher-student interaction, four, 45-minute lessons in each of the six classes were videotaped over a period of five months. Interaction acts seen as significant for the development of learning skills and socioemotional coping strategies were analyzed. The interaction modes of teachers were assessed within three categories in terms of how much the teacher's interactions (a) increased or reduced the students' coping strategies included in an ego-defensive orientation, (b) increased or reduced the students' coping strategies included in a social-dependence orientation, and (c) encouraged the adoption and elaboration of cognitive and metacognitive skills and the coherent construction of domain knowledge. Further, the assessment of interaction modes was made separately in terms of the teacher's interaction acts with students classified as having different orientation tendencies.

The results showed that the quality of the teaching-learning is strongly related to the students' orientation tendencies. In general, the teachers tended to reduce the ego-defensive coping tendencies but, at the same time, to increase students' tendency toward social dependence. Many interaction episodes were also observed in which the teacher systematically tried to improve students' task orientation by giving them a challenging task, by supporting their independent study processes, and by encouraging students toward higher levels of aspiration. However, this took place only in teachers' interactions with high achieving students who already had a strong tendency toward task orientation. The teachers' interactional patterns were distinctly different with the LD children. The teachers were prone to increase LD students' anxiety and tendency toward ego-defensive coping. They did this, for example, by giving overtly disapproving feedback and reinforcing negative self-attributions, by inducing social comparison, or by introducing tasks in an insufficiently explicit manner that made students feel a loss of control.

Fifth grader Sauli Lehmus (male) had the strongest tendency to ego-defensive orientation in the class. Almost always when the teacher was interacting with Sauli he used an interaction mode which he only rarely used with other students. The teacher almost never used Sauli's first name. Instead, he used nicknames which he formed from Sauli's surname and a title somehow connected to the content of the teaching. The nicknames typically conflicted with Sauli's incorrect or insufficient answer, and this frequently amused other students. For example "professor Lehmus" was not able to give a simple explanation of a biological phenomenon, "mathematician Lehmus" gave incorrect answers to simple arithmetic tasks, and "engineer Lehmus" failed to describe the function of an engine. The teacher typically gave very short comments on students' answers. He only indicated if the answer was right or wrong. In Sauli's case, however, he often used longer feedback expressions, which highlighted negative features of Sauli's person.

The teacher also used Sauli once as scapegoat when he wanted to punish the whole class. During a science lesson the teacher informed students of an upcoming exam. Several students (including Sauli) asked for a detailed description of what they should read for the exam. The teacher became frustrated because he had not yet planned the exam and was not able to give the students an adequate answer. "This discussion stops now, Mr. Lehmus. If you squeak once more, I give you a good hiding." This caused amusement among other students and a shame reaction in Sauli.

Likewise, teachers were strongly prone to increase LD students' tendency toward social-dependence, for example, by encouraging impulsive guessing, by introducing routine tasks without demands for intellectual responsibility, or by providing instant assistance.

In subsequent analyses, the association between a teacher's behavior and an individual student's socio-emotional coping tendencies was investigated in greater detail. The results indicated that two of the teachers varied their interaction in a therapeutic manner, by trying to reduce the maladaptive coping strategies of those students who had a very strong tendency toward ego-defensiveness or social dependence. But four teachers showed the opposite characteristic: they varied their behavior in such a way as to increase the maladaptive coping strategies in students. This was especially pronounced with highly ego-defensive children. The correlation between the teachers' tendency to increase ego-defensive orientation and the students' initial ego-defensive coping tendency was 0.48. In the case of social dependence, the correlation was 0.31. However, this low correlation was partly due to the fact that teachers tended to increase social-dependence behavior in all students, whatever their coping tendencies.

It is important to note that these results are based on videotaped lessons. This means that the teachers were aware that their behavior was to be analyzed by the researchers. Nevertheless, they frequently used interaction modes that tended to strengthen students' initial orientation tendencies and to increase interindividual motivational differences. The teachers seemed to vary their interaction from individual to individual. In most cases of LD children this leads to transactional changes in the teacher's behavior and in the student's coping which promote regression in the student's motivation and learning skills. We assume that weak stability and poor transfer of learning skills, which have been observed in many remedial programs, can be partially explained by the observed qualities of ordinary classroom interaction.

In our intervention studies (Lehtinen et al., 1995; Vauras, Lehtinen, Kinnunen, & Salonen, 1992; Vauras, Salonen, & Naskali, 1992), we have approached the problem of transaction of cognitive and motivational factors from a somewhat different angle. We were interested in seeing whether the defective learning of low-achieving students with a low task orientation could be enhanced in an environment in which both motivational coping strategies and cognitive and metacognitive skills are dealt with explicitly. We designed a 32-hour cognitive-motivational intervention program for fourth-grade, 10-11- year-old, learning disabled students. This cognitive-motivational treatment was compared to pure strategy and motivation treatments and to a control condition. We were also interested in finding how low-achieving children of contrasting motivational orientations responded to these treatments. Furthermore, after the intervention, an experimental classroom program was designed with the teachers to examine the post-intervention conditions for maintenance and transfer of the motivational orientations and cognitive skills which have been attained.

Although significant improvement was obtained both for the coping plus strategy training and for the strategy training alone, long-term transfer effects on learning processes were found only for the coping plus strategy intervention. As a result of the intervention in small groups, enhanced maintenance and transfer were found primarily among children with some initial task orientation. However, when extended classroom training was provided, even children with low initial task orientation and maladaptive coping tendencies began to show more lasting training effects and transfer of learning. Training was least successful in children who displayed dominantly strong and stable ego-defensive coping tendencies in various instructional and learning situations. Most of these children showed no progression, or very sporadic progression with high situational variation, as a result of training in small groups (Lehtinen et al., 1995).

Typical of the children who showed only short-term training effects were low task orientation and a strong tendency to cope in learning situations through social compliance. Transfer of learning skills could be achieved mostly among the children who consistently displayed some degree of task orientation and only low-to-moderate levels of ego-defensiveness or social compliance. However, when extended training was provided, even children with low initial task orientation and maladaptive coping tendencies showed more lasting training effects and transfer to learning. It is not enough to change individual students' orientation tendencies in small group training; rather the whole system of expectations and interaction patterns in the classroom should be changed to turn a regressive circle toward progressive motivational and cognitive development (Lehtinen et al., 1995).

CONCLUSIONS

We have proposed a theoretical model for explaining the different effects of teacher activities and expectations on students' development. Our model, however, only indirectly deals with the traditional Pygmalion study of Rosenthal and Jacobson (1968). We do not directly focus on teacher expectations but aim at understanding the systemic conditions of reciprocal adaptation between student and teacher. We claim that initial teacher expectations, whether accurate or biased, cannot explain the whole problem of school-reinforced diverging student careers. Teachers' interaction behavior in a classroom situation results from many institutional frame factors and situational features. Our results indicate that it originates from dynamic but gradually stabilizing interaction patterns between the teacher and individual students. Initial teacher expectations concerning the capacity of an individual student certainly play a role in the formation of these interaction modes, but this reciprocal process is also affected by each student's initial orientation in the classroom situation.

We believe that this leads to two conclusions that could partly explain the controversial results of teacher expectations research. First, teachers expectations do not have similar effects in students with different initial orientation tendencies. Students with high task orientation are not particularly sensitive to teachers' personally related expectations. They define school situations above all in terms of the cognitive demands of the tasks. Initial tendency toward task orientation makes students more resistant to teachers' expectations concerning their ability, whereas non-task-oriented students are much more prone to recognize teachers' expectations and beliefs related to their person. In the case of non-task-orientation, teacher expectations can decrease (positive expectations) or increase (negative expectations) a student's social dependence or ego-defensiveness. Second, initial teacher expectations will not necessarily have permanent, long-term effects on students' development. The effects of self-fulfilling prophecies may dissipate due to the interactive formation of students' and teachers' reciprocal orientations. This means that the interactive formation of students' situational orientations goes beyond teacher expectations and that the self-fulfilling prophecies hypothesis is insufficient to explain the problem of strongly divergent student careers.

The descriptions of three orientation types could be criticized as too general and inaccurate. However, as systemic concepts they have proved to be helpful in diagnosing different learning careers in classrooms and in designing effective remedial programs for learning disabled students. When dealing with possibilities to improve learning by using technologies, Gavriel Salomon (1996) has stressed the importance of better understanding of systemic changes in interaction and activity patterns in classrooms. Possible effects cannot be explained using only individual variables or mere statistical interactions between several factors. Our theoretical analyses and empirical results support this notion. Especially when we try to understand long-term development of differentiating learning careers or to change individual students' regressive developmental cycles into progressive learning processes, a more systemic or transactional view of classroom activities is needed.

It might be important for remedial educational programs to convince teachers of the positive effects of high expectations, but this can be of little consequence if teachers do not become aware of the interactive formation of divergent communication patterns with different students. High teacher expectations are not enough to counteract regressive developmental cycles of students. Our results suggest that more versatile and individually focused interventions are needed to change simultaneously students' learning and coping strategies and teachers' interaction modes with those students.

REFERENCES

Abraham, J. (1989). Testing Hargreaves' and Lacey's differentiation-polarisation theory in a setted comprehensive. *British Journal of Sociology, 40*, 46-81.
Abramson, L. Y., Seligman, M. E., & Teasdale, J. D. (1978). Learned helplessness in humans: Critique and reformulation. *Journal of Abnormal Psychology, 87*, 49-74.
Acker, S. (1995). Carry on caring: The work of women teachers. *British Journal of Sociology of Education, 16*, 21-36.
Adelman, H. S., & Taylor, L. (1983). Enhancing motivation for overcoming learning and behavior problems. *Journal of Learning Disabilities, 16*, 384-392.
Ames, C. (1990). Motivation: What teachers need to know. *Teachers College Record, 91*, 409-421.
Baker, L., & Brown, A. (1984). Metacognitive skills and reading. In P. Pearson, R. Barr, M. Kamil, & P. Mosenthal (Eds.), *Handbook of reading research* (pp. 353-394). New York: Longman.
Ball, S. J. (1981). *Beachside Comprehensive: A case-study of secondary schooling*. Cambridge: Cambridge University Press.
Bandura, A. (1989). Human agency in social cognitive theory. *American Psychologist, 44*, 1175-1184.
Barga, N. K. (1996). Students with learning disabilities in education: Managing a disability. *Journal of Learning Disabilities, 29*, 413-421.
Bereiter, C., & Scardamalia, M. (1993). *Surpassing ourselves: An inquiry into the nature and implications of expertise*. Chicago: Open Court.
Berends, M. (1995). Educational stratification and students' social bonding to school. *British Journal of Sociology of Education, 16*, 327-351.
von Bertalanffy, L. (1969). Chance or law. In A. Koestler & J. Smythies (Eds.), *Beyond reductionism: New perspectives in life sciences* (pp. 56-84). London: Hutchinson.
Black, F. W. (1974). Self-concept as related to achievement and age in learning disabled children. *Child Development, 45*, 1137-1140.
Blumenfeld, P. C., Pintrich. P. R., & Hamilton, V. L. (1987). Teacher talk and students' reasoning about morals, conventions, and achievement. *Child Development, 58*, 1389-1401.
Bronfenbrenner, U. (1979). *The ecology of human development: Experiments in nature and design*. Cambridge, MA: Harvard University Press.
Brophy, J. (1983). Research on the self-fulfilling prophecy and teacher expectations. *Journal of Educational Psychology, 75*, 631-661.
Brown, A., Campione, J., & Barclay, C. (1979). Training self-checking routines for estimating test-readiness: Generalization from list learning to prose recall. *Child Development, 50*, 501-512.
Brown, J. S., Collins, A., & Duguid, P. (1989). Situated cognition and the culture of learning. *Educational Researcher, 18*, 32-42.
Brown, A., & Ferrera, R. (1985). Diagnosing zones of proximal development. In J. Wertch (Ed.), *Culture, communication and cognition: Vygotskian perspectives* (pp. 273-305). New York: Cambridge University Press.
Brown, A., Palincsar, A., & Purcell, L. (1986). Poor readers: Teach, don't label. In U. Neisser (Ed.), *The school achievement of minority children: New perspectives (pp. 105-144)*. Hillsdale, NJ: Erlbaum.

Buckley, W. (1967). *Sociology and modern systems theory.* Englewood Cliffs, NJ: Prentice-Hall.
Burgess, R. L., & Conger, R. D. (1978). Family interaction in abusive, neglectful, and normal families. *Child Development, 49,* 1163-1173.
Butkowsky, I., & Willows, D. (1980). Cognitive-motivational characteristics of children varying in reading ability: Evidence for learned helplessness. *Journal of Educational Psychology, 72,* 408-422.
Carlberg, C., & Kavale, K. (1980). The efficacy of special versus regular class placement for exceptional children: A meta-analysis. *Journal of Special Education, 14,* 295-309.
Charlesworth, W. (1976). Human intelligence as adaptation. In L. Resnick (Ed.), *The nature of intelligence* (pp. 147-168). Hillsdale, NJ: Erlbaum.
Cicourel, A. V. (1973). *Cognitive Sociology: Language and meaning in social interaction.* Harmondsworth: Penguin.
Cicourel, A. V., Jennings, K., Jennings, S., Leiter, K., Mackay, R., Mehan, H., & Roth, D. (1974). *Language use and school performance.* New York: Academic Press.
Cole, M., & Bruner, J. S. (1974). Cultural differences and inferences about psychological processes. In J. S. Bruner (Ed.), *Beyond the information given* (pp 452-467). Old Woking, Surrey: Unwin.
Cooper, H., & Good, T. (1983). *Pygmalion grows up: Studies in the expectation communication process.* New York: Longman.
Covington, M. V., & Beery, R. (1976). *Self-worth and social learning.* New York: Holt, Rinehart & Winston.
Crittenden, P. M. (1981). Abusing, neglecting, problematic, and adequate dyads: Patterns of interaction. *Merrill-Palmer Quarterly, 27,* 201-218.
Crittenden, P. M., & Di Lalla, D. L. (1988). Compulsive compliance: The development of an inhibitory coping strategy in infancy. *Journal of Abnormal Child Psychology, 16,* 585-599.
Crombie, G., Pilon, D., & Xinaris, S. (1991). Children's problem-solving performance: The effects of compliance and cognitive factors. *Journal of Genetic Psychology, 152,* 359-369.
Crutchfield, R. (1962). Conformity and creative thinking. In H. Gruber, G. Terrell, & M. Wertheimer (Eds.), *Contemporary approaches to creative thinking* (pp. 120-140). New York: Prentice-Hall.
Durrant, J. E. (1993). Attributions for achievement outcomes among behavioral subgroups of children with learning disabilities. *Journal of Special Education, 27* (3), 306-320.
Dweck, C., & Leggett, E. (1988). A social-cognitive approach to motivation and personality. *Psychological Review, 95,* 256-273.
Edwards, C. H. (1994). Learning and control in the classroom. *Journal of Instructional Psychology, 21,* 340-346.
Elashoff, J. D., & Snow, R. E. (1971). *Pygmalion reconsidered.* Worthington, OH: Jones.
Elley, W. (1992). *How in the world do students read?* Hamburg: The International Association for the Evaluation of Educational Achievement.
Ericsson, K. A., & Charness, N. (1994). Expert performance. Its structure and acquisition. *American Psychologist, 49,* 725-747.
Eshel, Y., & Klein, Z. (1981). Development of academic self-concept of lower-class and middle-class primary school children. *Journal of Educational Psychology, 73,* 287-293.
Feffer, M. (1970). A developmental analysis of interpersonal behavior. *Psychological Review, 77,* 197-214.
Feffer, M., & Suchotliff, L. (1966). Decentering implications of social interaction. *Journal of Personality and Social Psychology, 4,* 415-422.
Feuerstein, R., Rand, Y., Hoffman, M. B., & Miller, R. (1980). *Instrumental enrichment.* Baltimore: University Park Press.
Frankel, A., & Snyder, M.L. (1978). Poor performance following unsolvable problems: Learned helplessness or egotism? *Journal of Personality and Social Psychology, 36,* 1415-1423.
Fuchs, D., & Fuchs, L. S. (1994). Inclusive schools movement and the radicalization of special education reform. *Exceptional Children, 60,* 294-309.

Gerber, M. M., & Semmel, M. I. (1984). Teacher as imperfect test: Reconceptualizing the referral process. *Educational Psychologist, 19*, 137-148.
Goffman, E. (1972). *Relations in public.* Harmondsworth: Penguin Books.
Gutierrez, K. D., & Larson, J. (1995). Cultural tensions in the scripted classroom: The value of the subjugated perspective. *Urban Education, 29*, 410-442.
Hamalainen, M., & Lehtinen, E. (1989, September). *The formation of learning skills in teacher student interaction.* Paper presented at the European Conference for Research on Learning and Instruction, Madrid.
Hargreaves, D. H. (1967). *Social relations in a secondary school.* London: Tinling.
Harter, S. (1974). Pleasure derived by children from cognitive challenge and mastery. *Child Development, 45*, 661-669.
Harter, S. (1975). Mastery motivation and the need for approval in older children and their relationship to social desirability tendencies. *Developmental Psychology, 11*, 186-196.
Harter, S., & Jackson, B. K. (1992). Trait vs. nontrait conceptualizations of intrinsic/extrinsic motivational orientation. *Motivation and Emotion, 16*, 209-229.
Haywood, H. C. (1968). Motivational orientation of overachieving and underachieving elementary school children. *American Journal of Mental Deficiency, 72*, 662-667.
Hepler, J. B. (1994). Mainstreaming children with learning disabilities: Have we improved their social environment. *Social Work in Education, 16*, 143-153.
Holt, J. (1964). *How children fail.* New York: Pitman.
Juel, C. (1988). Learning to read and write: A longitudinal study of 54 children from first through fourth grades. *Journal of Educational Psychology, 80*, 437-447.
Kauffman, J. M. (1989). The Regular Education Initiative as Reagan-Bush policy: A trickle-down theory of education of the hard-to-teach. *The Journal of Special Education, 3*, 256-278.
Kauffman, J. M., Gerber, M. M., & Semmel, M. I. (1988). Arguable assumptions underlying the Regular Education initiative. *Journal of Learning Disabilities, 21*, 6-11.
Kistner, J. Osborne, M., & LeVerrier, L. (1988). Causal attributions of learning-disabled children: Developmental patterns and relation to academic progress. *Journal of Educational Psychology, 80*, 82-89.
Kivinen, O., & Rinne, R. (1996). Higher education, mobility and inequality: The Finnish case. *European Journal of Education, 31*, 289-310.
Knapp, M. S., Shields, P. M., & Turnbull, B. J. (1995). Academic challenge in high-poverty classrooms. *Phi Delta Kappan, 76*, 770-776.
Lacey, C. (1970). *Hightown grammar.* Manchester: Manchester University Press.
Lave, J. (1988) *Cognition in practice.* Cambridge, MA: Cambridge University Press.
Lehtinen, E., Vauras, M., Salonen, P., Olkinuora, E., & Kinnunen, R. (1995). Long-term development of learning activity: motivational, cognitive, and social interaction. *Educational Psychologist, 30*, 21-35.
Licht, B. (1983). Cognitive-motivational factors that contribute to the achievement of learning-disabled children. *Journal of Learning Disabilities, 16*, 483-490.
Lundgren, U. (1981). *Model analysis of pedagogical processes.* Stockholm Institute of Education, Department of Educational Research, CWK Gleerup.
Macbeth, D. H. (1990). Classroom order as practical action: The making and un-making of a quiet reproach. *British Journal of Sociology of Education, 11*, 189-214.
Madden, N., & Slavin, R. (1983). Mainstreaming students with mild handicaps: Academic and social outcomes, *Review of Educational Research, 53*, 519-569.
Magnusson, D. (1985). Implications of an interactional paradigm for research on human development. *International Journal of Behavioral Development, 8*, 115-137.
Marcus, R. F. (1975). The child as elicitor of parental sanctions for independent and dependent behavior: A simulation of parent-child interaction. *Developmental Psychology, 11*, 443-452.
McLaughlin, H. J. (1991). Reconciling care and control: Authority in classroom relationships. *Journal of Teacher Education, 42*, 182-195.

Mc Neil, L. (1988). *Contradictions of control: School structure and school knowledge*. New York: Routledge.

Medway, F. J. (1979). Causal attribution for schoolrelated problems: Teacher perception and teacher feedback. *Journal of Educational Psychology, 71* (6), 809-818.

Mehan, H. (1979). *Learning lessons*. Cambridge, MA: Harvard University Press.

Mikkila, M., Olkinuora, E., & Laaksonen, E. (1997). *Teacher's textbook use and student's learning orientation*. Manuscript submitted for publication.

Miller, A. (1995). Teachers' attributions of causality, control and responsibility in respect of difficult pupil behavior and its successful management. *Educational Psychology, 15*, 457-471.

Minke, K. M., Bear, G. G., Deemer, S. A., & Griffin, S. M. (1995). Teachers' experiences with inclusive classrooms: Implications for special education reform. *Journal of Special Education, 30*, 152-186.

Mischel, W. (1979). On the interface of cognition and personality: Beyond the person-situation debate. *American Psychologist, 9*, 740-754.

Nelson-LeGall, S. (1981). Help-seeking: An understudied problem-solving skill in children. *Developmental Review, 1*, 224-246.

Nicholls, J. (1978). The development of the concepts of effort and ability, perception of academic attainment, and the understanding that difficult tasks require more ability. *Child Development, 49*, 800-814.

Nicholls, J., & Thorkildsen, T. (1988). Children's distinctions among matters of intellectual convention, logic, fact, and personal preference. *Child Development, 59*, 939-949.

Nicholls, J., & Thorkildsen, T. (1989). Intellectual conventions versus matters of substance: Elementary school students as curriculum theorists. *American Educational Research Journal, 26*, 533-544.

Oakes, J. (1985). *Keeping track: How schools structure inequality*. New Haven, CT: Yale University Press.

Oakes, J. (1992). Can tracking research inform practice? Technical, normative, and political considerations. *Educational Researcher, 21*, 12-21.

Oakes, J., & Lipton, M. (1992). Detracking schools: Early lessons from the field. *Phi Delta Kappan, 73*, 448-454.

Olkinuora, E., & Salonen, P. (1992). Adaptation, motivational orientation, and cognition in a subnormally-performing child: A systemic perspective for training. In B. Wong (Ed.), *Intervention research in learning disabilities: An international perspective* (pp. 190-213). New York: Springer-Verlag.

Osofsky, J. D. (1971). Children's influence upon parental behavior: An attempt to define the relationship with the use of laboratory tasks. *Genetic Psychology Monographs, 83*, 147-169.

Osofsky, J. D., & O'Connell, E. (1972). Parent-child interaction: Daughters' effects upon mothers' and fathers' behaviors. *Developmental Psychology, 7*, 157-168.

Palincsar, A. (1986). The role of dialogue in providing scaffolded instruction. *Educational Psychologist, 21*, 73-98.

Palinscar, A., & Brown, A. (1984). Reciprocal teaching of comprehension-fostering and monitoring activities. *Cognition and Instruction, 1*, 117-175.

Paris, S., & Jacobs, J. (1984). The benefits of informed instruction for childrens' reading and comprehension skills. *Child Development, 55*, 2083-2093.

Pearl, R. (1982). LD children's attributions for success and failure: A replication with labeled learning disabled sample. *Learning Disability Quarterly, 5*, 173-176.

Piaget, J. (1961). The genetic approach to the psychology of thought. *Journal of Educational Psychology, 52*, 275-281.

Piaget, J. (1971). *Science of education and the psychology of the child*. London: Longman.

Pressley, M., Levin, J., & Bryant, S. (1983). Memory strategy instruction during adolescence: When is explicit instruction needed. In M. Pressley & J. Levin (Eds.), *Cognitive strategy research: Psychological foundations* (pp. 25-49). New York: Springer.

Pugach, M. C. (1995). On the failure of imagination in inclusive schooling. *Journal of Special Education, 29*, 212-223.
Resnick, L. B. (1987). Learning in school and out. *Educational Researcher, 16*, 13-20.
Rogoff, B. (1990). *Apprenticeship in thinking: Cognitive development in social context*. New York: Oxford University Press.
Rosenthal, R. (1985). From unconscious experimenter bias to teacher expectancy effects. In J.B. Dusek (Ed.), *Teacher expectancies* (pp. 37-65). Hillsdale, NJ: Erlbaum.
Rosenthal, R., & Jacobson, L. (1968). *Pygmalion in the classroom: Teacher expectation and pupils' intellectual development*. New York: Holt, Rinehart & Winston.
Rueda, R., & Mehan, H. (1986). Metacognition and passing: Strategic interactions in the lives of students with learning disabilities. *Anthropology & Education Quarterly, 17*, 145-165.
Salomon, G. (1996). Studying novel learning environments as patterns of change. In S. Vosniadou, E. De Corte, R. Glaser, & H. Mandl (Eds.), *International perspectives on the design of technology-supported learning environments* (pp. 363-377). Mahwah, NJ: Lawrence Erlbaum.
Salonen, P., Lepola, J., & Niemi, P. (in press). The development of first-graders' reading skill as a function of pre-school motivational orientation and phonemic awareness. *European Journal of Psychology of Education*.
Sameroff, A. J. (1975). Early influences on development: Fact or fancy. *Merrill-Palmer Quarterly, 21*, 267-294.
Sameroff, A. J. (1983). Developmental systems: Contexts and evolution. In P.H. Mussen & W. Kessen (Eds.), *Handbook of child psychology* (Vol. 1, pp. 237-294). New York: Wiley.
Sarason, S., & Doris, J. (1979). *Educational handicap, public policy, and social history: A broadened perspective on mental retardion*. New York: Free Press.
Schofield, J. W. (1993). Promoting positive peer relations in desegregated schools. *Educational Policy, 7*, 297-317.
Schoggen, P. (1989). *Behavior settings: A revision and extension of Roger G. Barker's ecological psychology*. Stanford, CA: Stanford University Press.
Schultz, K. (1994). "I want to be good; I just don't get it." A fourth grader's entrance into a literacy community. *Written Communication, 11*, 381-413.
Schunk, D. (1989). Self-efficacy and cognitive development: Implications for students with learning problems. *Journal of Learning Disabilities, 22*, 14-22.
Seligman, M. (1975). *Helplessness: On depression, development, and death*. San Francisco: Freeman.
Semmel, M. I. Gerber, M. M., & Mac Millan, D. L. (1994). Twenty-five years after Dunn's articles: A legacy of policy analysis research in special education. *The Journal of Special Education, 27*, 481-495.
Sheer, V. C., & Weigold, M. F. (1995). Managing threats to identity: The accountability triangle and strategic accounting. *Communication Research, 22*, 592-611.
Sindelar, P., & Deno. S. L. (1979). The effectiveness of resource programming. *Journal of Special Education, 12*, 17-28.
Stanovich, K. E. (1986). Matthew effects in reading: Some consequences of individual differences in the acquisition of literacy. *Reading Research Quarterly, 21*, 360-406.
Stanovich, K. E. (1993). Dysrationalia: A new specific learning disability. *Journal of Learning Disabilities, 26*, 501-515.
Stipek, D., & Weisz, J. (1981). Perceived personal control and academic achievement. *Review of Educational Research, 51*, 101-137.
Stone, A. C., & Wertsch, J. V. (1984). A social interactional analysis of learning disabilities remediation. *Journal of Learning Disabilities, 17*, 194-199.
Thomas, A. (1979). Learned helplessness and expectancy factors: Implications for research in learning disabilities. *Review of Educational Research, 49*, 208-221.
Thompson, T. (1994). Self-worth protection: Review and implications for the classroom. *Educational Review, 46*, 259-274.

Tinbergen, N., & Tinbergen, E. (1985). *Autistic children: A new hope for a cure*. London: Allen & Unwin.
Tollefson, N., Melvin, J., & Thippavajjala, C. (1990). Teachers' attributions for students' low achievement: A validation of Cooper and Good's attributional categories. *Psychology in the Schools, 27*, 75-83.
Torgesen, J. (1977). The role of nonspecific factors in the task performance on memory task of learning disabled children: A theoretical assessment. *Journal of Learning Disabilities, 10*, 27-34.
Torgesen, J., & Licht, B. (1983). The learning disabled child as an inactive learner: Retrospect and prospects. In J. McKinney & L. Feagans (Eds.), *Current topics in learning disabilities* (Vol. 1, pp. 3-31). Norwood, NJ: Ablex.
Urdan, T., Midgley, C., & Wood, S. (1995). Special issues in reforming middle level schools. *Journal of Early Adolescence, 15*, 9-37.
Vauras, M., Kinnunen, R., & Kuusela, L. (1994). Development of text-processing skills in high-,average-, and low-achieving primary school children. *Journal of Reading Behavior, 26*, 361-389.
Vauras, M., Lehtinen, E., Kinnunen, R., & Salonen, P. (1992). Socio-emotional coping and cognitive processes in training learning-disabled children. In B. Wong (Ed.), *Intervention research in learning disabilities: An international perspective* (pp. 163-189). New York: Springer-Verlag.
Vauras, M., Salonen, P., & Naskali, T. (1992). *Interaction of learning-disabled children's coping and cognitive activity in training text processing skills.* Unpublished manuscript.
Vygotsky, L. S. (1978). *Mind in society. The development of higher psychological processes.* Cambridge, MA: Harvard University Press.
West, C., & Anderson, T. (1976). The question of preponderant causation in teacher expectancy research. *Review of Educational Research, 46*, 613-630.
Wine, J. D. (1979). Test anxiety and evaluation threat: Children's behavior in the classroom. *Journal of Abnormal Child Psychology, 7*, 45-59.
Wong, B. Y. L. (1985). Metacognition and learning disabilities. In D. Forrest-Presley, G. MacKinnon, & T. Waller (Eds.), *Metacognition, cognition, and human performance* (Vol. 2, pp. 137-180). Orlando, FL: Academic Press.
Wong. B. Y. L. (1991). The relevance of metacognition to learning disabilities. In B. Y. L: Wong (Ed.), *Learning about learning disabilities* (pp. 231-258). San Diego, CA: Academic Press.
Yarrow, M. R., Waxler, C. Z., & Scott, P. M. (1971). Child effects on adult behavior. *Developmental Psychology, 5*, 300-311.
Zigmond, N., Jenkins, J., Fuchs, L. S., Deno, S., & Fuchs, D. (1995). When students fail to achieve satisfactorily: A reply to McLeskey and Waldron. *Phi Delta Kappan, 77*, 303-306.

SHARED EXPECTATIONS:
CREATING A JOINT VISION FOR URBAN SCHOOLS

Catherine D. Ennis

When educators discuss expectations, they often focus on goals for student achievement or behavior articulated by professional educational organizations or school boards (e.g., American Association for the Advancement of Science, 1993; McCombs, 1992; National Council of Teachers of Mathematics, 1989). These documents articulate a top-down or outsiders' view of how students should approach the tasks of schooling and what knowledge they will learn. Recently, we have taken a more bottom-up or insiders' view, focusing on expectations that students have for school and for their own learning. Not surprisingly, these different perspectives reveal wide discrepancies between educators' and students' expectations for achievement and behavior in schools.

Nowhere are these differences more evident than in urban public schools. Although disagreements over expectations occur in rural and suburban schools, urban schools often include a variety of factors that make consensual expectations an essential component of successful school life. Even when variables of income level and location are disaggregated, urban students still face greater threats to the quality and integrity of their education than do students in rural or suburban schools. These unique educational challenges are associated with family background (i.e., single-parent families, school mobility), school experiences (i.e., difficulty hiring certified and effective teachers, low teacher control over curriculum, inadequate classroom discipline, more frequent weapons possession, student pregnancy), and student outcomes (i.e., lower school achievement and completion rates and poverty and unemployment of young adults) (Lippman, Burns, & MacArthur, 1996).

Divergent viewpoints about acceptable behaviors and the nature of content to be learned lead to serious confrontations between educators and students in urban schools (e.g., Ennis, 1996; Farrell, Peguero, Lindsey, & White, 1988). For educators, confrontations frequently result in feelings of unhappiness, discomfort, job dissatisfaction, disengagement, and, at times, fear when confrontations are vocal or violent. Administrators and teachers often respond by attempting to control expectations through a proliferation of rules and policies that are hard to enforce and that students perceive as unfair. Students may respond with a loss of interest in learning and an unwillingness to engage in the educational process, leading to detentions, suspensions, or expulsions. In each instance the membership bonds between the student, the educators, and the school are weakened or broken (Hirishi, 1969; Wehlage, Rutter, Smith, Lesko, & Fernandez, 1989).

Some teachers attempt to enhance student engagement and promote school membership by negotiating expectations with students (Ennis, 1995). The negotiation process preserves some of the expectations valued by teachers and students, while forcing each to give up or give in to a few of the others' wishes. Unfortunately, some students feel they must compromise their cultural values to earn a high school diploma (Farrell et al., 1988; Ladson-Billings, 1995). Similarly, teachers express frustration when they must compromise their content focus to maintain classroom order and motivate students to engage (Ennis, 1995). In each situation the zero-sum nature of the compromise leaves both educators and students frustrated and dissatisfied. Thus, although it may be possible to negotiate a few common goals that both teachers and students believe essential, other valued expectations are lost in the compromise process.

Rather than teacher-controlled or negotiated expectations, I believe greater success is found in a side-by-side integration of expectations as teachers and students create shared goals based on a joint vision for the future. The creation of shared expectations produces an environment where both teachers and students find meaning and value. Opening and establishing dialogues provides the foundation for a trusting relationship in which both students and teachers can acknowledge success and willingly engage in the educational process.

In this chapter I first will discuss motivational and social aspects of expectation formation within the frameworks of teacher-controlled and teacher-student negotiated perspectives. I will illustrate these using examples from my ethnographic, interpretive accounts of expectations in urban public schools. These sections will be followed by a proposal for a more side-by-side or integrated approach to expectation formation. In this process teachers and students socially construct a shared vision in which both acquire more than they hoped for or dreamed possible through educationally meaningful and personally satisfying experiences.

TEACHER-CONTROLLED EXPECTATIONS

Teachers' rights and responsibilities to control student expectations are founded in the moral authority traditionally given to them as community trustees (Metz, 1978). Recently, however, in loosely structured urban communities, increasingly impoverished and ethnically diverse parents gradually have lost direct influence, efficacy, and the ability to determine policy and expectations in public schools (Brantlinger, 1991; Kantor & Brenzel, 1992). One of the consequences of this loss of parental influence in schools has been a reduction of parents' perceptions of themselves as stakeholders in the success of the school and a lessening desire to contribute to the school's effectiveness. This can produce home-school rifts leading to reductions in parental support for the school's agenda. Without parental support, educators' find that students question their moral authority to create and enforce expectations. Students are less likely to acknowledge academic and behavioral expectations different from the values and aspirations learned at home and on the streets and more willing to confront teachers regarding curriculum and teaching methods (Ennis, 1994). Faced with students who are unprepared academically, some teachers lower their academic standards to avoid feelings of personal and professional failure (Page, 1990). Students in these classrooms find the curriculum increasingly irrelevant and meaningless, intensifying their sense of alienation and resistance to the educational process. In this section I will discuss how teachers' loss of moral authority, lowered academic and behavioral expectations, and teacher-student confrontations impact teachers' ability to control academic and social expectations in urban schools.

Moral Authority

Historically, adults have held the moral authority to articulate and enforce specific expectations for adolescents and children in schools (Burbules, 1995). Moral authority is the authority given to an individual to carry out the policies and rules of society (Metz, 1978). Authority entails the superordinates' right to command and the subordinates' duty to obey. Moral authority is intact in schools in which both teachers and students acknowledge the teachers' authority to select and teach socially sanctioned content and the students' obligations to learn that content. In schools where teachers and students agree on the relevance of academic content and a set of socially appropriate behaviors, disagreements and confrontations are minimal.

Teachers' moral authority to establish and enforce behavioral and academic expectations has a long history in public schools (Kliebard, 1987). For example, numerous public documents have been written articulating expectations for socially responsible behavior implemented as school rules and policies (Wentzel, 1991). Expectations for academic achievement are articulated as goals and objectives used to focus content and evaluative practices. Research repeatedly has supported the effectiveness of appropriate academic expectations to stretch and challenge students to perform above their current achievement levels (e.g., Good & Brophy, 1994).

Burbules (1995) argues, however, that legitimate teacher authority has clear boundaries. Teachers can exceed their authority by making arbitrary demands on students, allowing themselves privileges not available to students, or selecting content topics or examples consistent with the teacher's interest and expertise, but inconsistent with those of the students. Also researchers investigating schools on a more individual basis found extensive diversity in school contexts which contributed to varied expectations for achievement and behavior (e.g., McNeil, 1988). In urban high school classes I have examined, the moral authority of the teacher was confronted by students who were both noncompliant and disruptive (Ennis, 1995, 1996). Teachers responded by lowering expectations both for students and for themselves, contributing to an educational environment that students perceived as alienating and meaningless.

Expectations for Low-Income and Ethnically Diverse Students

Urban schools represent settings in which educators are increasingly asked to teach low-income students from diverse cultural and language backgrounds (Banks, 1994). Many of these students require remediation to work at acceptable grade-level standards. They depend on the school to provide services and emotional support that their families may be unable or unwilling to provide (Kantor & Brenzel, 1992). In this context, some teachers lower their expectations for students to compensate for the lack of readiness or previous academic preparation that constrains students' ability to meet external expectations (Ennis, 1995). These teachers may resent being assigned to teach lower ability students and the need to motivate, stimulate, and assist some students to focus on learning tasks. Additional problems with behavior in the form of disruption or noncompliance often leads to teacher dissatisfaction and disappointment with the class and school environments (Page, 1990).

In this atmosphere it is easy for educators to blame someone else for social problems, leaving themselves free to adjust expectations to avoid failure. Adjustments usually result in lowered expectations with concomitant increases in control, both in the classroom and in the school (Noguera, 1995). Students are given less responsibility, fewer intellectually stimulating tasks, and reduced opportunities to work collaboratively with peers (Page, 1990). This response may preserve the self-respect of some school personnel, but it does little to enhance the achievement levels of their students (Ennis, 1995).

The reality of poverty for many low-income students often leads to insufficient access to supportive adults (Fine & Mechling, 1993) and limited opportunities to experience innovative, challenging curricula (Page, 1990). Students may not have access to school environments with high expectations for academic success that provide future opportunities for legal, well-paying, and meaningful employment (Fine, 1991; Knapp, 1995). Over the years, lowered expectations for many low-income students and students of color has resulted in a sense of educational powerlessness and meaninglessness (Fine, 1991).

Although many students acknowledge the connection between a high school diploma and social mobility, they are less sure of the long-term value of academic knowledge and skills instrumental to graduation (Fine, 1991). The membership bonds of attachment, commitment, involvement, and belief in the value of schooling (Hirschi, 1969) that tie students to school and enhance the possibility of school engagement weaken or fail to form (Wehlage et al., 1989; Willis, 1977). School becomes a social gathering place in which adolescent desires for peer interaction and status are fulfilled (Peshkin & White, 1990).

Teachers in these schools observe a steady stream of students entering their classrooms who are increasingly more difficult to motivate and engage in the educational process (Solomon, 1992). Some teachers may become less interested and motivated to connect with students who appear distant and preoccupied with a social environment that includes gang affiliation (Vigil, 1993) and teen pregnancy (e.g., Wehlage et al., 1989). As teachers adjust to students who feel increasingly alienated and marginalized, they, in turn, may lower their expectations for their own teaching and the academic performance of their students (Schlosser, 1992). The consequences of this iterative process are manifest in an environment with decreasing standards for students who see little value in their schooling. In my urban school research, these factors often contributed to a context in which teachers' expectations changed from those associated with high to moderate academic achievement to expectations of student apathy, disengagement, noncompliance, disruption, and confrontation (Ennis, 1995, 1996). The research examined first in this paper (Ennis, 1995) was conducted in 10 urban high schools in a large school district (enrollment greater than 110,000) on the East Coast. The schools were selected because they enrolled a high percentage of African-American students (69%) who were taught by predominately European-American teachers (80%). The 10 teachers who participated averaged 25 years of teaching experience and represented an equal number of males and females. The district had been operating under a court-ordered desegregation plan since 1972. It used an extensive magnet school program k-12 to entice European-American and high-ability students to travel voluntarily to minority and low-income areas of the district. Field notes were collected in classes teachers considered their "best, average, and most difficult." Teachers were interviewed on three separate occasions: during the observation period, at the conclusion of the observation, and two months following the last observation. Concrete references to situations observed formed the basis of the semi-structured interview questions. Field notes and interviews were analyzed using constant comparison (LeCompte & Priessle, 1993). Themes were developed across schools and teaching situations. A detailed explanation of methods and findings from this study are reported in Ennis (1995).

A second study (Ennis, 1996) was conducted the following year in the same school district and focused on field notes and interview data from 10 teachers in 10 urban high schools different from those reported in the earlier Ennis (1995) research. Teachers represented five subject areas and ranged in teaching experience from 21 to 38 years. Field notes were collected over a six-week period. Each teacher was interviewed three times using protocols similar to the earlier Ennis (1995) research. Data were analyzed using constant comparison (LeCompte & Priessle, 1993). A detailed explanation of methods and findings from this study are reported in Ennis (1996).

In these schools teachers actively sought explanations for student lack of interest and low achievement in the students' backgrounds, ethnicity, and out-of-school affiliations, such as those that led to gang activity, late parties, or graffiti writing:

> [Teachers] believed that many students did not care about education in general, while others explained students' lack of motivation and low expectations as the product of dysfunctional families. They suggested that students who lacked emotional support were less likely to care about educational goals. Other teachers focused on the inconsistency between the schools' educational expectations for students and the students' own aspirations for the future (Ennis, 1995, p. 449).

McLaughlin and Health (1993) found that civic leaders held similar expectations for youth when discussing the difficulty of maintaining viable neighborhood youth organizations. One civic leader suggested:

> Most of the youth have no motivation. It's learned at home from parents who don't work hard or work only when it is convenient for them. They come home at night and drink beer. Or maybe they don't even come home, a lot of them just carouse at bars. The kids do poorly in school. We need to redirect the parent. Their main problem is the parents have no respect for education (p. 211).

By framing the problem outside the auspices of the school, civic leaders and teachers deflect the focus from educational solutions. They absolve themselves of responsibilities to persuade students to accept societies' academic expectations. With little hope of overcoming the students' background, teachers see no need to design and implement innovative plans to assist low-achieving students (Brophy, 1996; Ennis, 1995; Fine, 1991). One teacher explained:

> It's just that the general attitude of the students has changed to not caring for the teachers or the school or themselves. The ones who do care are a minority and they don't have any influence on the rest. We have a lot more negative [student] leaders they are willing to follow (Ennis, 1995, p. 450).

Teachers also revealed their traditional reliance on grades as a motivator and as a means of communicating their academic expectations. Historically, they depended on parents' and students' esteem for passing grades to stimulate positive behavior and achievement. From a behavioral standpoint, grades were used to coerce students to follow directions, accept and complete assignments, and project at least a neutral, if not a positive, attitude toward academic activities. One teacher in an urban high school explained:

> I think the major change [over the last five years] has been in my expectations for myself. Knowing that my young people are not the least bit interested in grades....They could care less if they get an A or a D as long as they get the credit. It matters not to them. If students come to class with their homework completed, even if it is not correct, I am very pleased; I praise that student. But most don't complete homework assignments, most do not participate in class. I teach only the most basic concepts in my classes. And I teach them over and over again to the same students. I try to use different tasks and different assignments, but they don't learn even the simplest concepts. They are not dumb; they just aren't interested (Ennis, 1996, p. 150).

As students become increasingly more alienated from the school environment or their expectations for a high school diploma become diminished, they are less willing to be controlled by grades. For some students the opportunity to pass the course with a D was satisfactory. Others attended school to pass the time and to socialize with friends, holding no expectation for passing grades. Because students did not hold educational expectations, they were often disruptive, leading to frequent suspensions. Many ultimately were expelled or dropped from the school rolls upon reaching age 17 (Ennis, 1995; Fine, 1991).

Teachers in these urban schools argued that the school's educational mission and purpose were different from the students' goals for success. School documents and teacher statements focused on the traditional subject area knowledge necessary to achieve a high school diploma, pass standardized tests, and further college aspirations. These goals were consistent with expectations from the community of taxpayers who supported the schools and from the administrators and teachers who had, themselves, benefited from this acquisition of formal knowledge. Educators expressed frustration and dissatisfaction when they were forced to admit that their students did not want the knowledge they had spent their professional career acquiring. This rejection of the knowledge base and lack of expectations for academic and community success also was perceived as a rejection of themselves as educators and individuals. Instead, students viewed school as a social setting where they enjoyed interactions with friends, received valued social support, and perhaps met their future spouse. One teacher expressed her frustration with this perspective:

> The main issue that we see here is a disinterest or a lack of understanding of what school is for. Some see it as totally a social environment. They are here for girlfriends and boyfriends, parties, and having a good time. School and classrooms are not a part of their lives except to provide a place to meet socially (Ennis, 1995, p. 451).

Kantor and Brenzel (1992) point out that lack of interest among low-income students or students from oppressed minorities (Ogbu, 1994) is rooted in decades of cultural and economic subservience. Peer norms associated with low achievement and underemployment of high school graduates appear to affect students' classroom performances adversely. When students do not expect their effort, hard work, and sacrifice to lead to a better future, they are less likely to expend energy on academic achievement. Page (1990) and Fine (1991) provide elaborate descriptions of schools in which low-income, low-achieving, minority students received irrelevant, fragmented curricula. Students were tracked into these classrooms in an effort to remediate their deficiencies. Oakes (1992) has provided ample evidence indicating that remedial or lower track curricula, such as those found in these programs, do not contribute substantially to student achievement gains. Further, these programs employed a rigid retention policy and disciplinary practices that made schools feel more like prisons than learning environments (Noguera, 1995). Not surprisingly, teachers in Page's and Fine's studies as well as those in my research reported low expectations for student success that blended negatively to decrease academic performance and students' perceptions of their own efficacy.

Confrontations to Moral Authority

Confrontations occur in schools when individuals disagree and are unwilling to negotiate, compromise, or work collaboratively to create shared understandings (Ennis, 1996). They range from angry, loud disagreements to subtle refusals, such as ignoring directions. Confrontations between educators and students can occur with one student or with an entire class or group of students. They are usually initiated when students perceive that administrators' or teachers' requests or expectations are unfair, unfeasible, or will lead to student failure. Students also may purposefully choose to confront educators when they are bored with the content and recognize that confrontations add excitement to an otherwise uninteresting class (Erickson & Shultz, 1992). As one student explained, "If nothing is happening in class, we will make something happen!" (Ennis & McCauley, 1996a).

Environments where teachers or students expect confrontation are stressful even when such confrontation does not occur. Expectations for confrontation create unpleasant situations in which individuals assume defensive postures that reduce their desire and ability to work collaboratively to enhance learning. Confrontation with students is an expectation for teachers in many urban high schools (Ennis, 1996; Schlosser, 1992). Both male and female teachers reported numerous instances where they had confronted students over class management or content issues. Disputes arose when students questioned why they were required to learn specific content or to participate in a particular activity. Expectations for confrontation were so alarming to some teachers that they reported altering the curriculum. Some even avoided content that they anticipated would lead to confrontation. A music teacher explained:

> I begin [in the fall] by talking about their music but I think they should also listen and become knowledgeable about music from other cultures as well. This year I tried to include Japanese,

Russian, and Austrian composers, but it was a battle from the start. The students were disruptive. It was clear that they were not interested in the characteristics of music from cultures [other than African American]. I finally just gave up. I simply skipped that part of our music curriculum... (Ennis, 1996, p. 150).

A mathematics teacher also expressed frustration and discouragement with his students' lack of interest in his subject:

> It has gotten to the point where I just don't want to come to my sixth-period class anymore. The students don't care about their grades, and they certainly don't care about geometry!...I am teaching; I am trying. But with these kids, it is very difficult. They have been turned off to math and turned off to school. I have stopped even trying to teach more than just the basics (Ennis, 1996, p. 150).

Students in these teachers' classes refused to comply with traditional school expectations for particular content and learning procedures. Other urban students, described by their teachers as disruptive and confrontational, reported they refused to complete work sheets as the primary learning task (Ennis & McCauley, 1996a). Students were particularly angry when teachers used the work sheets to avoid interactions with them and admitted they were insubordinate when they refused. Nevertheless, they felt confrontation was their only avenue to avoid the "boring" tasks associated with this teaching format.

Erickson and Shultz (1992) noted that teachers are constrained in their selection of content by the decrees of school boards and textbook committees. When neither teachers nor students are able to connect with a content that is meaningful in their lives, they cannot establish ownership. They are unable to become interested in the topic or engaged in the teaching and learning process (Hidi, 1990; Tobias, 1994). Erickson and Shultz explained:

> If students like and trust the teacher, they may do the work assigned even if they do not understand or own its purposes. But if students have not bonded with the teacher and the assigned work lacks intrinsic interest for them, they may withhold efforts on the tasks assigned or go through the mere motions of learning. Collusions develop, in which teachers do not press students to learn what may be meaningless or face-threatening for them and students do not press the teacher by disrupting the class (p. 471).

Teacher-control of expectations may be a thing of the past in many urban schools. Teachers report that students question and reject traditional expectations designed by outsiders and supported and implemented by teachers. Teachers, in turn, lowered expectations for disruptive and disengaged students and students from low-income or dysfunctional families. As school curricula became less stimulating and uninteresting, many students resisted even these expectations through confrontation and refusal to engage. Teachers responded by attempting to negotiate expectations, giving in to student demands to maintain classroom order.

NEGOTIATED EXPECTATIONS

Burbules (1995) explained that for teachers to enhance students' learning through positive educational experiences, they must have the students' respect and at least their provisional cooperation in activities. Teachers interviewed in my research often characterized their students as disengaged. Specifically, students were uninterested or reluctant to connect with the teacher, the subject matter, or the school (Ennis, 1995, 1996). Further, they were unwilling to complete assignments where they were requested to work independently or learn in isolation. Teachers' efforts to negotiate with students involved explicit tasks to motivate students, maintain class order, and adjust teaching methods to entice students to engage. Students responded positively to these attempts when teachers structured interesting and relevant tasks and cooperative activities.

Negotiating Motivation

Many urban teachers reported making substantial adjustments in recent years to their expectations for student behavior and learning (Ennis, 1995). They indicated that they began their professional careers teaching a prescribed content that was acceptable to them and to their students. The recent negative changes they observed in students' motivation, interest, and willingness to engage convinced them to compromise their expectations in an effort to enhance student motivation. They willingly compromised the "curriculum of knowledge and skills" to achieve "motivation and order" in their classrooms (Ennis, 1995, p. 453). They discussed the curriculum of knowledge as composed of traditional or historically acceptable content advocated by university professors and professional associations and formally approved by school boards. They complained, however, that these outside individuals usually had little contact with students in their schools. They resented the constant "academic preaching" and standardized tests designed by the "university and state department types" whom they viewed as totally disconnected from the problems they confronted daily in their classrooms. One English teacher explained the difficulty he experienced in motivating students to participate:

> The difficulty is that they really don't want to work as hard as you need to work to learn. They don't see the point of learning about "dead white men" anyway. It is really difficult to get them to concentrate and study in order to incorporate this information into their understanding. Our school's test scores in math and English are way below where they should be...but our students just aren't interested in most of this information, and they certainly don't care whether they do well on our state tests.

Teachers explained that they spent much of their class time and energy simply motivating students to participate in class activities. They reported using rewards, student choice activities, and delegation of responsibilities to help students focus on and complete class work. Although students in these classrooms often appeared orderly, a closer examination revealed they were not engaged in the learning tasks (Ennis, 1996; Knapp, 1995). Tasks appeared to be superficial exercises designed to keep students working at their desks. Activities were structured so that students worked independently with minimal peer interaction. Discussions with teachers and students revealed that students were working for the opportunity to participate in "Free Friday," when they could sit and talk quietly at their desks and not have to complete work sheets or test papers. On these days, teachers reported they breathed a sigh of relief that students valued these rewards, resulting in an orderly, uneventful day in class.

Expectations for Orderliness

"Orderliness" was described by these teachers as an essential expectation for classes in their urban schools. The need for order appeared to surpass the normal expectations for class management (Ennis, 1995, p. 454). An orderly classroom was predictable. Students were compliant and appeared to be engaged in academic activities. Knapp (1995) described these classrooms as orderly, restrictive environments in which assignments were completed, but the "spark" of interest and motivation often was missing. Administrators emphasized and reinforced the expectation for order in these schools. Teachers reported that their administrators had to focus on the most difficult problems in their schools, involving disruptive students, weapons, and unauthorized visitors. They had little time to serve as instructional leaders or to deal with noncompliant or disorderly students. A mathematics teacher explained this problem:

> Our administrators just don't back up the teachers' attempts to discipline students. I think it is because they have so many other more serious things to deal with. They feel [student-teacher confrontations] are a classroom management problem (Ennis, 1996, p. 158).

Discussions with principals also supported this view:

> I am constantly monitoring the halls and working with our security guards and assistant principals to keep students in class. We also ask teachers to patrol the halls during class changes. At times just before and after holidays and large school functions, we conduct hall sweeps to move students out of the halls and into classrooms. My assistant principal for instruction tries to assist teachers with curriculum, but that is not my major function at present (Ennis, 1996, p. 159).

It is not surprising that these urban principals felt preoccupied with school security measures and encouraged teachers to maintain order in their classrooms and hallway. The National School Boards Association (1993) reported that 82 percent of the 2,000 school districts surveyed indicated student violence had increased over the last five years. Most respondents (77%) identified the primary causes of the increase as "changing family situations." Ninety percent of urban school districts reported increases in violent acts initiated by a student against other students and in incidents of weapons brought to school. In addition, almost 50 percent of urban districts reported increases in student violence against teachers (see also, Guerra, Tolan, & Hammond, 1994).

In many urban schools expectations for violence are high among teachers and administrators (Ennis, 1995, 1996). These expectations lead them to include activities that keep students occupied and prevent or diffuse situations that occur when students congregate in groups in the halls or classrooms. Thus, order becomes a prized commodity that is valued by educators and compliant students. Teachers achieved expectations for order through avoidance of content that disruptive students considered "difficult or unpleasant." Teachers reported that these students were unwilling to work hard or risk failure attempting difficult work (Ennis, 1996).

Some students reported they became angry and disorderly when they were not permitted choices in how the class was to be managed and what content was to be learned (Ennis & McCauley, 1996a). Unfortunately, when administrators and teachers believed they must control these decisions, students reacted and resisted, requiring additional control measures, such as detentions and suspensions. Noguera (1995) insisted that increased efforts to control students trigger more disruptive and violent student behavior, leading to additional, highly visible and expensive controls such as security cameras, chained gates, and metal detectors. In these high schools educators' and students' willingness to engage in enjoyable, orderly, though educationally irrelevant activities, was a negotiated compromise that satisfied teachers but left some students irritated and angry.

Students' Expectations for Effective Teaching

Urban students interviewed in our most recent research complained most bitterly about poor teaching (Ennis & McCauley, 1996a). The data reported in the Ennis and McCauley (1996a, 1996b) papers are part of a large ethnographic, interpretive study examining the perspectives of students, teachers, and administrators in one urban high school with an enrollment of 1,200 students, 90 percent of whom were African American. The first Ennis and McCauley (1996a) paper reported the findings from interviews with 50 disruptive and disengaged students. In this high school student complaints moved beyond teacher personality or excessive work concerns to focus on some teachers' inability to convey content. We found that students held expectations for teachers to provide quality instruction within a caring climate. When asked to describe a favorite teacher, Tamika (all names are pseudonyms) chose her government teacher:

> She lets us work together, to talk to each other about what we think about issues. She brings in interesting topics, like how government people make decisions. You know, about welfare, and taking it away and all. We all know somebody that depends on that check every month. We know what it feels like to want a job, but we can't get one. These be important things to talk about. Maybe one day we'll be the ones to make those decisions. So she talk to us...she don't tell us what to think. She helps us to think for ourselves and gets us to ask questions to each other. I really have learned a lot in her class (Ennis & McCauley, 1996a).

Dexter described how a physical education teacher created this environment in a weight training class:

> You don't have to say anything and he come to you to help you out. He likes to talk to me. You know, he likes to let me know something. Some teachers don't care what you do—you know what I'm saying? But with him, he stands there to see what you doing...he tell you, "You can do better than that." He says, "Go, go, go, you can do it." That's somebody who cares for you; who wants you to do something for yourself. Sometimes I know I drive him crazy, but he's my favorite teacher you know (Ennis et al., 1996).

Other students were less positive in their descriptions of teachers' classes. They described situations that made them angry—so angry at times, they were willing to disrupt the class and confront their teachers. David explained:

> Like in Mr. A's class...he just teach. He just, like if you ask him questions—sometimes he don't stop [lecturing] he just keep on, you know teaching the class. And then if you ask him a question...its like he's just teaching to teach sometimes, not to teach you...just to teach (Ennis & McCauley, 1996a).

Donnell also expressed frustration with his teachers:

> Well, two of my teachers, they just make it seem as though if you pass, you pass, if you fail, you fail...anyway they gonna get paid whether you pass or not, and they don't really help you. Most of my teachers, they really could care less whether you pass or you fail (Ennis & McCauley, 1996a).

Jovell described her experiences in one of her classes:

> Well, like sometimes my history teacher, she'll just—well, every once in a while we'll have lectures, and we take notes on what she lectures on. And sometimes she just looks really bored—that's how she's talking to us. I'm like, she doesn't want to be here and neither do we. Why are either one of us here? (Ennis & McCauley, 1996a).

Other students explained that the content simply was not relevant to their lives and the futures they were creating for themselves. They explained that teachers were unaware of situations they faced daily in their neighborhoods and on the streets as they walked to and from school. Ryan explained his expectations for content also included job and family related topics that would help support himself and his future family:

> I want to learn about the real world. Because I seen kids that grow up much faster than they needed to as far as girls having babies and kids struggling to get jobs. We are growing up real quick and we have to mature a little bit more faster than the older generation. We need teachers that understand this and know what we be going through. We need classes and topics that connect to what we need to learn to get a good job (Ennis & McCauley, 1996a).

In theory the knowledge that Ryan and other students are learning is consistent with information that employers want their employees to know. In practice, however, many students report, and their test scores confirm, that they are not receiving this knowledge in a manner that connects to personal or professional aspirations. Although students such as Ryan appear to be "growing up much faster," this often does not translate into maturity, goal setting, reflective decision-making ability, and positive social skills that most adults and potential employers expect. It is unfortunate that Ryan's teachers were unaware of his concerns and the need to explain and demonstrate how school knowledge was consistent with the knowledge and skills necessary to be successful in a high paying, personally and professionally satisfying job.

Crystal expressed frustration with teachers' lack of interest in working with students. To her, teaching meant involvement with students, helping them to understand and interpret their readings. She was dissatisfied with teachers who lectured without finding ways to involve her in the lesson:

> I mean teachers really don't teach you nothin'. You teach yourself in school. I mean, you know, 'cause when they talk at you or give you the books [to read or study], you teaching yourself. They not explaining nothing. They'll tell you, you know, turn to subject...talk, talk, talk...I mean, it's not real teaching, you teaching your ownself! (Ennis & McCauley, 1996a).

These students, like those interviewed by Sheets (1996) wanted the opportunity to be heard and to be part of the learning process—but on their own terms. They were eager to work interactively with the teacher on projects and problems. Jamal expressed his frustration:

> How is this—filling out this [work sheet]—in history gonna help me get a job as a mechanic? I need to learn things I can use in a tech school like the one my brother goes to. But all we do is sit there and listen to him talk about wars that happened hundreds of years ago. Or we read about arguments that people get in that have nothing to do with me. I am just wasting my time here (Ennis & McCauley, 1996a).

Many students discussed the importance of working cooperatively with others to learn. They reported that they learned more when knowledge was connected to peer experiences and understandings. In these situations they could talk with their friends, ask questions of a trusted peer, and perhaps interact with the teacher in a more informal setting. Most students pointed out that they did not want to learn alone. They did not want to struggle to understand the passage in the book by themselves, when they could discuss it with a friend. They did not like to sit alone in their seat and work on work sheets when they did not understand or were not interested in the content. They felt they learned better when they worked in cooperative groups. Further, they pointed out that "having to work independently" was an example of the teacher not doing his or her job. Delonte echoed Crystal's concern when he explained:

> Well, they give us books. You know, they expect us to do it ourself! That's all they do is give us books and then they say well here, you know, turn on such and such page, and we'll just turn to it and do the work. But they don't be teaching us nothing. If we are gonna learn in this class, we have to teach ourself (Ennis & McCauley, 1996a).

Negotiation to avoid confrontation required teachers to work with students to identify a common set of expectations both could accept. At times, it meant that both teachers and students had to alter or compromise their expectations to accommodate the other. For students, it meant a willingness to follow school and class rules for behavior that teachers believed were essential to maintain order and control in the classroom. Within this negotiated compromise, teachers also had to make changes in their teaching practices. They had to vary the methods they used to address student concerns and to teach the content. For instance, they had to provide opportunities for students to interact in cooperative settings, discuss topics more informally, and plan classes with numerous "hands-on" experiences. Showing students and permitting them to discover answers rather than telling students is an essential component of active approaches to learning (Shuell, 1986). These approaches focus less on teaching using a standardized formula and more on the engagement of students in the learning experience. For teachers who had been using a lecture or a "teaching as telling" format for 20 to 30 years, this required a major compromise.

The value of a negotiated compromise is the shared or common agreements that bring the parties closer together and permit work to continue. In environments where compromise is the expectation, such as government, individuals leave the bargaining arena pleased with what they have gained, yet concerned for what was lost in the bargain. In schools, forfeiting issues that form the essence of one's philosophy creates deep disappointment. Educators and students often have difficulty remaining focused and motivated when their ideological core is compromised. As demonstrated through the participants' voices in these studies, giving up meaningful aspects of content, culture, or behavior constitutes a great loss.

For teachers and students to negotiate willingly, they need to accept the other as a partner in the educational process. Teachers acknowledge that students' personal and experiential knowledge is valuable and worthwhile in the classroom. Students, in turn, agree that teachers' formal knowledge is relevant and functional now and in the future. Negotiated expectations can lead to an environment where both parties have gained on issues of importance. Students win the opportunity to propose content or teaching formats they find meaningful, while teachers win student compliance and the opportunity to present the knowledge they have been contracted to teach.

Teachers, however, expressed frustration, dissatisfaction, and disappointment when they were unable to persuade students to compromise and comply. Further, they were angry when students did not make the effort to learn using traditional educational methods. They lamented the loss of their moral authority that precipitated the need for these conversations and compromises. In situations where compromise could not be reached, teachers disengaged and disconnected, while students confronted teachers in an effort to influence, bully, or persuade them to teach in a manner that conveyed content in meaningful ways. Nevertheless, many teachers simply were unwilling or unable to give up their valued content and teaching methods when faced with the magnitude of the compromise required to win student compliance.

Students, in turn, were willing to disrupt the class when their requests for meaningful content went unheard. They demonstrated negative, disrespectful behaviors with the express purpose of being dismissed from boring classes. They questioned why they should remain in ineffective teachers' classrooms when they were not learning. Other students simply disengaged, daydreaming in the back of the room, or moving slowly to comply with teachers' instructions. They were unwilling to expend the effort needed to connect and engage, preferring to fail the class rather than memorize content that lacked interest or meaning. Enticing these students to engage and to focus on academic goals requires more than punishments or threats of disciplinary action. Successful reform in urban schools begins when teachers make personal connections with students to pave the way for successful curricular and instructional innovations to reverse the stalemate that has kept teachers and students apart for too long.

CONSTRUCTING SHARED EXPECTATIONS

Ellsworth (1989) argued that teachers no longer have any legitimate claims to moral authority in the school or classroom. From her perspective, teachers have the obligation to explain "why" content is relevant and "why" students need to know and understand the material. Textbook committee members and teachers are obligated to select texts that convey content within a meaningful format. Teachers must earn students' respect by their willingness to work with them to construct shared expectations valued by both.

The joint construction of expectations promises to blend teachers' and students' goals for content, methods, and behavior into a shared vision each can attain and value. Educational psychologists, such as Resnick (1989), define learning as a process of knowledge construction, dependent on students' prior knowledge and attuned to the contexts in which it is situated. Social constructivists take this notion of a jointly constructed vision one step further to examine how expectations are formulated and implemented in practice. They emphasize the importance of the context in which the learner lives and learns, and the central role of others in assisting with the construction of shared understandings and meanings (Wertsch & Toma, 1995). This can be formulated through dialogue with peers or parents that challenges and stimulates teachers to listen and integrate students' understanding with their own. It can also occur in the context of a mentor-apprentice relationship as the mentor selects content and provides experiences that are relevant to the needs and interests of the apprentice (Hausfather, 1996; Vygotsky, 1978).

Vygotsky suggested that this instruction occurs within a "zone of proximal development" in which the mentor provides tasks and scaffolds opportunities consistent with the apprentices' current abilities to understand and perform. As the apprentice progresses, the zone moves to entice and challenge the learner to acquire increasingly complex knowledge. New knowledge is connected to prior knowledge within a context that is familiar and understandable to the learner (Resnick, 1989; Shuell, 1986).

The essence of this constructivist perspective in schools is the social nature of the learning process. Students learn most effectively in interactive, cooperative classrooms. Teachers provide information that is connected to the context in which it will be learned. Their expectations for learning are familiar and acceptable to the individuals who will be responsive and responsible. Knowledge is connected in meaningful ways to the individuals who will benefit most from the learning experience. A social constructivist focus on the development of expectations is most successful in a shared or insider perspective. Expectations are developed by teachers and students working together to structure and shape positive goals. They are socially constructed by a student who is connected with the issues and needs of the educational community. The scaffolding necessary to entice the student into an educational zone of proximal development is provided by a teacher in tune with the student's current ability level, interests, and needs.

The process of joint construction is more sophisticated and challenging than traditional negotiation and compromise. Negotiating expectations between teachers and students requires both to give up essential elements of their ideologies, often leaving the educational experience meaningless. For individuals to be willing to engage in the hard work of teaching and learning, they must believe that the benefits are substantial and attainable. Expectations, therefore, must be constructed jointly to reflect the shared meanings of educators and students. They must integrate the shared expectations of teachers and students for content and methods that connect them to each other and permit both to grow in their understanding of the experience. In this section I will focus on the expectations of teachers and students as a web of joint constructions. The essence of joint construction is teachers' and students' ability and willingness to trust the intentions of the other and work cooperatively to enhance achievement.

Developing Trusting Relationships

The trust essential to create shared expectations typically begins slowly and develops gradually over time. It is fragile and can be shaken easily by a single distrustful act. Once lost, it can be more difficult to rebuild than to create initially:

> Trust involves the belief that you can rely on someone (specifically, their beliefs, dispositions, motives, and good will) or something (an institution, or a piece of equipment) where there is a greater or lesser element of risk. One may or may not be conscious of the trust relationship and it will involve varying degrees of personal commitment (White, 1995, p. 233).

Because many urban students live in situations where they learn to distrust others, they come to school not having experienced caring relationships (Noddings,

1992). These students may have particular difficulties developing trusting relationships necessary to construct shared expectations. Teachers, too, have experienced numerous events with students that have resulted in deep feelings of distrust. Over their careers they have been confronted and challenged by disruptive students to the point that it is difficult to contemplate a shared experience. Yet, in our discussions with urban high school teachers (Ennis & McCauley, 1996b), we found many teachers who were working to build trusting relationships with students necessary for the development of shared expectations. This research (Ennis & McCauley, 1996b) reported interview data collected from 18 teachers, representing eight different subject areas and six administrators. We asked them to describe disruptive and disengaged student behaviors and strategies they used to entice students to participate.

Teachers used a variety of techniques and strategies to convince disruptive and disengaged students that they trusted, cared for, and believed in them. These teachers believed that it was their responsibility to make the first move to design or create a trusting environment that they hoped would lead to shared expectations. They acknowledged that disruptive students often entered their classrooms with shields of distrust and anger developed over years of school failure and thwarted expectations. Likewise, disengaged students, marginalized by failure and irrelevant curricula, typically wore a mantle of protective disinterest equally difficult to penetrate (Ennis & McCauley, 1996b). Teachers explained that the construction of shared expectations began with a series of small steps such as offering second chances, positive attention, and student ownership. As the bonds of trust grew, teachers and students began the gradual process of constructing a web of shared expectations in which both could find meaning:

> I want my students to know that they can get fairness from adults, regardless of color or gender, and that takes time. That's just a question of gaining their trust. When kids are ready to go off, and they think I'm treating them unfairly, or I'm going to do something without checking things out, I say, "Have I ever treated you unfairly before?" And sometimes they'll stop and think and say, "Well, no you haven't." And I'll say, "Why do you think I'm going to treat you unfairly now?" The answer is simply, you have to be fair, you have to be honest, and you have to treat them with the same human dignity that everybody wants to be treated with. That's one of the biggest problems we have, is that some teachers don't, not all teachers. There are some wonderful teachers here. But some teachers don't treat them with the dignity they deserve. And that's how I have been successful for twenty years in education, it is simply having respect for other persons and treating them like I would want to be treated (Ennis & McCauley, 1996a).

Giving Expectations a Second Chance

Teachers at this high school (Ennis & McCauley, 1996b) estimated that between 30 and 60 percent of the urban students who first entered their classrooms had not found past educational experiences meaningful and relevant in their lives. They reported that these students often were unwilling to complete homework, class work, or engage positively in class discussions. These teachers attempted to connect with

Shared Expectations

students and refused to let them fail because of a lack of school-related effort, interest, or initiative. Instead, they created alternative management strategies to take students "from where they are" to help them reconnect with the teacher and their school work. A science teacher described this process:

> But nine times out of ten if you shower them with praise when they do something right and tell them honestly when they do something wrong, they will respect the fact that you didn't tell them that they were dumb or infer that they can't. You just simply say, "Well, you didn't study very much for this one did you?" and they'll be very honest, "No, I was busy last night and I didn't study." And I'll say, "Well, then you are going to have to come in on Tuesday after school and put in study time that you missed last night." So you give them another out, and if the students feel you care about them, they will work so hard for you that your class will become a priority, and eventually your work becomes less and less. And students accept more ownership and responsibility for their work. So that what you put into it in the beginning feeds back at the end (Ennis & McCauley, 1996b).

Second chances are opportunities for students to revisit class material that they either were unable, unwilling, or uninterested in learning when the material was initially presented. It permits the student to reconsider his or her decision or actions and offers an opportunity to try again. Rather than giving students a failing grade when they did not complete an assignment or perform successfully on a test, second chances provided teachers with another chance to engage students in the content, while students received another opportunity to learn. Although the strategy of offering second chances might be viewed by some as an example of lowered expectations, teachers in our study disagreed, explaining that second chances were an important step in the process of building shared goals and expectations. They acknowledged that many of their students did not come from traditional, stable homes where parents were present to monitor homework and provide safe, quiet, and supportive places for students to complete assignments. They realized that if students were to have this safe place, teachers had to create it in the classroom. Thus, they made their classroom a safe place and gave second chances to nurture the seeds of school and content relevance. The science teacher continued to explain how she nurtured and encouraged students in her classroom:

> If a student isn't working I'll come over and sit down beside her and say, "What's the matter? I see you haven't started, do you understand what the assignment is? Do you know the page? Do you have a book? I don't see a pencil moving, do you have the materials that you need? You don't have a book? Well, you can rent one." You know we go through that routine. I try to move a lot in the classroom, and if I see a kid sitting there and not really getting something, I'll say, "Stuck? Do you need some help? Can I tell you what page to look on for the answer? And so we try to give them enough that they can get started (Ennis & McCauley, 1996b).

Other teachers were sensitive to disruptive students' reasons for their behavior and worked to diffuse the situation in a way that preserved the student's self-esteem and taught alternative responses to the situation. These teachers attempted to guide students to reconsider their behavior and focus on the needs of both the teacher and themselves.

This was rarely easy. A physical education teacher explained that when working with confrontational or disruptive students, her first concern was for the student:

> When a student exhibits some behavior that concerns me, the first thing I do, is to [help that student] save face. I am considering their reaction to the way that I will respond to their behavior, because I can really set off a chain of events that can lead to a real chaotic situation. So I try to get the student out of the environment immediately. I ask the student to wait for me out in the hallway. Or if I have a student who is very angry with me and they say, "I'm not getting up, or I'm not moving." I usually go on to explain, "I really need..." and I put my need out there. "I need you to go into the hallway so that you and I can talk, It's no big deal, I just want to talk and I would feel more comfortable, and I think it would serve us both if you would meet me out there." And I'm trying to remain as calm as possible, because my reaction to situations is often what triggers the reaction by the student. I need to diffuse it as quickly as possible and try to get them out of the environment. Often times, kids will say, "Okay, okay, I'm sorry," or they will use their slang, "My bad." And they sit quietly and they'll fall in line (Ennis & McCauley, 1996b).

Teachers explained that without second chances and a nurturing environment, students were likely to disengage from school, fail their classes, and ultimately drop out of school. Teachers understood the importance of caring for their students and letting them know in tangible ways that they cared for them (Byrd, Lundeberg, Hoffland, Couillard, & Lee, 1996). Noddings (1992) suggests that the development of a caring environment is essential to the creation of trust, interpersonal commitment, and joint goals. Modeling, dialogue, practice, and confirmation are essential to the development of a caring environment. These teachers created opportunities to respond to students' needs and to provide or invite personally and socially relevant interpretations of traditional disciplinary knowledge. Teachers attempted to restore a sense of effort optimism (Ogbu, 1994), conveying to students that, "at least in my classroom, your efforts will be rewarded" (Ennis & McCauley, 1996b).

Connecting Expectations to Personally Meaningful Outcomes

Teachers also used the strategy of positive attention to create a personal connection or relationship with students. Disruptive students, in particular, often satisfy needs for attention through negative social and educational behaviors (Finn, Pannozzo, & Voelkl, 1995). After years of successful negative attention seeking, some appeared unable to reverse these behaviors to take advantage of educational and social reward systems in high schools. Teachers reported asking knowledgeable disruptive or disengaged students to assist or demonstrate for other students. They also appointed them as group leaders to help with classroom duties and to run errands. Although these techniques are used frequently by many teachers to reward engaged and compliant students, the teachers we interviewed reported that these positive attention techniques were essential when connecting with disengaged, disruptive, or distrustful students. They constituted an additional, public way that teachers demonstrated their trust in the student. A social studies teacher explained how he assisted students to engage in class:

Shared Expectations

> Well, I usually try to get those people who appear to be disengaged or disruptive to take on more of a role in class. When students get in a group situation or a cooperative learning situation and everybody has an assigned role, you have not only the sense of autonomy with that particular student, but you also have the [student] group pressure of saying, "Hey, we all have to pull together on this and produce something here, so we need your cooperation." It's a different tactic to stop me from always doing the asking. It's their friends asking, and all of a sudden the idea of the grade takes on a little bit more importance, I think, than it might otherwise (Ennis & McCauley, 1996b).

Teachers also found ways to personally connect with students and to demonstrate care and concern for their learning and behavior. This sometimes included teasing, poking fun at themselves, and generally providing a "light" touch in an otherwise serious school environment. An art teacher described how she used humor in her class:

> There is one guy in my third-period art class who is tough. He would like to be disruptive, but I won't let him. I know he's tough, and I'll say stuff to him like, "If you don't settle down, I'm going to hurt you." And he looks down at me and says, "We'll I guess I better sit down..." It is a light moment when I say it to them, and it is not like embarrassing them. They know I can't hurt them. But they understand that I am communicating with them in terms they understand (Ennis & McCauley, 1996b).

Shared expectations involved instances when teachers and students agreed to accept and value school expectations and students' interpretations of meaning. Teachers who were most successful treated all students with respect and integrity, joked and teased, and gave them positive attention. Even disruptive students were more likely to pause to consider the requests from teachers who had established a positive, respectful rapport. The relationship between positive attention and meaning-making is important because it highlights the connections between feelings of self-worth and engagement in the challenging process of constructing shared expectations.

Student Ownership

Many teachers and administrators told us that students worked particularly well when they were allowed to create meaningful classroom policies or infuse their unique personalities or beliefs into the final product. Several teachers explained how they had abandoned drill-like methods because of their intense focus on teacher control and their meaninglessness to students. Students were uninterested and unlikely to engage in memorization or other forms of repetitive work. The teachers, however, were able to use cooperative and peer teaching strategies to entice disruptive or disengaged students to connect with others who were having difficulty learning the material:

> Cooperative learning usually works well, especially where they help each other. Take the students who already know something and let them, in turn, help others because it just helps

them learn it more by helping others. Most of our best work is done in hands-on, group projects (Ennis & McCauley, 1996b).

Teachers provided numerous examples to demonstrate their emphasis on student ownership. Many teachers were convinced that difficult students needed to feel in control and responsible for classroom tasks that were associated with their own learning. Ownership often led to a greater sense of responsibility for self and a willingness to control their own behaviors. Stinson (1993) emphasized that students respond positively when they are offered choices in matters that they consider important. She encouraged educators to design situations in which students know they have an active and direct role in their futures. Decision about where to sit, when to talk, whom to work with, and what and how to learn are very important to adolescents. Providing increasing opportunities for students to have ownership of the content and take responsibility for their own behavior helps disruptive and disengaged students connect to the educational process.

Educational communities built on shared expectations thrive on open-ended dialogue and careful consideration of others' needs. The process of creating shared expectations in urban schools requires that teachers first connect with students in personally relevant ways. Teachers begin by providing second chances to students who have experienced consistent failure in the educational system. They supplement this with small opportunities for students to take responsibility for classroom chores and for their own learning. Teachers attempt to provide a stable, fair environment where students know that they will be treated with care and respect. They work constantly to connect personally with students by teasing, joking, and creating light moments. They also ask important questions that begin with "why" (for example, "Why do you feel that way?, Why did you choose to do that?") instead of condemning the performance or the behavior without listening to the students' reasons for the action.

Building Shared Expectations

Teachers who are effective at building shared expectations connect with students in trusting relationships, treat students humanely, demonstrate that they care personally for students, and provide opportunities for student ownership for aspects of the class that students' value. This occurs early in the teacher-student relationship during the first days and weeks of school. Typically teachers must initiate the process, demonstrating to students that this class will be different from those that you have experienced in the past. With students who have been disruptive or disengaged for many years, teachers must be persistent when their students reject their initial efforts.

It is evident from our research that little learning occurs without teachers first making this connection with students (Ennis, 1995). Many urban African-American students willingly work and learn for the teacher rather than for them-

Shared Expectations 173

selves (Ennis, 1995; Ennis & McCauley, 1996a). Negative threats or punishments, such as low grades, parent conferences, retention, detention, suspension, or expulsion rarely have the same power to motivate found in a warm, caring teacher relationship. Building shared expectations in urban schools may start with very modest goals. The process requires that teachers make the first move to reach alienated students and to continue to convince students to engage and to increase their academic expectations (e.g., Wehlage et al., 1989).

One outstanding science teacher we interviewed had developed a special rapport with her students that permitted her to nurture and teach many students who were failing other subjects. The teacher was white and had been teaching for 15 years in several different high schools. She had worked with students in science and technology magnet programs in prestigious high schools, but found this to be "too easy and not challenging." Instead, she had experienced greater professional satisfaction working with low-income, disruptive and disengaged students in a comprehensive high school. Her principal suggested that her classes were the last chance for students who had failed in other high schools and in other classes in this school.

This science teacher pointed out that understanding her students' home environment and providing instructional accommodations that permitted school work to be a priority in their lives was critical to their success and the first step in sharing expectations:

> In the past, I have taught at high schools that routinely produce National Merit Scholars, but that is rarely the case at this high school. My students here, do not come to me motivated. Most of them have come from very low socioeconomic environments. Most of them are on reduced lunches. Most of them cannot afford to provide materials for themselves, even basic materials such as paper and pencils. They won't let you know that, but if you look into their backgrounds, that's the situation. Most of the students here have experienced a lot of failure and a lot of disappointment in their upbringing and all of them are skills deficient. Some more than others. They are not motivated. Many of them do not have parents at home or the parents at home have severe problems that disable them or prevent them from providing support. And in many cases the students take care of the parents, rather than the parents taking care of the students. Many of my students have children of their own and parenting responsibilities of their own. Most of them have jobs that are required in order for them to supplement family income and to provide the attire that they wish to have. Students here are very content if they can get a D, and a C is an extraordinary grade. There is absolutely no motivation for most of the students that I see to get As and Bs. So what I try to do is change their level of expectation so that not only do they experience success, but they expect a higher level from themselves (Ennis & McCauley, 1996b).

Part of the process of shared expectations is understanding the everyday complexity, stress, and demands of students' lives. It is impossible to build shared expectations without a realistic understanding of the contextual factors that constrain and prohibit learning in some students' lives. Shared expectations evolve from this understanding and are created in a process in which both teachers and students begin to believe that academic goals are possible and worthwhile. It is hard for students to aspire to Bs and As when they have not received these grades previously. They must also understand how the knowledge represented by these grades can be useful in their current lives and can be used to build a future to which they aspire. Providing the incentive, encouragement, and skills necessary to dream of achieving higher grades is a critical step in expectation formation. In these situations, most students are dependent on the teacher initially to help them set realistic goals and take the first steps to reach them:

> I begin with very simple tasks in September to find the level of skill that matches each student's ability. Most of the students who are placed in my classes use disruptive behaviors because they are skill deficient, and they exhibit behaviors to distract the teacher away from the real problem that they can't function academically. So I then try to establish the right learning environment. I teach the lowest level science classes for comprehensive students. These classes also include students who are severely emotionally impaired, the higher functioning special education students, and other students in special programs who have experienced failure in some way. That gives my classes a unique mix because I have some students who lack skills because they haven't attended school and others who are so emotionally consumed by their lifestyle that they are unable to focus on their school work. During the first month of school, I try to find the strength of each student. Someone in each class can draw, someone is good in math, someone is articulate; reading levels vary all over the place. When I plan my weekly plans I try to make certain that during the week we do an activity that will hit the strengths of each of my students through some activity. At some point everyone has to be successful. I do not have students raise their hands, and I do not allow students to call out the answers. I call on students, and by calling on students every student has to participate and every student is accountable for the material (Ennis & McCauley, 1996b).

Because these students were skill deficient, they had been consistently unsuccessful. For this teacher part of building a trusting relationship was based on helping students feel successful and praising their legitimate academic accomplishments. For many students this class was the only class in which they experienced success. Slowly they began to vest energy in the class and to trust the teacher to lead them toward learning. This teacher explained, however, that even as students were beginning to focus on academic work, many of the behaviors that were valued in the peer group or at home, but were detrimental to the learning process cast a shadow over the class. Behaviors such as making excuses, negative outbursts, and personal criticism of others for mistakes can stifle learning and feelings of positive self-esteem. This teacher worked persistently to eliminate these behaviors and replace them with shared expectations for supportive learning environments:

> I do not let anyone make any kind of a negative comment in my class; no snickering when a student makes a mistake. We don't laugh at mistakes. You would be surprised at how much of this can go on if you do not make it a rule and enforce it. By enforcing that rule so that there are no negative comments, then we establish a place where people can make mistakes without the fear of being laughed at. As students become more comfortable in class, it is much easier to take the risks necessary to study and learn science. This rule is critical to helping students understand their potential and to begin to set goals for themselves (Ennis & McCauley, 1996b).

One of the most challenging things that teachers must do is disrupt the cycle of failure that traps many urban students. Low or deficient skills combined with un-

Shared Expectations

productive school behaviors, lack of organization, and low motivation create a nonproductive spiral that is difficult to break. Concrete experiences that demonstrate to students realistic ways that behavior change can increase their skills and result in school success is an essential step in the process:

> In the beginning of the year, most of my students are in the F or D-F range in terms of their skills. Every week as a class, we average our grades for the week and calculate how this weeks' grades have helped or hurt our grades. I try to encourage the students to take ownership for their grades. If they're late to class and they get the zero, they have to see that zero; and what the impact of that zero is in their grade-point average. Most of my students don't have a clue of how a zero will take away from a good grade. We look at their grades and we average them two ways: with the zero included, which is the grade I will give them if they don't take action, and then we average it based only on the grades they submitted. So I can say, "Well, you have five As and two Bs, if I were to give you a grade based on only the work you turned in, you would have a high B average. That's where your functioning capability is. You are smart; you can work. However, you only turned in seven out of 15 assignments. So that means when I add these zeros in, your grades goes from this wonderful 88 down to this rotten 42. But you have control of this. Now you have one weekend to make that better, or you have one week to make that better. Anything you turn in within this grace period I will give you credit for without penalty, but I will not accept a single assignment after this day" (Ennis & McCauley, 1996b).

By providing a concrete message in which students saw clear results of completing work, this teacher began to reverse the culture of low motivation she found in her students. She adjusted her expectations to consider students' entry level of motivation and skills, and then she provided concrete experiences through which students understood the significance of their actions and the possibility for change. Together they created a class setting in which the teacher and her students could share a common set of expectations for improving academic skills and grades:

> Nine times out of ten, the students that I teach are very disorganized. They have started the work and may have almost finished it with only a question or two left to be completed, but they didn't have the initiative or the support at home to do that. But with the grace period, they will go home and do the work. They see concretely that their F can become a B. So for the first time, they are completing something at home and working independently (Ennis & McCauley, 1996b).

Students and the teacher were able to share the expectation of independent work for at least some of the class assignments. Students were permitted to fail and then receive a second chance to recover and perform at their level of ability. Although this process did not completely eliminate the problems and excuses that students used to explain their performance, it did provide a process that permitted students and the teacher to come closer together and to share some academic expectations. Nowhere is this more difficult than in teaching students to study and communicate their knowledge effectively on tests. This teacher provided clear test guidelines, expectations, and consequences that helped students to experience success:

> I also try to encourage a more confident atmosphere or attitude associated with test taking. I permit students to retake tests to learn that they can learn and improve their scores. The grade

doesn't have to stay an F, but students do have to come after school on their time and give me a minimum of one hour of study time in order for them to be eligible to retake the test. That way I can get them to come in and sit down, and we can look at their mistakes. I can show them where they made their mistake and maybe for the first time they will put in an hour of study. But then the two tests are averaged together. So it is not replaced. We don't say it is okay to fail it the first time and then you can retake it and get a higher grade. It doesn't work that way. You have to accept the consequences for that first grade. But I give them the opportunity to save face. Those that wish to do well will have the opportunity, and I try to eliminate as many excuses as I can in order for them to do well. And I feel that when I receive students, they are usually disorganized; they are not very structured; and so by providing them with structure, providing them with an opportunity to fail without making it a finality, that way my students respond better. My first-quarter grades are awful; they call my bluff; and I show them that I mean what I say. But second quarter their grades just leap and then by the end of the year their grades usually are pretty decent (Ennis & McCauley, 1996b).

This process of building shared expectations required time to develop the trusting relationship and to implement concrete experiences. Students gradually learned that they had sufficient control of the learning environment to begin to have higher expectations for themselves. The teacher fostered this and nurtured these expectations, but admitted that she had to "bend frequently" to accommodate her students' limitation. Gradually, she found strategies to entice most students to engage in the learning process:

> My students are like oysters. Life has not been good to them. Most of them have failed. Most of them are very skills deficient. Most of them have not come from the kind of environment that I would want to come from, and to protect themselves they build this shell around them. As a teacher what I have to do is wear away, literally by perseverance, wear away that shell to the point where I can get inside. I have to make them open it up. Sometimes they will open up because of their success or because they connected with the content in a unit or they just like the atmosphere in the class. Sometimes a few open up in a negative way. If I cannot get them to be responsible for materials or to complete their work, I will harass them to the point where they'll say, "I can't stand this anymore!" And I'll say, "Okay, then do something about it. Bring your book and I won't have to ask about it." But the point is, if they lash out at you in hostility, they've opened the shell. Because the only kid you can't reach is the kid who is so apathetic he won't react. But, if you can make them angry, they've opened their shell and then you've got to sneak in. Because once you can show the student that they can succeed, you've found that pearl that is hidden deep inside. And once the student feels comfortable enough to let you see the pearl that they have hidden, then they will work with you and for you, and you can create a sharing relationship that is valuable and meaningful for both of you (Ennis & McCauley, 1996b).

The Web of Shared Expectations

Sharing the goals of student learning provides a motivating atmosphere for both teachers and students. Expectations for content, achievement, and social behavior form a web of shared experiences that build bonds for continued growth. Building shared expectations can occur throughout the school as teachers and administrators work with students to create viable expectations for the future. Preservice and staff development opportunities that assist present and future teachers to create shared

Shared Expectations 177

expectations with students can enhance their professional satisfaction and facilitate student learning.

Teachers' understanding of the importance of shared expectations can begin early in their careers as preservice or novice teachers. Each university instructor and school-based resource teacher can emphasize several key factors in classes and consultations that can assist new teachers to connect positively with students and enhance their achievement:

1. Understand your students' backgrounds and the constraints and opportunities they provide for learning.
2. Provide numerous opportunities for legitimate academic success early in the school semester or year. Focus initially on students' strengths and use those strengths to build student confidence and scaffold later learning tasks.
3. Explain your grading system with concrete examples that demonstrate to students how they might enhance their grade. Confront the issue of low or failing grades by explaining the situation that led to the grade and providing an alternative or second chance for students to complete the work, thus learning important study, writing, or test-taking skills useful in future assignments.
4. Construct sequenced learning tasks that facilitate skills acquisition in your subject area and be willing to remediate skills both within and outside of your subject when necessary to promote student success.
5. Focus on the process of how you teach that includes positive, affirming ways of interacting with students and insist that they treat each other with the same level of courtesy.
6. Provide frequent opportunities for students to express their goals and aspirations for learning. Assist students to set positive, realistic goals and envision a successful future. Use these clues as you work with students to jointly create expectations that you both can share.

Teacher preparation programs can facilitate the process of creating shared expectations by providing multiple opportunities for university students to interact with urban public school students (Weiner, 1993). More than simply a semester practicum experience, college students need the opportunity to meet and work with adolescents and to vest in the success of a class or school. Professional preparation faculty who themselves facilitate the formation of shared expectations, model these behaviors, assist students to acknowledge their importance, and accept their joint responsibility in expectation development. For instance, university students in methods classes may benefit greatly from the opportunity to teach multiple lessons to urban students prior to student teaching. They are encouraged to create academic lessons within the realistic context of students' lives. Additional experiences that facilitate the development of personal bonds between preservice teachers and students also enhance this process. Future teachers should be encouraged to volunteer at urban high schools and to learn how to talk with adolescents. This learning experience is even more critical when students represent ethnic and racial backgrounds different from the teacher. When this process occurs early in the professional preparation process prior to student teaching, it plants the seeds of expectation sharing and facilitates the maturity of this thinking and planning process.

Historically some new teachers were warned, "Don't smile 'til Christmas!" as a first step in establishing discipline and a serious, businesslike attitude in their class. Unfortunately, this may be one of the most detrimental caveats that can be issued in today's urban schools. Adolescents are seeking adults with whom they can talk and who can advise them on serious issues in their lives. Young teachers, instead, need to be reminded to be friendly with their students. This does not been being a pal or peer, it does mean being polite and respectful with students, taking students' comments seriously, smiling and laughing in the classroom, and finding ways to make the subject matter and the learning process enjoyable.

Successful teachers with whom we have talked also emphasize the importance of treating students fairly. Concrete, nurturing relationships are built by providing each student with the same opportunities to be successful and giving attention to each as equally as is possible within the sometimes hectic classroom setting. Adolescents are vigilant for signs that teachers are playing favorites, leading to hostility and disengagement. I have emphasized in this section that creating shared expectations first requires developing a positive rapport with students. This rapport is based on mutual respect and concrete academic experiences through which students realize that they can be successful. The teacher often must take the lead in developing this relationship. As students connect with the learning process, they are more willing to accept academic expectations and to work with the teacher to share the responsibility for their creation.

Perhaps the most detrimental and difficult problem to address in urban schools is that of disengaged teachers. Teacher disengagement, whether because of extreme changes in the school context, lack of teaching skills, or inability to connect with students, can cause serious problems for students, parents, administrators, and the community. Because the process of building shared expectations usually begins with teacher effort and initiatives, disengaged teachers often never begin the process. Students remain disruptive and disconnected from the learning process further isolating themselves and confirming teachers' decisions to withdraw from these relationships. Staff development experiences to reverse this process need to begin with retraining to teach disengaged teachers classroom pedagogical skills that include all class members as viable members of the learning community. Opportunities to connect with students and community groups outside of class may also facilitate this process. Finally, teachers may need to work with their administrators to build shared expectations for their classrooms. The web of shared expectations extends to teachers and administrators as well as students. Educators are responsible for learning and implementing positive strategies that contribute to the development of shared expectations.

Shared Expectations

Because knowledge and learning are shared constructions, students and teachers working as partners in side-by-side or integrated relationships can jointly construct a web of expectations for behavior and learning. Attempts at top-down or outsider designations of academic and behavioral expectations are limited by the declining levels of moral authority in urban schools. Negotiated compromises also are constrained when passionate beliefs and ideologies are not permitted a valued place in the final agreement (Ladson-Billings, 1995). Loss of value for experiential knowledge and personal beliefs can lead to both student and teacher disengagement. Within these formal arrangements, individuals compromise the essence of their culture and their sense of community. Unfortunately, some students respond by refusing to engage and, instead, seek ways to disrupt the learning environment. They become negative participants who create debilitating environments for themselves and others. Attempts to control these individuals through punitive consequences have proven consistently unsuccessful. The sheer numbers of students who are responding with disengagement and disruption often overwhelm the urban school setting, drawing economic and personnel resources from the educational process (Newberg & Sims, 1996; Talbert, McLaughlin, & Rowan, 1993). Further, attempts to maintain order when students are intent on disorder appear to further escalate emotions, leading to violence.

When framed as a shared collaboration, expectations for academic success and appropriate social behaviors lead to valued outcomes for all participants. Although the shared vision does not evolve easily in urban public high schools, it is critical to creating a learning environment that is positive and leads to academic and social success. Efforts by caring teachers to create a shared vision often succeed with one student at a time. Each student is greatly in need of attention, ownership, and success. Each must be persuaded that the teacher and the content are worthy of their trust and their engagement (Ennis & McCauley, 1996b).

Shared expectations evolve from a dialogue between teachers and students. The dialogue must truly reflect the aspirations and visions of both. Shared expectations are integrated expectations. They focus on the culturally sensitive and personally meaningful constructions that shape the teaching and learning expectations for educators and students. Working within a web of shared expectations may require individuals to learn new knowledge and skills necessary for living out the vision. Creating a web of shared expectations is really creating a shared future in which both teachers and students focus on the academic skills and social behaviors that students need to be successful in the community and that teachers need to teach the knowledge base effectively. Each must connect with the other in the present while they prepare for the future. In this way, shared expectations become a joint vision to carry them beyond their current circumstances toward a positive future beneficial for all.

REFERENCES

American Association for the Advancement of Science. (1993). *Benchmarks for science literacy*. New York: Oxford University Press.

Banks, J. A. (1994). *Multiethnic education.* Boston: Allyn & Bacon.
Brantlinger, E. (1991). Low income adolescents' perceptions of social class related peer affiliations in school. *Interchange, 22,* 9-27.
Brophy, J. (1996). *Teaching problem students.* New York: Guilford Press.
Burbules, N. (1995). Authority and the tragic dimension of teaching In A.G. Rud & J. Garrison (Eds.), *The educational conversation: Closing the gap* (pp. 97-129). Albany, NY: State University of New York Press.
Byrd, J., Lundeberg, M. A., Hoffland, S. C., Couillard, E. L., & Lee, M. S. (1996). Caring, cognition and cultural pluralism: Case studies of urban teachers. *Urban Education, 31,* 432-452.
Ellsworth, E. (1989). Why doesn't this feel empowering? Working through the repressive myths of critical pedagogy. *Harvard Educational Review, 59,* 297-324.
Ennis, C. D. (1994). Urban secondary teachers' value orientations: Social goals for teaching. *Teaching and Teacher Education, 10,* 109-120.
Ennis, C. D. (1995). Teachers' responses to noncompliant students: The realities and consequences of a negotiated curriculum. *Teaching and Teacher Education, 11,* 445-460.
Ennis, C. D. (1996). When avoiding confrontation leads to avoiding content: Disruptive students' impact on curriculum. *Journal of Curriculum and Supervision, 11,* 145-162.
Ennis, C. D., Cothran, D. J., Davidson, K. S., Loftus, S. J., Owens, L. M., & Hopsicker, P. (1996). Implementing curriculum within a context of fear and disengagement. *Journal of Teaching in Physical Education, 17,* 58-81.
Ennis, C. D., & McCauley, M. T. (1996a). *Disruptive students' rationales for their class behavior.* Paper presented at the annual meeting of the American Educational Research Association, New York.
Ennis, C. D., & McCauley, M. T. (1996b). *Enticing disruptive and disengaged students to learn.* Paper presented at the annual meeting of the American Educational Research Association, New York.
Erickson, F., & Shultz, J. (1992). Students' experiences of the curriculum. In P. W. Jackson (Ed.), *Handbook of research on curriculum* (pp. 465-485). New York: Macmillian.
Farrell, E., Peguero, G., Lindsey, R., & White, R. (1988). Giving voice to high school students: Pressures and boredom, Ya know what I'm sayin'? *American Educational Research Journal, 25,* 489-502.
Fine, G. A., & Mechling, J. (1993). Child saving and children's cultures at century's end. In S. B. Heath & M. W. McLaughlin (Eds.), *Identity and inner city youth* (pp. 120-146). New York: Teachers College Press.
Fine, M. (1991). *Framing dropouts: Notes on the politics of an urban public high school.* New York: State University of New York Press.
Finn, J. D., Pannozzo, G. M., & Voelkl, K. E. (1995). Disruptive and inattentive-withdrawn behavior and achievement among fourth graders. *The Elementary School Journal, 95,* 421-434.
Good, T. L., & Brophy, J. E. (1994). *Looking in classrooms.* New York: Harper & Row.
Guerra, N. G., Tolan, P. H., & Hammond, W. R. (1994). Prevention and treatment of adolescent violence. In L. D. Eron, J. H. Gentry, & P. Schlegel (Eds.), *Reason for hope: A psychosocial perspective on violence and youth* (pp. 383-404). Washington, DC: American Psychological Association.
Hausfather, S. J. (1996). Vygotsky and schooling: Creating a social context for learning. *Action in Teacher Education, 18* (2), 1-10.
Hidi, S. (1990). Interest and its contribution as a mental resource for learning. *Review of Educational Research, 60,* 549-571.
Hirschi, T. (1969). *Causes of delinquency.* Berkeley, CA: University of California Press.
Kantor, H., & Brenzel, B. (1992). Urban education and the "truly disadvantaged": The historical roots of the contemporary crisis, 1945-1990. *Teachers College Record, 94,* 278-314.
Kliebard, H. M. (1987). *The struggle for the American curriculum: 1893-1953.* New York: Routledge.

Knapp, S. (1995). *Teaching for meaning in high-poverty classrooms.* New York: Teachers College Press.
Ladson-Billings, G. (1995). Toward a theory of culturally relevant pedagogy. *American Educational Research Journal, 32,* 465-491.
LeCompte, M., & Priessle, J. (1993). *Ethnography and qualitative design in educational research.* New York: Academic Press.
Lippman, L., Burns, S., & MacArthur, E. (1996). *Urban schools: The challenge of location and poverty* (NCES-184). Washington, DC: United States Department of Education. National Center for Educational Statistics.
McCombs, B. L. (1992). *Learner-centered psychological principles: Guidelines for school redesign and reform.* Washington, DC: American Psychological Association.
McLaughlin, M. W., & Heath, S. B. (1993). Casting the self: Frames for identity and dilemmas for policy. In S. B. Heath & M. W. McLaughlin (Eds.), *Identity and inner city youth* (pp. 210-239). New York: Teachers College Press.
McNeil, L. M. (1988). *Contradictions of control.* New York: Routledge.
Metz, M. H. (1978). *Classrooms and corridors: The crisis of authority in desegregated secondary schools.* Berkeley: University of California Press.
National Council of Teachers of Mathematics. (1989). *Curriculum and evaluation standards for school mathematics.* Reston, VA: Author.
National School Boards Association. (1993). *Violence in the schools: How America's schoolboards are safeguarding your children.* Alexandria, VA: Author.
Newberg, N.A., & Sims, R.A. (1996). Contexts that promote success for inner-city students. *Urban Education, 31,* 149-176.
Noddings, N. (1992). *The challenge to care in schools: An alternative approach to education.* New York: Teachers College Press.
Noguera, P.A. (1995). Preventing and producing violence: A critical analysis of responses to school violence. *Harvard Educational Review, 65,* 189-212.
Oakes, J. (1992). Can tracking research inform practice? Technical, normative, and political considerations. *Educational Researcher, 21* (5), 12-21.
Ogbu, J. U. (1994). Racial stratification and education in the United States: Why inequality persists. *Teachers' College Record, 96,* 264-298.
Page, R. N. (1990). Games of chance: The lower track curriculum in a college-preparatory high school. *Curriculum Inquiry, 20,* 249-281.
Peshkin, A., & White, C.J. (1990). Four black American students: Coming of age in a multiethnic high school. *Teachers College Record, 92,* 21-38.
Resnick, L. (1989). Introduction. In L. Resnick (Ed.). *Knowing, learning, and instruction: Essays in honor of Robert Glaser* (pp. 1-24). Hillsdale, NJ: Erlbaum.
Schlosser, L. K. (1992). Teacher distance and student disengagement: School lives on the margin. *Journal of Teacher Education, 43,* 128-140.
Sheets, R. H. (1996). Urban classroom conflict: Student-teacher perception: Ethnic integrity, solidarity and resistance. *The Urban Review, 28,* 165-183.
Shuell, (1986). Cognitive conceptions of learning. *Review of Educational Research, 56,* 411-436.
Solomon, R. P. (1992). *Black resistance in high school: Forging a separatist culture.* Albany, NY: State University of New York Press.
Stinson, S. W. (1993). Meaning and value: Reflections on what students say about school. *Journal of Curriculum and Supervision, 8,* 216-238.
Talbert, J. E., McLaughlin, M. W., & Rowan, B. (1993). Understanding context effects on secondary school teaching. *Teachers College Record, 95,* 45-68.
Tobias, S. (1994). Interest, prior knowledge and learning. *Review of Educational Research, 64,* 37-54.
Virgil, J. D. (1993). Gangs, social control and ethnicity: Ways to redirect. In S. B. Heath & M. W. McLaughlin (Eds.), *Identity and inner city youth* (pp. 94-119). New York: Teachers College Press.

Vygotsky, L. S. (1978). *Mind in society: The development of higher psychological processes.* Cambridge, MA: Harvard University Press.

Wehlage, G. G., Rutter, R. A., Smith, G. A., Lesko, N., & Fernandez, R. R. (1989). *Reducing the risk: Schools as communities of support.* New York: Falmer Press.

Weiner, L. (1993). *Preparing teachers for urban schools: Thirty years of school reform.* New York: Teachers College Press.

Wentzel, K. R. (1991). Social competence at school: Relation between social responsibility and academic achievement. *Review of Educational Research, 61,* 1-24.

Wertsch, J. V., & Toma, C. (1995). Discourse and learning in the classroom: A sociocultural approach. In L. P. Steffe, & J. Gale (Eds.), *Constructivism in education* (pp. 159-174). Hillsdale, NJ: Erlbaum.

White, P. (1995). Education for citizenship: Obstacles and opportunities. In W. Kohli (Ed.), *Critical conversations in philosophy of education* (pp. 229-240). New York: Routledge.

Willis, P. (1977). *Learning to labor.* Farnborough: Saxon House.

PREFERENTIAL AFFECT:
THE CRUX OF THE TEACHER EXPECTANCY ISSUE

Elisha Babad

INTRODUCTION

This chapter deals with several teacher expectancy phenomena and their implications for researchers and educators. Hundreds of studies on teacher expectancies were conducted in the 1970s and into the 1980s in the wake of Rosenthal and Jacobson's (1968) classic *Pygmalion in the Classroom*, but the rate of publications has decreased considerably in recent years. Perhaps some issues (such as the very existence and intensity of teacher expectancy effects) had been satisfactorily resolved over the years, but many issues remained unresolved despite their potentially important implications for teaching in heterogeneous classrooms.

This chapter is focused on what I consider the crux of the teacher expectancy issue, namely teachers' differential *emotional* behavior in the classroom. Differential affect is a highly loaded and sensitive issue, probably the hardest to detect, control, and improve in teachers' conduct. The chapter is not intended to be a comprehensive review of the vast literature on teacher expectancies (interested readers are referred to reviews and books by Dusek, 1985; Blanck, 1993; Brophy, 1983, 1985;

Advances in Research on Teaching,
Volume 7, pages 183-214.
Copyright © 1998 by JAI Press Inc.
All rights of reproduction in any form reserved.
ISBN: 0-7623-0261-5

Babad, 1993a, 1993b; and many others). Rather, the relevant literature has been sampled selectively to focus on studies illuminating the issue of teachers' differential affect.

The first part of this chapter deals with teachers' differential behavior in the classroom—listing the most common specific behaviors enacted differentially by teachers; offering a comprehensive conceptualization of this phenomenon; examining student perceptions of teachers' differential behavior; and investigating systematic individual differences among teachers in proneness to demonstrate expectancy-related differential behavior. The second part of this chapter presents recent ideas and empirical findings on the teacher's pet phenomenon—a commonly occurring classroom situation in which extreme positive teacher affect is directed at particular favorite students. The third part examines classroom correlates of teachers' differential behavior and the pet phenomenon which are the presumed social consequences (or "costs") of teachers' preferential affect.

TEACHERS' DIFFERENTIAL BEHAVIOR IN THE HETEROGENEOUS CLASSROOM

Expression of Teacher Expectancies in Differential Behavior

Pygmalion in the Classroom (Rosenthal & Jacobson, 1968) created an intense storm, which still reverberates in educational circles, and opened a vast research domain focusing on teacher expectancies. Essentially, the Pygmalion research showed that if prophecies of potential late blooming for certain students are implanted in teachers' minds, there is a certain probability that some of these students may subsequently demonstrate improved intellectual performance in the absence of any other causal factor. I think that the controversy was really caused by the untested "Golem" implication (not investigated for obvious ethical reasons) that teachers' *negative* expectations may harm low achievers and depress their performance below their intellectual potential (Babad, Inbar, & Rosenthal, 1982a), and by educators' apprehension about potential "teacher bashing" in the wake of Pygmalion (Rist, 1987; Wineburg, 1987).

For teacher beliefs and expectancies (formed by experimental manipulation or by naturally available information about students) to influence student performance, a mediation process must exist (e.g., Darley & Fazio, 1980; Salomon, 1981), in which the expectations are expressed in differential behavior, which in turn is perceived by students, influencing their self-image and their academic behavior. Indeed, most of the teacher expectancy literature was focused on the behavioral mediation of expectancies through differential behavior toward different students (mostly toward low- versus high-expectancy students, but often tracing other partitions based on race, gender, etc.). The growing trend in Western educational systems to integrate different types of students in heterogeneous classrooms increased the importance of this issue, because teachers must face a wider array of students in the classroom.

When a teacher behaves differently toward different students (such as by varying length of eye contact, or wait time, toward low- and high-expectancy students when they fail to respond to a question), we can assume that the variation in her behavior reflects her differential beliefs and expectations. In the 1970s *all* differential behaviors were considered negative and dangerous: The overriding norm called for equal treatment of all students, and it seemed that whenever differential behavior was observed, the "low" students received the short end of the stick (more criticism, fewer instructional opportunities, less warmth, etc.).

In recent years, emerging approaches to instruction in heterogeneous classrooms (as well as progress in teacher expectancy research) changed educators' value orientation toward teachers' differential behavior. It is now recognized that in order to best satisfy the needs of all students and to promote educational equity, some (corrective) differentiality is legitimate and even desirable. It is also recognized that most differential expectations are not arbitrary (as is the case in Pygmalion-type experimental manipulations), but instead reflect real differences among students. If teacher expectancies are "accurate, reality-based, and open to corrective feedback" (Brophy, 1983, 1985)—then differential behavior may well be an effective educational practice.

First I shall list the most frequent differential behaviors reported in the literature on the basis of observational studies. Brophy (1983, 1985) and Harris and Rosenthal (1985, 1986a) presented detailed lists of the most common behaviors. Harris and Rosenthal conducted meta-analyses on specific behaviors which had been investigated in at least four independent studies each. (Meta-analysis is a statistical technique for the assessment of effect magnitude across studies, in this case, the intensity of the differential enactment of each behavior.) They examined separately experimental manipulations of bogus information and field studies on naturally existing teacher expectations, and distinguished between studies examining the relationships between teacher expectancies and their subsequent differential behavior and studies examing the relationships between differential behavior and subsequent student performance.

These two groups of studies represent the B-C and C-D links in Rosenthal's 10-Arrow model of expectancy mediation (Rosenthal, 1985; Harris & Rosenthal, 1985). This model presents five classes of variables: A—Distal independent variables (stable attributes of the expecter or expectee such as gender, race, status, ability, etc.); B—Proximal independent variables, which are the actual expectations (based on real distal variables or fabricated as in the case of the Pygmalion study); C—The mediating process by which teacher expectations are transmitted to students in verbal or nonverbal differential behavior; D—Proximal dependent variables, which are the immediate outcomes of the transmission of expectancies (short-term expectancy effects); and E—Distal dependent variables, or long-term outcomes of teacher expectancies. The "10

Arrows" are the various combinations, or "types" of expectancy research, such as A-B, B-C, B-D, C-D, and so on.

The following lists are organized according to the magnitude of the effects of the various differential behaviors and the consistency between the B-C and C-D links (see Babad, 1993a for greater detail).

(A) Differential behaviors with substantial and consistent effect magnitudes: Negative emotional climate; positive emotional climate; physical distance; off-task behavior; instructional input; and duration of interaction.

(B) Differential behaviors with moderate and/or inconsistent effect magnitudes: Frequency of interaction; asking questions; smiling; eye contact; encouragement; indirect influence; direct influence; giving directions; corrective feedback; head nods; wait time; lecturing; ignoring student; accepting student ideas; speech rate; criticism; praise.

(C) Differential behaviors with negligible effects (that is, the hypothesis that these behaviors are differentially enacted as a function of teacher expectancies was disconfirmed): Leaning toward student; touching student; gesturing; praise following correct answer; criticism following wrong answer; persistence; work-related contact; procedural contact; task orientation.

Brophy (1983, 1985) presented lists of differential behaviors with no empirical estimation of their effect magnitudes. Most were similar to behaviors listed above, but several were sufficiently different to warrant an additional list, specifying teachers' behaviors toward low-expectancy students: Not improving their answers; praising incorrect answers or inappropriate behaviors; demanding less; more private and more structured interaction with them; less credit in grading; shorter and less informative feedback; less intrusive instruction; less use of time-consuming instructional methods.

As early as 1973 Rosenthal presented the "four-factor theory" in which he grouped various differential behaviors in four meaningful clusters:

1. *Climate.* Teachers tend to create a warmer social/emotional climate for their high-expectancy students, transmitting warmth and support through verbal and nonverbal behavior.
2. *Feedback.* Teachers provide more positive feedback to high-expectancy students, and more negative feedback to low-expectancy students. In later formulations, teachers were described as giving more informative and instructive feedback to high-expectancy students.
3. *Input.* Teaching high-expectancy students more, and more difficult, material.
4. *Output.* Giving high-expectancy students more opportunities to respond, maximizing their chances of learning.

Harris and Rosenthal (1985, 1986a; see also Rosenthal, 1987, 1991) grouped the behaviors in their meta-analyses into the four factors and summarized the effect magnitudes for each factor. Substantial effects were found for differential enactment of affective climate ($r = .29$) and instructional input ($r = .27$), a smaller effect size was found for output ($r = .17$), and the accumulated effect size of feedback behaviors was practically negligible ($r = .10$). Thus, contrary to empirical findings and theorizing of the early 1970s, teachers in the 1980s demonstrated differentiality in affective behaviors and direct instructional behaviors, but they divided their feedback in an equal manner among students. In light of these findings, Rosenthal (1989, 1991) recently modified the four-factor theory into a two-factor theory: *Affect* (similar to the original climate factor) and *effort*, an instructional dimension combining input and output. In the remainder of this chapter I refer to these factors in a somewhat broader meaning as emotional support and learning support.

What happened to the feedback factor over the past two decades? Informal inquiries that I often conduct among groups of educators in in-service training repeatedly indicate that most practitioners think that expectancy-related differential behavior is expressed *most strongly* in differential feedback practices, and early investigators shared this notion. I think that the shift into a more equal distribution of feedback reflects the accumulated influence of the teacher expectancy literature, especially in the framework of progress in theory and implementation of teaching methods for integrated, heterogeneous classrooms. I think that teachers are aware of the expectancy issue and try very hard to dispense their feedback equally. However, as Brophy (personal communication) suggested, it is also quite conceivable that the results of the meta-analyses depended considerably on how the measures were defined and aggregated. Perhaps subtle and qualitative aspects of feedback are more differentiated and more related to expectancy differences than the *quantity* of feedback which was usually measured in these studies.

A Proposed Conceptualization of Teachers' Differential Behavior

On the basis of the accumulated observational literature on the behavioral mediation of teacher expectancies and collaborative investigations with several colleagues, I proposed recently a conceptualization of teachers' differential behavior (Babad, 1993a, 1993b). The three components in this conceptualization are:

1. The desirability and ideological legitimacy of each differential behavior;
2. The (presumable) "real" state of teachers' emotions toward different students;
3. Teachers' ability to control their verbal and nonverbal behavior.

The dominant educational ideology today holds that underprivileged students must be compensated for their low starting point by investment of extra resources ("corrective preference") in order to enrich their education and lead them toward an equal state with their more privileged peers. Therefore, differential instructional behavior is condoned, and teachers can and should provide more learning support to low achievers and low-expectancy students. (In my opinion, a trace of this ideology could be found in empirical findings about the input factor mentioned above, reflected in gaps between the reported levels of quantity and quality of instructional input: Low-expectancy students were often reported to receive *more* instruction, but of *lower quality*. Thus, teachers often compensated low-expectancy students, but the quality of that instruction followed their expectations.)

The ideologically legitimate corrective preference is strictly and explicitly limited to the instructional domain. In affective behavior and manner of relating to students, teachers are required to treat all students equally. Teachers (and parents) are not supposed to demonstrate emotional preferences even if their inner emotions toward various children are differential. In affective displays, total equality is demanded regardless of one's true feelings. We can presume that high-expectancy students are usually better liked by their teachers: They achieve the academic objectives successfully and give teachers much satisfaction; they are usually more obedient and do not violate the norms of the "good student" role; and teachers can more readily control their behavior and teach them effectively (Cooper, 1979, 1985; Cooper & Good, 1983). Therefore, in order to treat all students equally, teachers must exercise a great deal of control over their emotional behavior and conceal their differential affect.

The ability to control affective displays and to conceal negative emotions is not only a dimension of individual differences between teachers, but varies greatly *within* the person. Some channels (e.g., words and verbal content) can be readily controlled to demonstrate equality, whereas other channels (body, voice intonation, and to a lesser degree facial expressions) cannot be controlled as easily. Therefore, negative affect cannot be concealed and "leaks" through these channels. The concept of nonverbal leakage was first introduced by Ekman and Friesen (1969, 1974), and a great deal is known today about the hierarchy of channels in terms of leakage potential (see Rosenthal & DePaulo, 1979; Zuckerman, DePaulo, & Rosenthal, 1981, 1986; Babad, Bernieri, & Rosenthal, 1989a). With regard to teachers' differential affective behavior, there is also a danger that attempts to control emotional displays may lead to excessive sweetness toward disliked students which may well be interpreted by them as reflecting *negative* affect.

The various combinations of the components of this conceptualization yield the following patterns:

1. Teachers actively compensate low-achieving students and invest extra instructional efforts and resources in providing learning support to them.
2. Teachers exercise self-control in trying to dispense verbal feedback to all students in an equal manner.
3. Despite possible attempts to conceal negative affect, teachers dispense their emotional support differentially, showing more warmth and supportiveness (perhaps even love) toward high-expectancy students. Thus, teachers' feedback is dispensed in an equal manner, whereas patterns of differential behavior for learning support and emotional support are reversed, lows

receiving more instructional support but highs receiving a more positive emotional treatment. The crux of the expectancy issue lies in the preferential affect, but teachers may not be cognizant of that, believing that they are as successful in controlling their emotional support as they are in controlling their feedback behaviors.

Babad and his associates (Babad, Bernieri, & Rosenthal, 1987, 1989a, 1989b, 1991; Babad & Taylor, 1992; and a summary chapter, Babad, 1993a) conducted a series of context-minimal studies of teachers' expectancy-related nonverbal behavior when talking about and talking to low- and high-expectancy students. Context was minimized by separating verbal and nonverbal channels (face, body, voice intonation, verbal content, etc.), by showing judges very brief 10-second instances depicting teacher behavior, and by showing only the teacher in these clips. Clear expectancy effects, that is, differential treatment of low- and high-expectancy students, were evident in fine nuances of nonverbal behavior, even when teachers were just talking *about* different students (Babad, Bernieri, & Rosenthal, 1989b). When talking to individual students and interacting with them, facially communicated expectancy differences were found in judges' ratings of negative affect and active teaching behavior, confirming the proposed conceptualization: Low-expectancy students were treated with more vigorous teaching activity (learning support compensation), but with more negative affect. Students (as young as fourth graders) who served as judges in subsequent studies were acutely aware of teachers' differential nonverbal behavior, and could detect well beyond chance when teachers were interacting with (unseen) low- or high-expectancy students (Babad, Bernieri, & Rosenthal, 1991; Babad & Taylor, 1992). This last finding leads into the next section, in which students' perceptions of teachers' differential behavior are discussed.

Students' Perceptions of Expectancy-Related Teachers' Differential Behavior

Subjective perceptions and attributions have been central foci of theory and research in social psychology for a long time, and they have become more important in educational research in recent years. Low-inference measures (mostly systematic behavioral observations) have been replaced by high-inference techniques based on perceptions and judgments of students and teachers. High inference measures are less costly and easier to administer, and it is argued (e.g., Chavez, 1984; Fraser, 1987, 1989) that they reflect more directly the reality of the classroom as it is experienced by its participants.

Student perceptions are particularly important in teacher expectancy research, because all sequential models describing the stages leading to the realization of a teacher's expectation in a student's behavior (e.g., Brophy, 1983, 1985; Cooper, 1985; Darley & Fazio, 1980; Jussim, 1986), emphasize that students must perceive teachers' expectancy-related differential behavior, internalize its meaning, and accommodate their self-image and behavior to that meaning. Indeed, when Marshall and Weinstein (1984; see also Weinstein, 1986) interviewed young students to find out how they knew how smart they were in school, perceived teacher behavior toward them was the major source of students' self-knowledge. However, despite the significance of student perceptions in the expectancy confirmation process, and contrary to the abundance of other types of expectancy research, relatively little research has focused on student perceptions of teachers' differential behavior. I shall refer here mostly to a series of studies conducted by Weinstein, Marshall, and their colleagues in Berkeley (Weinstein, 1983, 1985, 1986, 1989; Weinstein, Marshall, Brattesani, & Middlestadt, 1982; Weinstein, Marshall, Sharp, & Botkin, 1987; Weinstein & Middlestadt, 1979; Brattesani, Weinstein, & Marshall, 1984; Marshall & Weinstein, 1984, 1986), by Babad in Israel (Babad, 1990a, 1990b, 1992, 1993a, 1993b, 1995, 1996), and, to a lesser extent, by Cooper and Good (1983), then in Missouri.

In general, all investigators found that students are highly sensitive to teachers' differential behavior and can describe it in systematic terms which are quite consistent with observational findings. However, students often interpret teacher behavior differently from the standard interpretations given by (adult) trained observers, by researchers, or by the teachers themselves. For instance, the behavior of "calling on students" which supposedly carries the universal meaning of providing instructional support in the "output" category (i.e., providing response opportunities), was found (Babad, 1990b) to have different meanings in students' perceptions, depending on *who* is the target of the teacher's questions: If the teacher calls on a high achiever, the students perceive this behavior as reflecting learning support, but calling on a low achiever is perceived within the cluster of "putting pressure." There may well be other behaviors that the teachers enact with a positive and helpful intent but which students consider as negative and as reflecting low teacher expectations.

Weinstein and Babad independently discovered that the most effective method of tracing students' perceptions of teachers' differential behavior involves asking about the teacher's behavior toward a hypothetical low-expectancy and high-expectancy student "that she had taught in the past." This method allows students to freely project their current perceptions of their teacher. Direct questions about how your teacher treats you or how she distributes each behavior in the classroom are too obtrusive, evoking defensive and/or protective reactions denying teacher differentiality. Cooper and Good (1983) asked directly about teachers' feedback behaviors, and indeed reported lesser differentiality (althogh in any case, the feedback factor does not show substantial effect magnitudes in meta-analyses of teachers' differential behavior).

Weinstein and Middlestadt (1979) found that students' perceptions of teacher interactions with high-expectancy students reflected high expectations, academic demand, and special privileges, whereas low-expectancy students were perceived as receiving fewer chances but greater teacher concern and vigilance. Weinstein and colleagues (1982) found that low-expectancy students were perceived as receiving more negative feedback, more direction, and more work- and rule-oriented treatment, whereas high-expectancy students were perceived as receiving higher expectations and more opportunity and choice. These findings are well in line with the proposed conceptualization of teachers' differential behavior, although Weinstein and her colleagues did not put a special emphasis on teachers' affective behaviors in their investigations.

In Babad's (1990a, 1990b, 1995, 1996) investigations of students' perceptions of teachers' differential behavior, the list of behaviors included three (factor-based) clusters: *Learning Support* (e.g., teacher approaches to observe student's work, gives student opportunity to think long enough before answering, explains student's mistakes and how to correct them, etc.); *Emotional Support* (teacher is warm and supportive of student, loves the student, gives student much attention, etc.); and *Pressure* (teacher addresses difficult questions at student, is very demanding of student). In large samples of Israeli fifth- to seventh-grade classrooms, students' perceptions of teachers' differential behavior were very clear and consistent: Teachers were systematically perceived giving low-expectancy students more learning support and putting less pressure on them compared to high-expectancy students; on the other hand, teachers were systematically perceived as giving more emotional support to high-expectancy students, being more attentive, warm, supportive, and loving toward them. These effects of teacher differentiality were of substantial magnitudes, supporting the conceptualization of teachers' differential behavior presented above. As mentioned earlier, the differences in learning support and pressure are ideologically legitimate and even desirable, but the differentiality in emotional support is illegitimate and very problematic, and it may well spoil the intended positive effects of the learning support differentiality.

The reader should note that strong statistical effects do not necessarily reflect strongly expressed behaviors, and the actual enactment of teacher affective differentiality may be very subtle, expressed in fine and often almost undetectable nuances of verbal and nonverbal behavior. Student perceptions are based on an accumulation of such nuances over a long period of time. Weinstein (1983) pointed out that sometimes strong student impressions (mostly negative, I presume) might be caused by a rare teacher behavior or utterance that deviates from her regular conduct. In the Babad, Bernieri, and Rosenthal (1989b) investigation of teachers' nonverbal behavior, some differentiality in teachers' negative affect was picked up by adult judges from actual teacher facial expression and body language. However, I think that the clear detection of whether the teacher was interacting with an unseen high- or low-expectancy student made by young students serving as judges in Israel (Babad, Bernieri, & Rosenthal, 1991) and in New Zealand (Babad & Taylor, 1992) was not necessarily based on clearly *negative* affective behavior. Rather, in scrutinizing the clips my impression was that the teachers seemed very intent to treat low-expectancy students in a warm and supportive manner. Perhaps the exaggeration and the implied hypocrisy were the signals leading students to correctly identify teacher interactions with low-expectancy students.

Babad (1990a, 1995) compared the teachers' own reports of their differential behavior to the perceptions of their students. The patterns of agreement and disagreement between teachers and students confirmed the proposed conceptualization of teachers' differential behavior, emphasizing again the grave significance of the teacher affect issue. With regard to learning support and pressure, both teachers and students agreed about the intensity and direction of teacher differentiality, namely, that teachers give low-expectancy students more instructional support but put less pressure on them compared to high-expectancy students. With regard to emotional support, the perceptions were directly opposed: Whereas the students perceived their teachers as giving more affective support to high-expectancy students, the teachers reported giving more emotional support to low-expectancy students! The patterns of agreement and disagreement between students and teachers were very systematic in both studies, characterizing almost every classroom. Teachers report a consistent and desirable pattern of compensating low achievers in both the instructional and the affective domains, whereas students recognize the instructional compensation but perceive their teachers as giving preferential affect to high-expectancy students.

Both the 1990 and 1995 studies included an intervention in which teachers received empirically based feedback on the gaps between their reports and their students' perceptions. In the earlier study (Babad, 1990a) we characterized each teacher as open or resistant to the feedback. Subsequent measurement of differential behavior showed a small change in the classrooms of the teachers characterized earlier as open to feedback. In these classrooms, the differences in emotional support were somewhat reduced (yet not eliminated) in the posttest. Paradoxically, the teachers and their students reported on the change differently, following their initial, diametrically opposed perceptions—the students reported a small reduction of the excess emotional support to high-expectancy students, whereas the teachers reported increasing emotional support to high-expectancy students (since they reported initially giving less emotional support to the highs). In the second study (Babad, 1995) the feedback intervention had no effect whatsoever.

Elsewhere (Babad, 1993b) I discussed the difficulties and obstacles in conducting applied interventions to change teachers' differential behavior, and reviewed some of the interventions reported in the expectancy literature. Changes in teacher behavior on the basis of empirical feedback are never readily attainable, but in this particular situation it is even more unlikely because of the opposing perceptions of students and teachers. Given that differential learning support and pressure are desirable, the only change required is in emotional support, and teachers should be requested to balance their affective behavior by increasing emotional support to low-expectancy students and perhaps even decreasing excess emotional support to high-expectancy students. Facing this demand, the teachers would argue (as they did in our intervention studies) that giving more emotional support to the lows is exactly what they are doing!

A decade ago I tended to "believe" the students with regard to differential emotional support and to "disbelieve" the teachers, viewing them as defensive and hypocritical. Today I see the teachers in a different light, and I believe that most teachers genuinely try to be warm and supportive to low-expectancy students. They fail to create this impression in students' minds because they are often over-intent and overdo their affective displays to the lows, and sometimes their hidden affection to high-expectancy students leaks out despite the attempted control.

Individual Differences Among Teachers in Tendency to Show Expectancy Effects and Differential Behavior

The search for systematic individual differences in proneness to expectancy effects (personality correlates research) is quite different from the "mainstream" research on the existence and characteristics of expectancy effects (experimental research). To examine if expectancies are indeed expressed in expector differential behavior and if they can influence the behavior or performance of expectees, the investigator constructs an experimental design (in the laboratory or in the field) where experimental and control groups are compared (e.g., comparing the IQ of the "late blooming" Galatea group to the no-information control group in the original Pygmalion study). An expectancy effect would then be manifested in a statistically significant difference between means. Such an effect is calculated by dividing the difference between the means (numerator) by some index of the variance (denominator). The expectancy effect will be more significant and of substantial magnitude (a) the larger the difference between the means and (b) the smaller the variance among the subjects in each group. Therefore, in experimental research, the variance (that is, individual differences) is considered as *noise* to be reduced as much as possible. On the other hand, personality research on individual differences stems from interest in the variance, and the investigator tries to trace systematic patterns in the variance. The goal is to identify common attributes characterizing individuals showing a common behavioral pattern. In this case, the main question is whether we can identify and characterize teachers who are prone to demonstrate expectancy effects in their classrooms.

Expectancy effects represent a probabilistic phenomenon, even if their existence has been proven beyond chance and their effect magnitude is substantial. Not all investigated experimenters show effects of self-fulfilling prophecies, and not all teachers demonstrate expectancy effects in their classrooms (although *all* teachers must have differential expectations for different students). The objective of the individual difference research is to trace and identify (and subsequently predict) the teachers prone to demonstrate expectancy effects through salient differential behavior in their classrooms. Such investigation is of both theoretical and applied significance. If a corrective intervention is ever planned to reduce the negative manifestations of teacher expectancies (Kerman & Martin, 1980; Proctor, 1984), we must be able to identify the target audience who actually needs the intervention so as not to waste resources on those who do not need the treatment. Despite the importance of investigating individual differences in proneness to expectancy effects, this area has remained rather neglected over the years.

An expectancy effect combines cognitive and interactive aspects. The cognitive aspect involves the formation of the expectancy on the basis of available (or experimentally provided) information, and the interactive aspect involves subsequent differential behavior which expresses and transmits this expectation to the expectee. Finn (1972) labeled these aspects suggestibility and communicability, and Cooper and Hazelrigg (1988) discussed the need for social influence, expressiveness, and likability of the expector.

Conceptually, individuals prone to demonstrate expectancy effects are hypothesized to be rigid and inflexible. The rigidity would characterize both their cognitive style and social behavior, a combination of stereotypic perception causing the formation of bias, and rigid expectations and subsequent stereotypic behavior which would transmit those expectations in systematic and unchangeable manner. This description fits the well-known personality syndrome of *dogmatism* (Rokeach, 1960) or *authoritarianism* (Adorno, Frenkel-Brunswick, Levinson, & Sanford, 1950), with its salient characteristics of rigidity, intolerance of ambiguity, and resistance to disconfirming information. Indeed, the central hypotheses and the major findings in the investigations of proneness to demonstrate expectancy effects involved dogmatism-related constructs.

A typical study of this sort examined self-fulfilling expectancy effects in an experimental situation in which experimenters were given bogus biasing information. Subsequently, experimenters were classified according to the magnitude of their expectancy effects as measured in the behavior of their subjects, and potential demographic and personality correlates of proneness to expectancy effects were examined. Dogmatism-related variables reported to be correlates of expectancy effects included need for social approval (Perlmutter, 1972), social desirability (Blake & Heslin, 1971), repression-sensitization (Dana & Dana, 1969), locus of control (Clarke, Michie, Andreasen, Viney, & Rosenthal, 1976), cognitive complexity and intolerance of ambiguity (Tom & Cooper, 1983; Tom, Cooper, & McGraw, 1984), and, in a different experimental situation involving biasing effects in peer counseling, dogmatism and rigidity (Harris & Rosenthal, 1986b; Harris, 1989).

Studies conducted on teachers showed conceptually similar correlates of expectancy proneness and teachers' differential behavior. Brattesani, Weinstein, and Marshall (1984) reported differences in teachers' level of differentiation, Cooper and his associates (Cooper, 1979, 1985; Cooper & Good, 1983; Cooper & Tom, 1984) reported on the moderating influence of teachers' sense of control over students' performance, and Brophy and Good (1974; Brophy, 1983, 1985) conceptualized (but did not provide empirical data supporting their claim) that teacher expectancy effects are likely to occur with "over-reactive" teachers who hold rigid expectations harmful to low-expectancy students.

Babad (1979) devised a performance measure of susceptibility to biasing information, and used it to identify teachers prone to demonstrate expectancy effects in their classrooms. In this measure, teachers and university education students were asked to score two drawings in a Draw-A-Person test that they had practiced, allegedly drawn by a high-status and a low-status child (according to demographic, ethnic, and socioeconomic information provided about each child). The drawings were actually reproduced from the test manual (the drawing attributed to the high-status child had a manual score three points higher than the drawing attributed to the low-status child), and the differences between the scores given to the two children (minus the three-point objective difference) reflected the scorer's level of susceptibility to biasing information.

In a series of administrations of this measure to hundreds of teachers and education students over several years, stable distributions of bias scores were observed. About one-sixth of the subjects scored the drawings objectively and were uninfluenced by the biasing information, about one-half were mildly biased, about one-fourth were highly biased and attributed a much higher score to the high-status child. Also, 3 to 5 percent of the scorers showed reversed bias, giving a higher score to the low-status child. (For years, I wondered whether the reversed bias persons should be characterized as similar to the highly biased or the unbiased individuals. I finally examined this question in a four-year follow-up (unpublished) study of teachers-in-training, and discovered that the small reversed-bias group resembled the unbiased more than the biased group in almost all characteristics. My cynical conclusion was that it sometimes pays to be hypocritical, because although this group demonstrated bias in scoring, they "compensated" the low-status child by giving him a higher score than objectively warranted!)

In a series of subsequent studies, substantial differences were found between unbiased and highly biased individuals, all consistent with the hypothesized conceptual links between dogmatism, bias, and expectancy proneness. Highly biased individuals held political views (Babad, 1979) and educational views (Babad, 1985) in a more extreme fashion than unbiased individuals, in line with Rokeach's (1960) conception of dogmatism as representing extremity of style independent of content. Highly biased teachers showed strong halo effects in nominating high- and low-expectancy students, while unbiased teachers showed no halo effect (Babad, Inbar, & Rosenthal, 1982b). Babad and Inbar (1981) found that highly biased teachers used more dogmatic statements than unbiased teachers in written analyses of educational events. Babad (1979) also found that highly biased individuals (but not the unbiased ones) described themselves on an adjective checklist as over-reasonable, highly objective, logical, reasoned, and unbiased! The dogmatism scale itself did not distinguish between the bias groups. I suspect this is due to the transparent nature of the scale, which consists only of highly dogmatic statements (in addition to my personal mistrust of self-report personality questionnaires in general). In a four-year (unpublished) follow-up study of biased and unbiased teachers-in-training throughout their college years, I found that the highly biased group was lower than the unbiased group in intellectual level, academic performance, and professional level in their fieldwork. The biased group also showed a higher drop-out rate throughout the college years.

The crucial connection between susceptibility to biasing information and teachers' classroom conduct was investigated in several studies. Babad and Inbar (1981) conducted behavioral observations in the classrooms of highly biased and unbiased teachers-in-training, and found very substantial differences on almost all variables, indicating that the unbiased group consisted of better teachers. These differences were subsequently confirmed by judgments of these teachers' immediate supervisors, who were totally unaware of the measurement of susceptibility to biasing information or of any partition among the investigated teachers. The measurement of teachers' nonverbal behavior in the context-minimal studies by Babad, Bernieri, and Rosenthal (1989a, 1989b) showed that highly biased teachers demonstrated more nonverbal leakage (indicative of unsuccessful attempts to conceal negative affect) toward their classrooms than unbiased teachers.

The most important study examined directly the relationship between teachers' susceptibility to biasing information and their tendency to demonstrate positive ("Pygmalion") and negative ("Golem") expectancy effects in their classrooms (Babad, Inbar, & Rosenthal, 1982a; also Rosenthal & Babad, 1985). In this study highly biased and unbiased physical education teachers nominated three high-expectancy and three low-expectancy students (presumably for observations by the Sports Authority of the Israeli Ministry of Education), following the methodology of experimental field studies of teacher expectancies. A Pygmalion-type experimental intervention was added, in which each teacher was provided by a person of professional authority the (randomly selected) names of two "physical education late bloomers" in the classroom, allegedly chosen according to the ministry's newest testing instrument. The dependent variables for measuring the expectancy effects included both teachers' differential behavior and students' actual performance for these teachers. The results were strong and highly consistent: Highly biased teachers demonstrated substantial expectancy effects (evident in their own differential behavior *and* in students' performance), whereas no expectancy effects were found in the classrooms of the unbiased teachers. The Golem effect in the classrooms of the highly biased teachers was particularly salient: These teachers treated their low-expectancy students in a negative manner, and the students, in turn, responded with particularly low levels of performance. In contrast, unbiased teachers did not treat their low-expectancy students differently than the other students, and the performance of the low-expectancy students was almost equal to that of the high-expectancy students.

THE "TEACHER'S PET PHENOMENON:" AN EXTREME CASE OF PREFERENTIAL AFFECT

We all know about teachers' pets and recognize this phenomenon in the classroom. Some of us may remember nostalgically how we might have been teachers' pets ourselves in the foregone past or, as teachers, how we might have loved a particular student. All of us probably remember our emotional reactions toward students who were considered teachers' pets and toward teachers who had special relationships with their favorites. Doris Day sang some decades ago "I want to be teacher's pet," but educators probably do not hold this phenomenon in a particularly positive light.

Despite the accumulation of empirical knowledge on other aspects of teacher expectancies and teacher-student interaction and despite the high recognition factor of the teacher's pet phenomenon, very little empirical research has focused directly on this phenomenon. The issues discussed next include the existence and measurement of this phenomenon in actual classrooms, its rate of occurrence, the potential verification of "folk knowledge" about aspects of this phenomenon (such as favoritism), and the characteristics of pets and of teachers who have pets.

The teacher's pet phenomenon might be considered either as a teacher attitude or as a teacher expectancy issue. It deals with the teacher's affective response to a student and it involves differential, even preferential, treatment, but expectations are not necessarily focused on students' achievement. Teachers' positive expectations and love toward their particularly chosen pets (usually one student, never more than three) are believed to be expressed in conspicuous favoritism and preferential affect which can strongly influence classroom climate and students' reactions to their teachers.

Silberman's Conceptualization of Teachers' Emotional Attitudes

Silberman (1969, 1971) came close to defining and measuring the teachers' pet phenomenon. He defined four emotional attitudes teachers hold for particular students—attachment, concern, indifference, and rejection—and examined teachers' behaviors toward students who were objects of these four attitudes. Silberman defined "attachment" as an affectionate relation to students that derives from the pleasure they bring to the teacher's work. In Silberman's questionnnaire, teachers nominated those students "they would like to keep for another year for the sheer joy of it." Given this emotional bond, numerous attachment students could probably be considered as teachers' pets, but the attachment group could also have included excellent students academically who are not necessarily teachers' pets. The inclusion of "academically best students" in the attachment group was most salient in Good and Brophy's (1972) investigation.

Silberman's conceptualization evoked much interest, and a number of studies focused on teachers' emotional attitudes and their manifestations in the classroom in the next dozen years. Silberman (1971) and Good and Brophy (1972) examined the behaviors of students defined as objects of the four teacher attitudes, and found that attachment students showed high conformity and achievement orientation which were rewarding for their teachers. Brophy and Everston (1981) found attachment students to be most compliant and participative, judged by their teachers as closest to the ideal student role image. Brophy and Good (1974) reported that teachers described attachment students in the most positive terms along a spectrum of attributes and characteristics.

However, as far as actually observed teacher behavior toward attachment students was concerned, the empirical findings were weak. In summarizing this line of research, Good and Brophy (1974) and Brophy and Everston (1981) concluded that, whereas differential patterns of teacher behavior toward concern and rejection students were found, very few behavioral distinctions were observed for attachment students, despite the fact that teachers' affection for these students was clear from the very fact that they nominated them. I believe that the investigated teachers were careful to monitor their behavior and to treat all students equally in front of the observers after they themselves had nominated the attachment students. Given that the existence of teachers' pets might be a sensitive and loaded issue (especially since the term "pet" carries more negative connotation then the term "attachment"), it would have made more sense if pets would not have been nominated directly by their own teachers. It must be pointed out, however, that the Brophy and Everston work marked a shift from purely quantitative aspects to more qualitative and subtle aspects of teacher student interaction. Perhaps a more complete shift to qualitative measures might have yielded clearer findings regarding attachment students.

Next, I shall review a series of studies on the teacher's pet phenomenon conducted by my colleagues and me in Israel during the last decade. To the best of my knowledge, no other empirical examinations focusing directly on this phenomenon have been published.

Students' and Teachers' Thoughts about the Teacher's Pet Phenomenon

Our initial study (Tal, 1987; Tal & Babad, 1989) investigated students' and teachers' perceptions and attributions about the teacher's pet phenomenon. In the absence of published research or theorizing on this phenomenon, it seemed appropriate to examine first students' and teachers' familiarity with it and their attitudes about it. A sample of students of varied grade levels, teachers-in-training, and experienced teachers participated in this investigation.

The recognition factor of the teacher's pet phenomenon was extremely high. Over 90 percent of the respondents recognized the phenomenon and could describe the special emotional relationship between the teacher and her pet and the mutual rewards they provide for each other. The phenomenon was characterized mostly in negative terms, emphasizing teachers' preferential treatment of their pets. Whenever positive aspects were mentioned in respondents' open-ended descriptions, they referred mostly to the personality and positive social attributes of children who become teachers' pets.

In this study the reactions of elementary and high school students were more negative than those of the teachers, and the attitudes of teachers-in-training (still being students, but soon to be teachers) fell between these extremes. Teachers related to the pet phenomenon in a somewhat lenient and tolerant manner, believing more strongly than the students that they can conceal from the other students their special attachment to their pets. (This belief was disconfirmed by the high recognition factor and by numerous other findings in subsequent studies.) The teachers also expressed apprehension toward the investigators for prying into this topic, whereas students were quite satisfied to be asked about this phenomenon.

In another part of this study three student "types" were presented—"teacher's pet," "academically best student," and "leader"—and all respondents rated the perceived social distance between each type and (a) the teacher, and (b) the other students. The findings showed high consensus among all respondents. Academically best students were perceived as close to both the teacher and the other students; leaders were seen as close to the students but remote from the teacher; and teachers' pets were considered very close to the teachers but remote from the other students. These perceptions confirmed the existence of an implicit attributional typology in which the teacher's pet is a clearly distinct "type." However, when perceived social distances were measured again in a field study where specific students were actually identified as representing these types in their classrooms (Tal, 1987), the abovementioned patterns were not observed with the same level of clarity. The main reason for the undifferentiated pattern was that a considerable number of salient students were identified in their classrooms as representing *two* types (both teacher's pet and leader, or both teacher's pet and academically best student) or even as representing *all three* types. When student types are considered hypothetically, they can be distinguished from each other more sharply.

Actual Measurement of the Pet Phenomenon and it s Rate of Occurrence in the Classroom

Next (Tal, 1987; Tal & Babad, 1990) a method was developed to identify teachers' pets and to assess the intensity of the teacher's pet phenomenon in schools. The method was based on students' sociometric judgments—which differ from conventional sociometric attractions in that the respondent is not asked to nominate affective choices of potential friends, but rather to judge "cognitively" who are the students who occupy particular social roles in the classroom. Students were asked to name the two students "best liked" by their homeroom teacher, and later to nominate the student "most loved" by the teacher, as well as nominating academically best students, leaders of the boys or the girls, quiet students, and eventually also the "scapegoats." When the rate of consensus in nominating a particular child (or two-three children) as best liked and most loved by the teacher was very high and most students nominated the same children, that classroom was designated as a pet classroom. Tal and Babad distinguished between "exclusive pet classrooms" (i.e., extremely high consensus about a single pet); "non-exclusive pet classrooms" (i.e., high consensus in identifying one, or two, or three students as teachers' pets); and "no pet classrooms" (where there was no consensus in identifying pets). Non-exclusive pet classrooms were considered to reflect a pet phenomenon of lower intensity than exclusive pet classrooms.

In a sample of 80 Israeli fifth- and sixth-grade classrooms, Tal and Babad (1990) found that 80 percent were characterized as pet classrooms—26 percent exclusive and 54 percent non-exclusive—and only 20 percent were no pet classrooms. In a more recent study on another sample of 80 Israeli fifth- and sixth-grade classrooms, Babad (1995) found only 49 percent pet classrooms (13% exclusive and 36% non-exclusive). However, using somewhat less strict sociometric consensus criteria, the phenomenon could be clearly identified in over three-quarters of the classrooms. Thus, the rate of occurrence of the teacher's pet phenomenon is very high

Characteristics of Teachers' Pets

What characterizes students who are more likely to become teachers' pets? Attributions made by students and teachers (Tal, 1987; Tal & Babad, 1989) and examinations of the measured characteristics of actually identified pets (Tal, 1987; Tal & Babad, 1990; Babad, 1995) indicated that pets are very positive, likable, and attractive children. They tend to be girls more frequently than boys, perhaps because at this age, girls are less rebellious and more likely to accommodate themselves to school norms than boys. Pets are almost always students of high academic standing, although not necessarily the very best students in their classrooms. It seems that their intellectual ability is complemented by charm, beauty, and social skills (which include good manners, compliance, and flattery, although the teacher's pet is clearly distinguished from the classroom flatterer). In the 1990 study pets were found more frequently to be of Israeli, European, or American ("Ashkenazi") origin than of African or Asian ("Sephardi") origin, but this finding was not replicated in the 1995 study. Finally, although research on teachers' differential behavior shows that teachers often demonstrate excessively positive behavior toward low-expectancy and disadvantaged students, the empirical evidence indicated that such students were *not* consensually held by their classmates to be teachers' pets.

Are Teachers Aware of the Pet Phenomenon in their Classrooms?

Having pets is a delicate issue for teachers, and their potential awareness of its occurrence in their own classrooms is loaded with presumed defensiveness and denial. In the attributional study (Tal & Babad, 1989) we found that teachers were well aware of the teacher's pet phenomenon in hypothetical terms, but they were also more confident than students and teachers-in-training of their ability to conceal from their students their special attachments to their pets.

When asked to guess the consensual sociometric nominations actually made by their students (Tal & Babad, 1990), the investigated teachers did a very poor job. Either they failed to answer this question (some pointing out that there was no such consensus among their students) or they named another student, *not* consensually viewed as a pet by their students. Only 19 percent of the teachers named the pets correctly. In comparison, 50 percent of the teachers were accurate in guessing the nominated leader and the most popular student, and 68 percent correctly identified the students nominated by their peers as best students academically. It was concluded long ago in the sociometric literature (e.g., Evans, 1962) that teachers are not acutely aware of students' sociometric status, but the low rate of correct guessing of teacher pet nominations fell far below the rates of teacher accuracy reported in the literature for sociometric status, popularity, and so on.

Favoritism and Preferential Treatment

When people think about teachers' pets, they think almost automatically about favoritism and preferential treatment, and that is probably the main reason for the negative connotation of the teacher's pet phenomenon. In the Tal and Babad (1989) attributional study, preferential treatment was mentioned very frequently in respondents' open-ended descriptions of this phenomenon, and pets were described as taking personal advantage of their special status. Many respondents also thought that the other students would choose to send the pets as their representatives to the teacher to attain privileges for the entire class or to prevent collective punishment for wrongdoing.

In the 1990 field study, Tal and I examined whether exclusive pets actually received overly generous math and language grades compared to their relative position among their peers on their objective achievement scores in these areas. For this analysis, the achievements of the exclusive pets in teacher grades and objective achievement scores in math and language were transformed to standard scores within their classrooms. Higher standard scores in teacher grades compared to objective achievement were conceived to reflect teacher favoritism. No overall favoritism was found, although a subgroup of teachers who expressed more authoritarian educational attitudes showed a trace of favoritism in assigning higher grades to their pets. However, the power of these analyses was constrained by the fact that all pets were students of high academic standing, so that all comparisons were limited to the positive end of the achievement distribution. Unfortunately, no behavioral observations were conducted in this series of studies, and therefore no empirical data are available on behavioral manifestations of preferential treatment which presumably accompany the preferential affect inherent in the nature and very existence of the teacher's pet phenomenon.

Having Pets: A Teacher's Trait or a Situational Phenomenon?

When a phenomenon of this type is investigated, one of the first emerging questions is whether teachers' tendency to have or to not have pets reflects their inherent personality traits and emotional needs or arises out of situational factors only. Social-psychological literature on the "fundamental attribution error" (Ross, 1977) and the "correspondence bias" (Jones, 1990) illuminates a generalized human preference to interpret behavioral phenomena in terms of fixed traits rather than changeable situational factors. In that light, our initial tendency would be to view the teacher's pet phenomenon as reflecting teachers' fixed traits and needs.

The empirical findings bearing on this issue are equivocal. In 1990 Zohar Tal and I found that teachers who had pets (and particularly exclusive pets) demonstrated more authoritarianism in educational and classroom-related attitudes on a questionnaire adapted from the Minnesota Teacher Attitude Inventory. That finding supported a trait interpretation. Five years later (1995) I returned to the classrooms of 34 teachers from the previous research to examine whether each teacher's pet status category in the first study (no pet, non-exclusive, or exclusive pet) could predict the teacher's pet status in her new classroom. The assumption was that consonance between the teachers' pet status in the two measurements would have supported a trait interpretation. No such relationship was found, and the trait interpretation was not supported.

However, given our earlier findings on teacher authoritarianism (supported by the previous findings on susceptibility to biasing information and its power to predict teacher expectancy effects); given the systematic relationships between teacher's differential behavior and the intensity of the teacher's pet phenomenon; and given the influence of our intuitive belief in trait explanations, it is conceivable that particular combinations of traits and states may be identified in future research. In other words, perhaps in examination of other samples in future research, teachers possessing particular traits and needs may be shown to adopt pets in particular classroom compositions and settings whereas other teachers may not.

The Teacher's Pet Phenomenon and Students' Perceptions of Teachers' Differential Behavior

Another avenue of research, connecting the two parts of this chapter, consisted of the investigation of the relationship between the teacher's pet phenomenon and students' perceptions of teachers' differential behavior (Babad, 1995). The two constructs were measured via different methods and were separated from each other. Teachers' pets and the intensity of the pet phenomenon were derived from analyses of students' sociometric judgments in which students nominated the names of peers best liked and most loved by their teacher. Student perceptions of teachers' differential behavior were derived from a questionnaire in which students rated the presumed behaviors of their teacher toward hypothetical high- and low-expectancy students that "she had taught in the past." The high-expectancy student was described in the questionnaire only in terms of academic excellence, with no hint of any special relationship with the teacher which might have characterized a teacher's pet. Tal and Babad (1990) and Babad and Ezer (1993) demonstrated the differentiation between the student types "academically best student" and "teacher's pet."

For these analyses of the relationship between the pet phenomenon and student perceptions of teachers' differential behavior, the pet status of every teacher was operationally defined as a continuous variable, expressed by the intensity of students' consensus in identifying the first pet in her classroom. This variable was correlated to students' perceptions of their teacher's differential behavior toward the hypothetical high- and low-expectancy students in the instructional and affective domains. The question was whether the intensity of having pets would be related to the degree of perceived teacher differentiality. In research of this type which examines classroom phenomena, the methodological requirement is that the unit of analysis will be the classroom rather than individual students. Therefore, the results reported next were computed for $N = 80$ classrooms means.

The results were clear and highly consistent: The higher the intensity of the pet phenomenon, the more the teachers were perceived by their students as giving differential, and preferential, treatment to high-expectancy students. The specific patterns for emotional support and learning support were reversed: In emotional support behaviors—where teachers are generally perceived as giving more positive treatment to high-expectancy than to low-expectancy students—the differential in favor of high-expectancy students grew with the growing intensity of the pet phenomenon; in learning support—where teachers are generally perceived as investing more instructional effort in helping low-expectancy students—the growing intensity of the pet phenomenon was related to decreased differentiality, that is, the teachers were perceived as giving *less* advantage and extra support to low-expectancy students. Thus, the tendency to have pets was found related to negative and potentially damaging manifestations of differential behavior, favoring high-expectancy students in emotional support (warmth, love, supportiveness) and hindering low-expectancy students in learning support.

PSYCHOLOGICAL COSTS OF TEACHER EXPECTANCIES

Do Negative Teacher Expectancies Hinder Students' Academic Achievement?

The central question in expectancy research is whether the expectations of person A (expector) can in themselves, in the absence of any other causal factor, influence the behavior and performance of person B (expectee). Throughout the decades of expectancy research, this question was broken into several sub-questions, such as: How are expectations formed (bias research)? How are expectations expressed in the expector's differential behavior and how are they transmitted to expectees? How are expectations perceived, interpreted, and "internalized" by expectees? And can experimenters' or teachers' expectations actually modify expectees' behavior and performance? The last question is, of course, the most critical. The findings from the experimental studies are very clear, indicating that experimenter's expectations can indeed influence subjects' behavior in short experimental situations. In education, the potential impact of teacher expectancies on students' long-term academic performance remains an open issue.

The great impact of *Pygmalion in the Classroom* was caused in large part by Rosenthal and Jacobson's explicit claim that positive teacher expectancies can indeed influence students' IQ. As mentioned earlier, the really nagging question following *Pygmalion* was whether teachers' *negative* expectations can hinder the educational performance of low-expectancy students, increasing rather than decreasing educational gaps and defeating educational ideals and meta-goals. Following great debates, hundreds of studies, and numerous reviews and meta-analyses, the evidence on teachers' expectancy-related differential behavior is solid, but the impact on students' scholastic achievement is disputable. Dusek's (1985) edited book on teacher expectancies was dedicated in large part to this issue, and the editors (Dusek, also Hall and Meyer) and some of the authors concluded that teacher expectancies do *not* have substantial and lasting impact on students' achievement. Rosenthal (1985; Harris & Rosenthal, 1985, 1986a) classified the variables in teacher expectancy studies according to a "10-Arrow" model, in which student achievement would be located in "E"—distal dependent variables. In the various meta-analyses the links connecting independent and mediating variables to E were the weakest. They were investigated less frequently (researchers preferring to examine short-term effects and neglecting long-term ones) and their results were less conclusive.

We have sufficient evidence that teacher expectations *can* potentially influence student performance and achievement. In addition to the abovementioned meta-analyses, examples include Jussim's (1989) careful statistical analyses, the Golem study (Babad, Inbar, & Rosenthal, 1982a), and Marshall and Weinstein's (1984) study, and many educators believe that the accumulated effects of negative teacher expectations along years of schooling are indeed detrimental to low-achieving students. Still, the main point is that teacher expectancies have not been demonstrated empirically beyond any doubt to have systematic and lasting effects on student achievement independent of any other causal factor.

The problems in determining the effects of teacher expectancies on student achievement can be summarized briefly as follows:

1. For ethical reasons, a manipulation of negative expectancies in a field study conducted in real classrooms is impermissible. Therefore, Golem effects cannot be examined directly in actual classrooms via powerful research designs.
2. In naturalistic studies, there is no known method to assess students' "true potential" accurately. Therefore, the extent to which negative expectancies might have hindered a student's true potential cannot be determined directly.
3. In natural educational settings, most teacher expectancies are "sustaining expectancies" (Cooper, 1985) or "self-maintaining expectancies"

(Salomon, 1981) which reflect real differences among students. Therefore, most differential behavior would reflect "student effects on teachers" rather than "teacher effects on students" (Brophy, 1983, 1985) and could not be considered "self-fulfilling prophecies." It is not easy to separate the net effect of teacher expectancies from the actual differential attributes and abilities of the students in a heterogeneous classroom.

4. Brophy (1983, 1985) argued that not only do teacher expectations reflect most strongly "student effects on teachers," but these expectations are "accurate, reality-based, and open to corrective feedback," and therefore could not be harmful to students. Similarly, Mitman and Snow (1985) argued that it is possible that differential instruction simply reflects teacher adaptation to long established performance differences among students that serves to reinforce but not to change historical trends.

5. Brophy (1983, 1985) further argued that some forms of differential treatment are appropriate. Hall and Merkel (1985) concluded that differential treatments are actually beneficial, and certainly not harmful to low-expectancy students. This view is compatible with the conceptualization of teachers' differential behavior presented in this chapter and with the empirical findings showing that teachers provide extra learning support to low-expectancy students.

Presumed Social and Affective "Price" of Teachers' Differential Behavior

The potential impact of teacher expectancies was generally treated in the literature in a unidimensional perspective. The overall question concerned the extent to which teacher expectancies advance the highs and hinder the lows, and insufficient attention was given to distinctions between cognitive and noncognitive outcomes and between individual versus whole classroom consequences. The difference between learning support and emotional support in the conceptualization of teachers' differential behavior in this chapter leads to differential hypotheses: Perhaps expectancy-related differential behavior in learning support can *promote* lows rather than hindering them, whereas differential emotional support may be harmful to low-expectancy students; and perhaps expectancy-related affective behavior influences the entire classroom more than individual students.

Summarizing Dusek's edited book on teacher expectancies, Meyer (1985) concluded that the real issue is not the influence of teacher expectancies on student achievement, but rather the influence on noncognitive factors and on the psychological functioning of children in school. I believe that the social consequences of teacher expectancies, teachers' differential behavior, and the teacher's pet phenomenon must be evaluated in terms of classroom climate and students' reactions to their teachers. Given the evidence that students are sensitive to teacher behavior and perceive differential behavior and the existence of pets clearly, we must discover if the students resent teachers' differential behavior and the existence of favorites and pets, and whether their hypothesized anger is related to particular parameters and manifestations of expectancy-related behavior.

Correlations between variables representing student perceptions of teachers' differential behavior and variables representing classroom climate and students' affective reactions to their teachers were reported in Babad's 1995 study. It should be noted that a correlational analysis cannot prove causation, even if the theoretical hypotheses underlying the research *are* causal. In correlational research it is possible that a third causal factor may influence both sets of variables, causing a correlation even when there is no direct causation between the variables. Educational field studies often consist of correlational data only, and the possibilities of experimental manipulation and control are quite limited. And yet the underlying theory may well include propositions about causal relationships.

In the 1995 study, students in all classrooms (80 classrooms of fifth and sixth graders) filled two self-report questionnaires measuring classroom climate (with scores derived for students' morale, students' affect, social climate, and instructional climate), and students' affective reactions to their teachers (including items about how much they love their teacher, their satisfaction when a substitute teacher replaces their teacher, and their wish to continue with the same teacher next year). These outcome variables were correlated with the students' perceptions in the three domains of differential behavior: learning support, emotional support, and pressure.

The correlations differed for each of the three factors: Pressure behaviors were unrelated to any of the classroom climate variables; some correlations with learning support variables were significant but not very high; whereas the correlations between emotional support variables and classroom climate and reactions to teachers variables were very substantial. Considering the fact that the correlations were computed on classroom means (i.e., $N = 80$ classrooms), a correlation coefficient exceeding .40 represents a very strong relationship.

As mentioned, perceptions of differential teacher pressure were unrelated to any of the outcome variables. Thus, in causal terms, the fact that teachers are perceived to put more pressure on high-expectancy students probably is accepted by students and does not evoke any reaction. Differential teacher behavior in the instructional domain was related to some affective reactions to the teachers, the significant correlations ranging from $r = .23$ to $r = .27$. The pattern showed more positive reactions to teachers who provide more learning support to the lows. A smaller differential in favor of the low-expectancy students was accompanied by less positive reactions to the teacher, greater happiness with a substitute teacher, and a stronger wish to not continue with the teacher. Thus, in causal terms, the students in these classrooms reacted in a mildly positive manner to teachers' efforts to provide more instructional support to low achievers.

The relationships between teachers' differential affect and the social outcome variables were stronger and more substantial, with significant correlations ranging from $r = .26$ to $r = .41$. It must be remembered that the teachers were uniformly perceived by their students as giving more emotional support to the high-expectancy students. A stronger differential in favor of the highs was accompanied by lower student morale, more negative reactions to the teachers, a stronger wish to not continue with the teacher next year, and greater happiness with a substitute teacher. Thus, in causal terms, the students seemed to be angry at teachers perceived to favor the highs in their affective displays.

Because of the powerful implications of the latter findings, additional analyses were conducted to examine the relationships separately for perceived affective teacher treatment of high-expectancy students and of low-expectancy students. The emerging pattern of correlations with the social outcomes variables showed that the most intense student reactions were related to their perceptions of teacher affective behavior toward the *lows*. The teachers were uniformly perceived as giving less emotional support to lows, but classes were more satisfied with their teachers the more they gave affection and love to low-expectancy students. Perceptions of the teacher as providing love to the low-expectancy students were related to higher student morale ($r = .35$) and student positive affect in the classroom ($r = .39$), to more positive reactions to the teacher ($r = .42$), loving the teacher ($r = .46$), wishing to continue with the teacher next year ($r = .35$), and with less happiness when the teacher is replaced by a substitute teacher ($r = .53$).

These correlations support the claim that teachers' differential affect is the crux of the student perception aspects of the teacher expectancy issue. The excess learning support given differentially to the lows is recognized by the students and to some extent appreciated, but students' (presumed) reactions to teachers' differential affect are fierce, with particular sensitivity (that is, psychological cost in the form of low morale, negative affect, and negative reactions to the teacher) to teachers' affective treatment of weak students.

Unfortunately, illuminating the problem is easier than offering a viable solution, and a number of obstacles may prevent corrective efforts to reduce differential behavior in the affective domain. First, it was mentioned earlier that teachers deny and negate students' perceptions of excessive emotional support given to high-expectancy students, claiming that they give more emotional support to the low-expectancy students, together with the extra investment in learning support. Second, strong student *perceptions* of negative affect do not mean necessarily that this affect is expressed in strong and salient negative behavior as had been conceived in expectancy research in the 1970s. Rather, these behaviors are very subtle and almost undetectable, expressed in fine nuances of verbal and nonverbal behavior, or even inferred from excessive niceness to weak students which is interpreted as deceitful and hypocritical. Some of these behaviors are uncontrollable, or at least not readily controllable. Third, it was mentioned earlier that the teachers who are more prone to demonstrate expectancy effects in their classrooms seem to possess personality dispositions which would make them more resistant to change. Fourth, perhaps students are so suspicious of teacher intentions that they might overinterpret innocent teacher behaviors as reflecting negative affect. A trend in that direction was reported by Babad (1996) concerning the perceptions of affective differential behavior by low-achieving students.

No easy solutions can be offered to correct teachers' differential behavior in the affective domain. As Meyer (1985) has argued, low-aptitude and low-achieving students would have low self-esteem and bad feelings in school even if teachers demonstrated no differential behavior at all. But the findings cited in this section demonstrate low morale, negative climate, and negative reactions to the teacher as a function of affective differential behaviors which characterize *entire* classrooms. I am hesitant to offer specific corrective steps for reducing differential behavior as numerous authors in the expectancy literature have done. However, my experience as a basic and applied educational psychologist guides me to recommend at least that teachers should not express their affect freely in the classroom, and refrain from excessive (negative *and* positive) displays. Keeping more distance and avoiding affective overcompensation are the best inoculation against involuntary leakage.

Negative Consequences of the Teacher's Pet Phenomenon

The same social variables reported above (classroom climate, students' affect, and their reactions to their teachers) were examined in the Tal and Babad (1990) and Babad (1995) studies on two independent samples of 80 classrooms as correlates of the occurrence of the teacher's pet phenomenon in particular classrooms. In the first study (Tal, 1987; Tal & Babad, 1990), one-way ANOVAs of the categories of pet classrooms yielded significant differences in social outcome variables, no pet classrooms showing the most positive climate and positive reactions to their teachers; exclusive pet classrooms showing the most negative climate and negative reactions to their teachers; and non-exclusive pet classrooms falling between these extremes, with reactions more similar to those of students in the exclusive pet classrooms.

These findings supported the hypothesis about the presumed social consequences of the teacher's pet phenomenon in terms of reduced student morale and presumed anger at teachers who have pets. However, these findings were not replicated in the recent Babad (1995) study. ANOVAs similar to those of the previous study did not yield significant differences between the three groups of classrooms. Analyses utilizing the intensity of the pet phenomenon as a continuous variable instead of the trichotomy of classrooms yielded nonsignificant correlations between the teacher's pet phenomenon and climate and morale variables. These latter findings did not support the proposed hypothesis and were quite surprising. Ideas emerging in an additional study conducted by Babad and Ezer (1993) led to an improved, more differentiated conceptualization of the teacher's pet phenomenon, which eventually illuminated the issue of the psychological cost of this phenomenon in a clearer and sharper way.

Babad and Ezer (1993) investigated the classroom seating locations of several student types (teacher's pet, leader, popular student, academically best student, flatterer, etc.) nominated by the sociometric consensus of their peers. (Seating locations in conventional rows-and-columns classrooms are sometimes determined by the teachers and sometimes by the students themselves, and these locations keep changing throughout the school year, reflecting students' needs and wishes and teachers' various concerns. The overall consensus in the seating location literature is that seating arrangements [determined in whichever way or combination of ways] are highly meaningful and reflect underlying social phenomena.) Babad and Ezer found that consensually identified leaders tended to sit more often in the back of the classroom, while flatterers tended to sit in the front, close to the teacher. Among academically best students, those nominated by the highest consensus showed a tendency to sit in the back, while those nominated by a weaker consensus sat closer to the teachers. Babad and Ezer thought that the former may be the free and more independent thinkers, whereas the latter may be the more conforming and conventional "best students" in its classroom connotation. As to the teachers' pets, in the initial analysis Babad and Ezer failed to discover a systematic tendency to prefer a given seating location. In subsequent analyses they distinguished between a "pure type" (a student nominated consensually as best loved and better liked by the teacher, but not nominated as holder of any other classroom role except for the frequent academic success) and a "mixed type" (a student identified as both a teacher's pet and a leader and/or most popular student). Systematic seating locations were found for pure and mixed types: Pure types tended more often to sit closer to the teacher in the front of the classroom, whereas mixed pets tended more often to sit in the back of the classroom, presumably reflecting a stronger influence of their leadership role than their pet status. The conceptual implication of this distinction between pure and mixed pets was that the teacher's love and special relationship with her pet may not be the only important aspect of the pet phenomenon, and different pet situations might be created when the attitudes of the other children in the classroom toward the pet are considered as well.

Applying this idea to the 1995 sample, the classrooms were reclassified into three new groups, crossing students' sociometric status as pets (that is, best liked and most loved by the teacher) with popularity and/or leadership among classmates. The groups were: (1) no pet classrooms (26%); (2) popular pet classrooms (where the same students were nominated as consensual pets and also as most popular and/or leader—46%); (3) unpopular pet classrooms (where the consensual pets were not nominated highly for popularity and/or leadership—29%). The reformulated hypothesis was that a teacher's love of a popular pet should not anger the other students nor cause negative classroom climate, because the students also like and appreciate the teacher's pet, and they accept the special status of that child.

The results confirmed these ideas. Negative classroom climate, lower morale, more negative affective reactions to the teacher, happiness about substitute teachers, and a stronger wish to not continue with the teacher next year were found only in the unpopular pet classrooms, where the teacher's pet did not enjoy popularity and leadership status among his or her peers. The students in these classrooms might have felt that the special status given to the pet by the teacher was unjustified and unfair. On the other hand, the popular pet classrooms did not differ from the no pet classrooms in classroom climate and students' morale. Thus, the negative connotation of the teacher's pet phenomenon and its presumed social consequences may actually be limited in a justifiable manner only to classrooms where the students do not share in the teacher's love and appreciation for her pet.

CONCLUSION

Research on teachers' differential behavior and on teacher expectancies in general is declining, but research on the teacher's pet phenomenon is still in its early stages, and my hope is that this chapter will serve as impetus for further work. Theoretical and applied research is needed to examine other aspects of this phenomenon (especially actual measurement of teacher and student behavior); its manifestations in different age levels, types, and compositions of classrooms, and teacher types; and its relations with a whole range of other psychological phenomena in the classroom. It would also seem quite promising to investigate deeply the students who are teachers' pets, to trace their needs, motives, and behavior, and the intended and unintended outcomes of their per status. Several of the Brophy and Everston (1981) studies on student characteristics gave the impression that part of the explanation for very positive teacher reactions to certain students resides in the ways these students treat their teachers—showing that they like and want to please the teacher, beaming when the teacher praises them, offering to do classroom chores, and so on.

With regard to teacher expectancies in general, more research is needed on noncognitive aspects of teacher behavior and influences of expectancy-related behavior on social and affective outcomes for entire classrooms as well as for individual students. Further down the road, perhaps more action research will be conducted in attempts to change teacher behavior and reduce the negative manifestations of teachers' preferential affect.

REFERENCES

Adorno, T., Frenkel-Brunswick, E., Levinson, D., & Sanford, N. (1950). *The authoritarian personality.* New York: Harper & Row.

Babad, E. (1979). Personality correlates of susceptibility to biasing information. *Journal of Personality and Social Psychology, 37,* 195-202.

Babad, E. (1985). Some correlates of teachers' expectancy bias. *American Educational Research Journal, 22,* 175-183.

Babad, E. (1990a). Measuring and changing teachers' differential behavior as perceived by students and teachers. *Journal of Educational Psychology, 82,* 683-690.

Babad, E. (1990b). Calling on students: How a teacher's behavior can acquire disparate meanings in students' minds. *Journal of Classroom Interaction, 25,* 1-4.

Babad, E. (1992). Teacher expectancies and nonverbal behavior. In R. Feldman (Ed.), *Applications of nonverbal behavioral theories and research* (pp. 167-190). Hillsdale, NJ: Erlbaum.

Babad, E. (1993a). Teachers' differential behavior. *Educational Psychology Review, 5*, 347-376.
Babad, E. (1993b). Pygmalion-25 years later: Interpersonal expectations in the classroom. In P. Blanck (Ed.), *Interpersonal expectations: Theory, research and application* (pp. 125-153). London: Cambridge University Press.
Babad, E. (1995). The "teacher's pet" phenomenon, teachers' differential behavior, and students' morale. *Journal of Educational Psychology, 87*, 361-374.
Babad, E. (1996). How high is "high inference"? Within classroom differences in students' perceptions of classroom interaction. *Journal of Classroom Interaction, 31*, 1-9.
Babad, E., Bernieri, F., & Rosenthal, R. (1987). Nonverbal and verbal behavior of preschool, remedial, and elementary school teachers. *American Educational Research Journal, 24*, 405-415.
Babad, E., Bernieri, F., & Rosenthal, R. (1989a). Nonverbal communication and leakage in the behavior of biased and unbiased teachers. *Journal of Personality and Social Psychology, 56*, 89-94.
Babad, E., Bernieri, F., & Rosenthal, R. (1989b). When less information is more informative: Diagnosing teacher expectations from brief samples of behaviour. *British Journal of Educational Psychology, 59*, 281-295.
Babad, E., Bernieri, F., & Rosenthal, R. (1991). Students as judges of teachers' verbal and nonverbal behavior. *American Educational Research Journal, 28*, 211-234.
Babad, E., & Ezer, H. (1993). Seating locations of sociometrically identified student types: Methodological and substantive issues. *British Journal of Educational Psychology, 63*, 75-87.
Babad, E., & Inbar, J. (1981). Performance and personality correlates of teachers' susceptibility to biasing information. *Journal of Personality and Social Psychology, 40*, 553-561.
Babad, E., Inbar, J., & Rosenthal, R. (1982a). Pygmalion, Galatea, and the Golem: Investigations of biased and unbiased teachers. *Journal of Educational Psychology, 74*, 459-474.
Babad, E., Inbar, J., & Rosenthal, R. (1982b). Teachers' judgments of students' potential as a function of teachers' susceptibility to biasing information. *Journal of Personality and Social Psychology, 42*, 541-547.
Babad, E., & Taylor, P. (1992). The transparency of teacher expectancies across language, cultural boundaries. *Journal of Educational Research, 86*, 120-125.
Blake, B., & Heslin, R. (1971). Evaluation apprehension and subject bias in experiments. *Journal of Experimental Research in Personality, 5*, 57-63.
Blanck, P. (Ed.). (1993). *Interpersonal expectations: Theory, research, and application.* London: Cambridge University Press.
Brattesani, K., Weinstein, R., & Marshall, H. (1984). Student perceptions of differential teacher treatment as moderators of teacher expectancy effects. *Journal of Educational Psychology, 76*, 236-247.
Brophy, J. (1983). Research on the self-fulfilling prophecy and teacher expectations. *Journal of Educational Psychology, 75*, 631-661.
Brohpy, J. (1985). Teacher-student interaction. In J. Dusek (Ed.), *Teacher expectancies* (pp. 303-328). Hillsdale, NJ: Erlbaum.
Brophy, J., & Everston, C. (1981). *Student characteristics and teaching.* New York: Longman.
Brophy, J., & Good, T. (1974). *Teacher-student relationships: Causes and consequences.* New York: Holt, Rinehart & Winston.
Chavez, R. (1984). The use of high inference measures to study classroom climates: A review. *Review of Educational Research, 54*, 237-261.
Clarke, A., Michie, P., Andreasen, A., Viney, L., & Rosenthal, R. (1976). Expectancy effects in a psychological experiment. *Physiological Psychology, 4*, 137-144.
Cooper, H. (1979). Pygmalion grows up: A model for teacher expectation communication and performance influence. *Review of Educational Research, 49*, 389-410.
Cooper, H. (1985). Models for teacher expectation communication. In J. Dusek (Ed.), *Teacher expectancies* (pp. 135-158). Hillsdale, NJ: Erlbaum.
Cooper, H., & Good, T. (1983). *Pygmalion grows up.* New York: Longman.

Cooper, H., & Hazelrigg, P. (1988). Personality moderators of interpersonal expectancy effects: An integrative research review. *Journal of Personality and Social Psychology, 55*, 937-949.

Cooper, H., & Tom, D. (1984). Teacher expectation research: A review with implications for classroom instruction. *Elementary School Journal, 85*, 77-89.

Dana, J., & Dana, R. (1969). Experimenter bias and the WAIS. *Perceptual and Motor Skills, 28*, 634.

Darley, J., & Fazio, R. (1980). Expectancy confirmation process arising in the social interaction sequence. *American Psychologist, 35*, 867-881.

Dusek, J. (Ed.) (1985). *Teacher expectancies.* Hillsdale, NJ: Erlbaum.

Ekman, P., & Friesen, W. (1969). Nonverbal leakage and cues to deception. *Psychiatry, 32*, 88-106.

Ekman, P., & Friesen. W. (1974). Detecting deception from the body and face. *Journal of Personality and Social Psychology, 29*, 288-298.

Evans, J. (1962). *Sociometry and education.* London: Routledge & Kegan Paul.

Finn, J. (1972). Expectations and educational environment. *Review of Educational Research, 42*, 389-409.

Fraser, B. (1987). Use of classroom environment assessment in school psychology. *School Psychology International, 8*, 205-219.

Fraser, B. (1989). Twenty years of classroom environment research: Progress and prospect. *Journal of Curriculum Studies, 21*, 307-327.

Good, T., & Brophy, J. (1972). Behavioral expression of teacher attitudes. *Journal of Educational Psychology, 63*, 617-624.

Good, T., & Brophy, J. (1974). Changing student and teacher behavior: An empirical examination. *Journal of Educational Psychology, 66*, 390-405.

Hall, V., & Merkel, S. (1985). Teacher expectancy effects and educational psychology. In J. Dusek (Ed.), *Teacher expectancies* (pp. 67-92). Hillsdale, NJ: Erlbaum.

Harris, M. (1989). Personality moderators of interpersonal expectancy effects: Replication of Harris and Rosenthal (1986). *Journal of Research in Personality, 23*, 381-397.

Harris, M., & Rosenthal, R. (1985). Mediation of interpersonal expectancy effects: 31 meta-analyses. *Psychological Bulletin, 97*, 363-386.

Harris, M., & Rosenthal, R. (1986a). Four factors in the mediation of teacher expectancy effects. In R. Feldman (Ed.), *The social psychology of education* (pp. 91-114). London: Cambridge University Press.

Harris, M., & Rosenthal, R. (1986b). Counselor and client personality as determinants of counselor expectancy effects. *Journal of Personality and Social Psychology, 50*, 362-369.

Jones, E. (1990). *Interpersonal perception.* New York: W.H. Freeman.

Jussim, L. (1986). Self-fulfilling prophecies: A theoretical and integrative review. *Psychological Review, 93*, 429-445.

Jussim, L. (1989). Teacher expectations: Self-fulfilling prophecies, perceptual biases, and accuracy. *Journal of Personality and Social Psychology, 57*, 469-480.

Kerman, S., & Martin, M. (1980). *Teacher expectations and the student achievement* (Teacher handbook). Los Angeles: County Superintendent of Schools.

Marshall, H., & Weinstein, R. (1984). Classroom factors affecting students' self-evaluations: An interactional model. *Review of Educational Research, 54*, 301-325.

Marshall, H., & Weinstein, R. (1986). Classroom context of student-perceived differential teacher treatment. *Journal of Educational Psychology, 78*, 441-453.

Meyer, W. (1985). Summary, integration, and prospective. In J. Dusek (Ed.), *Teacher expectancies* (pp. 353-370). Hillsdale, NJ: Erlbaum.

Mitman, A., & Snow, R. (1985). Logical and methodological problems in teacher expectancy research. In J. Dusek (Ed.), *Teacher expectancies* (pp. 93-131). Hillsdale, NJ: Erlbaum.

Perlmutter, L. (1972). Experimenter-subject needs for social approval and task interactiveness as factors in experimenter expectancy effects. *Dissertations Abstracts International, 32*, 6692-6693.

Proctor, C. (1984). Teacher expectations: A model for school improvement. *Elementary School Journal, 84*, 469-481.
Rist, R. (1987, December). Do teachers count in the lives of children? *Educational Researcher, 16*, 41-42.
Rokeach, M. (1960). *The open and closed mind.* New York: Basic Books.
Rosenthal, R. (1973). The mediation of Pygmalion effects: A four factor "theory." *Papua New Guinea Journal of Education, 9*, 1-12.
Rosenthal, R. (1985). From unconscious experimenter bias to teacher expectancy effects. In J. Dusek (Ed.), *Teacher expectancies* (pp. 37-65). Hillsdale, NJ: Erlbaum.
Rosenthal, R. (1987, December). Pygmalion effects: Existence, magnitude, and social importance. *Educational Researcher, 16*, 37-41.
Rosenthal, R. (1989). *The affect/effort theory of the mediation of interpersonal expectancy effects.* Donald T. Campbell Award Address, Annual convention of the American Psychological Association, New Orleans.
Rosenthal, R. (1991). Teacher expectancy effects: A brief update 25 years after the Pygmalion experiment. *Journal of Research in Education, 1*, 3-12.
Rosenthal, R., & Babad, E. (1985, September). Pygmalion in the gymnasium. *Educational Leadership*, 36-39.
Rosenthal, R., & DePaulo, B. (1979). Sex differences in eavesdropping of nonverbal cues. *Journal of Personality and Social Psychology, 37*, 273-285.
Rosenthal, R., & Jacobson, L. (1968). *Pygmalion in the classroom.* New York: Holt, Rinehart and Winston.
Ross, L. (1977). The intuitive psychologist and his shortcomings: Distortions in the attribution process. In L. Berkowitz (Ed.), *Advances in experimental social psychology* (Vol. 10). New York: Academic Press.
Salomon, G. (1981). Self-fulfilling and self-maintaining prophecies and behaviors that realize them. *American Psychologist, 36*, 1452-1453.
Silberman, M. (1969). Behavioral expression of teachers' attitudes toward elementary school students. *Journal of Educational Psychology, 60*, 402-407.
Silberman, M. (1971). Teachers' actions and attitudes toward their students. In M. Silberman (Ed.), *The experience of schooling.* New York: Holt, Rinehart and Winston.
Tal, Z. (1987). *Teachers' differential behavior toward their students: Investigation of the "teacher's pet phenomenon."* Doctoral dissertation, Hebrew University of Jerusalem, Jerusalem, Israel.
Tal, Z., & Babad, E. (1989). The "teacher's pet" phenomenon as viewed by Israeli teachers and students. *Elementary School Journal, 90*, 99-110.
Tal, Z., & Babad, E. (1990). The teacher's pet phenomenon: Rate of occurrence, correlates, and psychological costs. *Journal of Educational Psychology, 82*, 637-645.
Tom, D., & Cooper, H. (1983). Teacher cognitive style, expectations, and attributions for student performance. In J. Staussner (Chair), *Adult cognitive functioning and the expectancy phenomenon.* Symposium presented at the annual meeting of the American Psychological Association, New Orleans.
Tom, D., Cooper, H., & McGraw, M. (1984). Influences of student background and teacher authoritarianism on teacher expectations. *Journal of Educational Psychology, 76*, 259-265.
Weinstein, R. (1983). Student perceptions of schooling. *Elementary School Journal, 83*, 287-312.
Weinstein, R. (1985). Student mediation of classroom expectancy effects. In J. Dusek (Ed.), *Teacher expectancies* (pp. 329-350). Hillsdale, NJ: Erlbaum.
Weinstein, R. (1986). The teaching of reading and children's awareness of teacher expectations. In T. Rapael (Ed.), *The contexts of school-based literacy.* New York: Random House.
Weinstein, R. (1989). Perceptions of classroom processes and student motivation: Children's views of self-fulfilling prophecies. In C. Ames & R. Ames (Eds.), *Research on motivation in education: Volume 3. Goals and cognitions* (pp. 187-221). New York: Academic Press.

Weinstein, R., Marshall, H., Brattesani, K., & Middlestadt, S. (1982). Student perceptions of differential teacher treatment in open and traditional classrooms. *Journal of Educational Psychology, 75,* 678-692.

Weinstein, R., Marshall, H., Sharp, L., & Botkin, M. (1987). Pygmalion and the students: Age and classroom differences in children's awareness of teacher expectations. *Child Development, 58,* 1079-1093.

Weinstein, R., & Middlestadt, S. (1979). Student perceptions of teacher interactions with male high and low achievers. *Journal of Educational Psychology, 71,* 421-431.

Wineburg, S. (1987, December). The self-fulfillment of the self-fulfilling prophecy. *Educational Researcher, 16,* 28-44.

Zuckerman, M., DePaulo, B., & Rosenthal, R. (1981). Verbal and nonverbal communication of deception. In L. Berkowitz (Ed.), *Advances in experimental social psychology* (Vol. 14). New York: Academic Press.

Zuckerman, M., DePaulo, B., & Rosenthal, R. (1986). Humans as deceivers and lie detectors. In P. Blanck, R. Buck, & R. Rosenthal (Eds.), *Nonverbal communication in the clinical context.* University Park, PA: Pennsylvania State University Press.

EXPECTANCY EFFECTS IN "CONTEXT": LISTENING TO THE VOICES OF STUDENTS AND TEACHERS

Rhona S. Weinstein and Clark McKown

INTRODUCTION

The formative role of teacher expectations in the development of children's academic ability has been a topic of intense scrutiny and equally intense debate in the educational research literature for almost 30 years. In recent years the tone of researchers has been skeptical and the fervor of research activity has abated. This apparent loss of interest in the topic stands in stark contrast to the world of practice where the call for high expectations continues to be heralded in staff development, in the standards movement, and in recent school reform efforts (Good & Weinstein, 1986). However, prescribed interventions for practice (such as equalizing praise and criticism, raising standards, and promoting high expectations) have not been closely tied to the findings of research. Further, relatively little evaluative research regarding the effectiveness of these interventions has been conducted.

Much of the debate among researchers has centered upon the all-or-none question of whether the phenomenon exists, that is, do teachers' expectations for students become self-fulfilling prophecies? Largely ignored is the critical role of context, that is, "the interrelated conditions in which something exists or occurs" (Webster's). This coherence within or variation among individuals and environments is what predicts where, when, for whom, and how the expectations of teachers are likely to result in the most potent impact on children's outcomes. The more fruitful question to address is not "do expectancy effects exist?" but rather "under what conditions are such effects magnified or minimized or likely to be changed?" (Babad, 1993; Raudenbush, 1984; Weinstein, 1993).

This chapter highlights some findings from our research program on the dynamics of expectancy effects as seen from the perspective of students and from our applied, collaborative research with teachers and principals on expectancy-enhancement interventions for students. The work illuminates the role of *contextual* factors in magnifying or diminishing expectancy effects: critical characteristics of teachers and the classroom environments they create, students and their susceptibility to expectancy effects, and schools and their educational philosophy and practices. Contextual variation along these dimensions plays a role in determining the potential for and magnitude of expectancy effects. Understanding the role of these contextual features will help shape the scope and direction of needed changes in educational practice.

Without attention to the social and psychological contexts in which achievement expectations are formed, expressed, understood, and acted upon, the importance of expectancy effects in schooling is likely to be underestimated. Further, without this knowledge base, our efforts to create schooling experiences that promote the development of ability in *all* children will likely be less effective.

We begin with an explication of the background for our work. We highlight findings from three domains of research: (1) explorations of children's awareness of teacher expectations and its implications for classroom differences in expectancy practices and for a broadened model of expectancy communication; (2) studies of moderating factors, associated processes, and consequences of expectancy effects; and (3) studies of interventions with teachers and principals to create or change expectancy-relevant practices and policies. Based on this work, we identify principles critical to teacher education and efforts to reform schools and we make recommendations for the direction of future research.

THE BACKGROUND

Lessening Discrepancies Between Potential and Performance

In his autobiography, Seymour Sarason (1988) underscored that all of his work has been directed toward one problem, that is, "how does one account for variation in the discrepancy between potential and performance?" (p. 45) His work as a staff psychologist in 1942 at the newly opened Southbury Training School for the mentally retarded opened his eyes to the ways that social settings can limit the unfolding of human potential. As noted at his festschrift, Sarason recognized early on "that the vantage point from which we view human capacity sharply determines what we see" (Weinstein, 1990, p. 362). His entire career has been directed toward understanding schooling—its behavioral and programmatic regularities, that is, its culture—and toward elucidating the universe of alternatives, in order to help optimize the development of children's abilities. Among his many contributions was the attention he drew to the need to create stimulating conditions for teachers if we have any hopes of engaging and challenging students.

The late Sam Rabinovitch, who founded the McGill University-Montreal Children's Hospital Learning Centre, spearheaded a movement in Canadian schools to recognize that children with learning disabilities, despite their difficulties, are indeed of normal intelligence and can learn, given appropriate instructional techniques. His methods, developed over a career cut short by his early death in 1977, underscored the importance of teachers' taking responsibility for children's learning and of listening to and learning from students, capitalizing on their strengths to enable success.

As a student of Sarason and Rabinovitch (Weinstein) and as a student of the student (McKown), we, too, came to the field with an enduring interest in the ways in which educational settings can lessen the gap between potential and performance. Our own experiences in working with children who have special needs in a myriad of classrooms have sensitized us to the promise of and also the obstacles to creating environments that develop rather than undermine talent.

It is not surprising then that the seminal work of Merton (1948), of Rosenthal and Jacobson in *Pygmalion in the Classroom* (1968), and of Brophy and Good (1970) captivated our interest in the phenomenon of teacher expectations and how such expectations might function as self-fulfilling prophecies—perhaps, providing part of the explanation for the gap between potential and performance. This work offered respectively, a theory (that a false belief about a target could become confirmed in the target's behavior), an experimental demonstration (albeit controversial) of such effects on elementary school children's achievement scores, and empirical evidence of observed differential teacher behavior reflecting teachers' natural expectations that might account for changes in student achievement.

Armed with this operationalization of the phenomenon, a flurry of research activity followed, with ensuing and continuing controversy. But what was missing in the early research was an understanding of children's experiences of expectations in schooling and of the role that children might play in expectancy effects. It was here that our studies at Berkeley began.

Evolving Models of Teacher Expectancy Effects

The work of Robert Rosenthal and Lenore Jacobson in the classic *Pygmalion in the Classroom* (1968) study focused attention on the outcomes of experimentally manipulated teacher expectations, that is, the relationship between the induction of false positive expectations about "unfolding" student ability and demonstrated changes in students' achievement scores. This study and subsequent replications demonstrated the existence of expectancy effects: however, these studies did not explain the process by which teacher expectations led to changes in student achievement. Here, the research by Brophy and Good (1970) provided the earliest model of the mediation of expectancy effects in classrooms. Their theory addressed the formation of teacher expectations (for example, based on information from school records), the behavioral expression of these expectations in the differential treatment of students, and the responses of students through behaviors that complemented and reinforced teacher expectations.

Braun (1973, 1976), Weinstein and colleagues (1976, 1979), and Darley and Fazio (1980) extended this model, to incorporate the role of children's awareness of differential treatment. It was argued that in order for teacher expectations to lead to behavioral confirmation in students, children had to notice those expectations. Brophy and Good (1974) made the important distinction between two possible mediating pathways through which differential teacher treatment might impact student behavioral outcomes: direct effects (through exposure to materials and opportunities for practice) and indirect effects (through student awareness of the teacher's expectations, which could stimulate changes in motivation, self-concept, and level of aspiration).

These increasingly refined models of the mediation of expectancy effects reflected a paradigm shift from simpler behavioral models to more complex sociocognitive perspectives, with a clear role for student awareness and understanding and with consideration of a broader set of student processes and outcomes (beyond achievement). While children's perceptions and cognitions have been increasingly emphasized (e.g., Weinstein, 1983; McCaslin & Good, 1996), surprisingly little research within the expectancy paradigm has explored this perspective, with a few exceptions (Babad, 1993; Cooper & Good, 1983; Mitman & Lash, 1988). And the lessons derived from a mapping of children's understanding have not been well utilized. A parallel progression occurred in the consideration of teachers' cognitions and individual differences (Babad, Inbar, & Rosenthal, 1982; Cooper, 1979). Yet in much of the research, still evident today, teachers and students appear almost as "black boxes" into which expectations are induced or measured and outputs (teacher behaviors and student achievement) are assessed.

CHILDREN'S AWARENESS OF TEACHER EXPECTATIONS

Children Get It

Measuring Student Views of Teacher Behavior

Our research program began with the goal of addressing the missing student link in studies of expectancy effects. Did children even notice teacher behaviors that communicate academic expectations and what teacher behaviors are particularly salient to children? Our interest in the measurement of children's perceptions of teacher treatment grew in part from the need to clarify some puzzling findings of a longitudinal study of teacher expectancy effects in first-grade reading groups (Weinstein, 1976). Here, the documented widening gap in cognitive performance and social status between groups did not fit the pattern of observed differential teacher behaviors (which favored low-expectancy students with more praise and less criticism). We began to suspect that children made sense of teacher behavior in different ways than we did. We wondered if the coded critical comments signalled high expectations and the coded high rates of praise (for less than perfect answers, as pointed out by observers) conveyed "an indiscriminant 'fine, fine, fine' to those from whom less was expected" (Weinstein, 1976, p. 115). Turning to the student perspective appeared to be a critical missing link in the expectancy research. We and others (such as Babad, Bernieri, & Rosenthal, 1989; Babad & Taylor, 1992) did so, and developed evidence that children picked up subtle teacher cues that were not captured adequately by observer ratings of classroom dynamics.

At that time, considerable work had been done to measure aspects of classroom climate through a survey of children's perceptions of various aspects of classroom life (Moos, 1979; Walberg, 1976). However, in their emphasis on average levels, for example, of teacher warmth, these instruments were not sensitive to the possibility that different classroom environments for high- and low-achieving students might exist. Our challenge was to devise a measure of children's perceptions of subenvironments *within* classrooms for high- and low-achieving students.

To fill this gap, we developed an instrument, the Teacher Treatment Inventory (TTI), in which children independently report on the frequency of a variety of teacher behaviors directed toward a hypothetical high- and low-achieving student in their classroom. Thus, perceptions of differential teacher treatment are obtained indirectly, reflected in the magnitude of the difference between the ratings for hypothetical high- and low-target students. Teacher behavior items were selected based on a review of the literature on expectancy processes. Through refinement, items with maximum within-scale internal consistency and maximum discriminating power were retained in a 30-item, three-scale version of the TTI (see Weinstein et al., 1987 for the most recent version). An eight-item short version and a teacher treatment "toward self" version of the TTI were also created.

What Children See

In a series of studies of elementary school-aged children across a large number of classrooms (Weinstein and colleagues, 1979, 1982, 1987), children reported a consistent set of differences in the treatment by teachers of high and low achievers. First, children perceive more frequent negative feedback and teacher direction targeted toward low-achieving students. This is expressed through such behaviors as scolding a student for not trying or not listening and prescribing unilaterally and in detail how a student will spend his or her time. Second, children perceive teachers to be more concerned with low-achieving students' adherence to work and rules. Teachers communicate this concern, for example, by telling a student what to do, explaining the rules, or punishing the student for breaking the rules. Finally, children perceive teachers to express higher expectations and to provide more opportunity and choice to high achievers, such as in asking a student to lead activities or to answer questions, showing interest in the student, and making the student feel good about academic successes.

This awareness of differential teacher treatment of high and low achievers was documented in children as young as first graders and appeared to be shared by both high- and low-achieving students. Thus, across a number of replications, elementary school children demonstrated their knowledge that teacher behavior differs toward different students, typically favoring the high over the low achiever with more positive treatment.

Children Identify Differences Between Classrooms

Measuring children's perceptions of teacher behavior also gives us some purchase on a critical aspect of the social milieu in which teacher expectations are communicated, perceived, understood, and acted upon. Aggregated by classroom, children's perceptions of teacher treatment provide an index of perceived classroom environment that is a powerful tool with which to estimate the likelihood that teacher expectancies will result in self-fulfilling prophecies. Classrooms may be characterized by the degree to which teachers are perceived to differentiate their behavior. Indeed, our studies show variability across classrooms in the degree of differential teacher treatment reported by students. Consistent with the observational studies, children describe marked differential treatment in some classrooms but more equitable treatment in other classrooms.

This variability in perceived climate is validated by observational studies of teacher behavior using narrative records and qualitative analyses (Marshall & Weinstein, 1986, 1988) and by predictive studies, linking teacher expectations to student outcomes, such as self-expectations and achievement (Brattesani, Weinstein, & Marshall, 1984). Student-identified high and low differential treatment classrooms both look different and predict different patterns of student outcomes. Specifically, high levels of perceived differential treatment by teachers were associated with stronger relationships between teacher expectations and subsequent academic, social, and emotional outcomes among students.

Differential and Equitable Treatment Environments

Interviewing Children

How do classrooms that children identify as highly differentiated in teacher treatment toward high and low achievers differ from classrooms in which children perceive more equitable treatment? For a deeper look at children's thinking in "context," targeting the nuances of differential teacher treatment and the inferences drawn about teachers' expectations for self and for peers, we supplemented our Teacher Treatment Inventory with a semi-structured interview, "Learning about Smartness," which was taperecorded and transcribed. We interviewed a carefully chosen sample of 133 high- and low-achieving fourth graders (both male and female) from student-identified high and low differential treatment classrooms. These interviews allowed us to explore in children's own words the ways in which they learned about their relative smartness in their classrooms. Coupled with classroom observations, these interviews have informed us about the complexity of children's knowledge of differential teacher expectations and have led to a broadened model of expectancy communication in the classroom (Marshall & Weinstein, 1984; Weinstein, 1989, 1993, 1998).

Sophisticated Observers

There is a great deal to learn from what children describe as clues to their own relative position in the classroom achievement hierarchy. Children's awareness of teacher expectations rests on subtle distinctions in teacher behavior. Children draw from nonverbal as well as verbal cues and from the larger, historical context for interpretation, which can soften or accentuate the clues about ability contained in a single interaction. Children are highly sensitive observers, for example, in distinguishing between different types of "call on" and between different qualities of "praise." Often, a single critical incident is reported by many students in a classroom and is sufficient to communicate clear expectations. These findings underscore the limitations of behavioral categories and of the counting of teacher-student interactions and direct us to think, instead, about a classroom culture of expectations about ability. These findings also suggest that children are sensitive to underlying nuances of teacher behavior.

Insights into children's thinking have led to a broadened model of expectancy communication in the classroom. The model moves beyond dyadic interactions between teachers and students to include the structural or institutional organization of classroom and school life in which such exchanges are embedded and in which opportunities to learn and to demonstrate that learning are allocated.

Eight Interactive Features of Instruction

We have identified eight features of classroom and school practice which in their interrelationships create a culture that provides information to children about expected ability and also shapes their exposure to and experiences with learning. These features reflect and elaborate the dimensions of Rosenthal's (1973) original four-factor model of expectancy communication.

Inputs to the educational environment include: (1) the ways in which students are grouped for instruction ("and so you know they're smart cause they're in the highest group"); (2) the tasks and materials through which the curriculum is enacted ("they read more books like thick books"); (3) the motivational strategies that teachers use to engage students ("like today, the teacher gave me an award saying I was the second top in the class"); and (4) the role that students play in directing their own learning ("the teacher doesn't actually work with [the smart kids] because they know how to do their stuff").

Educational outputs and *feedback* speak to (5) how students are evaluated ("a very soft voice lets you know you're doing well"), which includes beliefs about ability, the response opportunities accorded (and hence the achievement products obtained), and the assessments provided. Finally, *climate* factors reflect relationships at multiple levels, including (6) the quality of classroom relationships ("not-so-smart girls can't play with the smart girls because smart girls just act like they ignore 'em"); (7) the quality of parent-classroom relationships ("They mother and father don't teach 'em anything at home"); and (8) the quality of classroom-school relationships ("the way you know a person is smart, Miss _ always picks on them to go different places").

Thus, the instructional choices that teachers make with regard to each of these eight features are informative to children about relative ability differences (that is, such choices can accentuate or minimize such information). In addition, these choices create, expand, or constrain opportunities for all children to learn and demonstrate that learning, thus resulting in comparable or differential educational experiences for different students.

Toward Positive Expectancy Practices: Equity in Treatment

In contrast to classroom environments where ability differences between students are made highly salient, teacher practices in equitable treatment classrooms systematically seek to expand motivational, learning, and performance opportunities for *all* students and to minimize information about *comparative* ability differences. One fourth grader described such a classroom as one in which "people who used to not be so smart, they're smart now." In such classrooms the following types of beliefs and practices are likely to be found. Further, there is a growing body of research that has linked these kinds of practices to more positive outcomes for a greater number of students (Weinstein, 1998, pp. 93-94):

> Heterogeneous grouping practices with challenging curricula invite more students to meet the challenge (Gamoran, 1987). Motivational climates that stress intrinsic motivation, "learning" rather than "performance" goals, and cooperative rather than competitive reward focus student attention on "what I am doing" rather than on "how I am doing" (Aronson, 1978; Ames, 1992; Covington, 1992; Slavin, 1983). Increasing student choice and responsibility for learning serves to uncover new talent and to increase student motivation (Corno, 1993). Beliefs that intelligence is malleable (Dweck & Leggett, 1988), that there are multiple abilities (Gardner, 1983), and that all can meet a specified standard (Goodlad, 1990) shift responsibility for failure from students to teachers, broaden the performance opportunities available, and offer absolute rather than relative criteria for accomplishment. Finally, classrooms and schools with more diverse and

differentiated opportunities for participation (with the demand for involvement in excess of the personpower to meet it) create the contexts for warmer, more concerned relationships between teachers, students, and parents (Barker & Gump, 1964; Butterworth & Weinstein, 1996).

This vision of equitable treatment in the classroom, derived from children's perspectives, both shaped our continuing interest in the kinds of processes and outcomes differentially associated with these two types of classrooms and also provided the direction for our future intervention work in schools.

CHILDREN'S RESPONSES TO TEACHERS' EXPECTATIONS: CONTEXTS AND ASSOCIATED OUTCOMES

Drawing upon the findings from a number of Berkeley doctoral dissertations and masters' theses (in preparation for publication in monograph form), we highlight what we have learned to date about children's responses to teachers' expectations. That children identified differences *between classrooms* in the degree of differential treatment expressed by teachers underscores the importance of looking for expectancy-related processes and outcomes in "context." That is, we sought to understand how these child-identified differences in classroom expectancy climates were linked predictively to, indeed moderated, a variety of child processes and outcomes. Here, we looked especially to broaden the kinds of outcomes explored beyond achievement, to encompass the social-emotional as well as the academic world of schooling, and to explore longer time periods for such effects, searching for carryover of effects beyond a single school year.

In addition to exploring the sequelae of classroom and thereby teacher differences in degree of differential treatment, we also investigated differences *between children* in their reponses to teacher expectations. Such differences could result from (1) development, that is, differing cognitive-developmental capacities to use expectancy cues in framing and internalizing self-perceptions, and/or from (2) the individual circumstances of children, that is, either differing demographic or personality characteristics or differing exposure to varying environmental stresses or supports.

Not Only Academic But Also Social-Emotional Responses

Motivational and Achievement Outcomes

In three studies we were able to uncover links between the salience of achievement cues in classrooms and children's awareness of their own differential treatment, self-reports of expectations, and ultimately, achievement. In a study of third through fifth graders (Brattesani, Weinstein, & Marshall, 1984), we demonstrated that in classrooms where students reported a great deal of differential treatment toward high and low achievers (in contrast to more equitable treatment classrooms), students were more likely to perceive their *own* treatment from the teacher as *congruent with* the expectation that their teacher held for them. That is, reports of more positive treatment were associated with higher expectations from teachers.

In a second study of fourth through sixth graders, in classrooms where children reported a great deal of differential treatment by teachers, children's expectations of themselves (how well they expect to do relative to their classmates) more closely matched teachers' expectations than was the case in classrooms where teachers treated students equitably (Brattesani, Weinstein, & Marshall, 1984). Thus, the gap in self-expectations between students from whom much is expected and students from whom little is expected is accentuated and more closely follows teacher expectations. Similarly, in classrooms with greater perceived differential treatment, after controlling for initial student achievement differences, teacher expectations of reading early in the year predicted an additional 9 to 18 percent of the variance in year-end reading achievement. This was much more than the 1 to 5 percent figures obtained from classrooms with little perceived differential teacher treatment. High levels of differential treatment increased the predictive link between teacher expectations and child motivational and academic outcomes, with initial student achievement controlled.

Additional evidence for the link between classroom differences in the salience of achievement cues and students' motivational outcomes has been found in other cultural contexts. An honors thesis conducted by Louise Foo (1991) contrasted the experiences of Chinese students in three fifth-grade classrooms that reflected different cultural contexts (a Hong Kong classroom, a San Francisco Chinatown classroom, and a suburban ethnically mixed classroom). Greater perceived differential teacher treatment (evident in the Hong Kong classroom) was associated with stronger beliefs on the part of the students that low achievers would be less likely than high achievers to try hard in their school work.

Peer Relationships

Our studies also provide evidence for spillover effects of expectancy effects to nonacademic outcomes. Botkin (1985) examined the relationship between the classroom level of perceived differential treatment and children's perceived as well as actual social competence in a sample of 92 fourth-grade students in seven classrooms. She found that in identified high differential treatment classrooms, children who were the target of low teacher expectations saw themselves as less socially competent and were in fact less often chosen as a work- or playmate by their peers than were children for whom teachers held high expectations. In contrast, in classrooms with identified equitable teacher treatment, no such differences in the social domain were documented between the targets of high and low academic expectations. Thus, in classrooms where achievement cues are highly salient, how a child performs academically has implications for peer relationships, thus narrowing the opportunties for the "perceived" low achievers to develop other competencies.

Defensive Responses

Botkin's doctoral dissertation (1990) explored children's defensive functioning—the unconscious strategies children use in warding off anxiety—in child-identified high and low differential treatment classrooms. Defenses were assessed using a projective method, involving analysis of children's reported thoughts, feelings, and behaviors in response to pictures of two school scenes that depicted a student's public failure to answer correctly and peer rejection at the lunch hour.

In a sample of third through fifth graders from 14 classrooms, Botkin found that after controlling for achievement differences, children demonstrated greater defensiveness (such as the use of the distancing defenses of devaluation, omnipotence, denial, and regression to ward off feelings of worthlessness and powerlessness) in classrooms where the achievement differentiation was more salient than in classrooms with more equitable treatment. This was true for children who were the target of high as well as of low expectations, suggestive of high anxiety for both types of students in such classrooms. In classrooms where teachers provide very clear messages about who is a "high" and who is a "low," children who are the targets of low expectations may use defensive distancing as a strategy to maintain self-esteem in the face of clear messages of failure. Children who are the targets of high expectations may feel that their performance carries with it a constant threat of losing face and status.

Longitudinal Relationships

If teachers' expectations for students indeed shape the students' achievement and self-views, what may be small differences each year can become larger if associated consequences (such as missed curricular exposure or lowered self-views, expectations, and achievement) are carried over to subsequent school years. Two theses have explored aspects of longitudinal predictive relationships.

Carryover to a Second School Year

In a sample of 103 first and second graders (younger cohort) and 75 third and fourth graders (older cohort), Kuklinkski (1992) examined the carryover effects, over two school years, of membership in child-identified high versus low differential treatment classrooms. She found that both younger and older children and both high and low teacher expectancy students from high differential treatment classrooms at Year 1 perceived more differential treatment in the next school year (after controlling for differences in achievement and in the Year 2 classroom environment), than did children from low differential treatment classrooms. For older children only, such classroom membership at Year 1 also predicted greater congruence between teacher and self-perceptions of competence in the subsequent year. These findings suggest that membership in classrooms where achievement cues are heightened carries with it an increased vigilence to differential treatment and teacher expectations on the part of students—a sensitivity that children carry with them to the next school year and apply, regardless of the current classroom environment. Ironically, such training by teachers to pay attention to ability differences is not to students' benefit given research findings that students are better off concentrating on the task than on their own abilities (e.g., Ames, 1992).

Predictions Across 14 Years

In another study, using a longitudinal sample of 110 children, Alvidrez (1994) charted the relationship between early teacher expectations for children at age four and subsequent high school achievement, assessed in terms of GPA at age 18. Evidence for expectancy effects was examined in the cases where teacher expectations for children did not match measured child IQ scores. Alvidrez found that teacher "underestimation" of student ability was greater the lower the socioeconomic status of the child and was associated with teacher ratings of the child as immature and insecure.

Four-year-old children for whom teacher expectations were lower than predicted from IQ scores ("underestimated") had significantly lower GPAs at age 18 than children whose ability was either accurately estimated or overestimated by teachers. However, most importantly, this predictive relationship depended upon qualities of the home environment as rated by trained observers. Lowered achievement was not documented in those cases where mothers were rated as significantly more career-oriented and active outside the home and showed a tendency to be more intellectually oriented. In interpreting these results, it is certainly possible that teachers were more sensitive than IQ scores in predicting children's achievement trajectories 18 years later, perhaps by identifying child behaviors that might stand in the way of achievement. It is also possible that teachers' under-predictions of children's ability reflected biased perceptions (where less was expected of poorer children even after IQ differences had been controlled) and may have influenced subsequent achievement. That mothers' more positive beliefs and/or actions may have buffered the effects of lowered expectations underscores the variability among the "underestimated" children. That is, not all children whose abilities were underestimated subsequently underperformed. While we lean toward an influential rather than predictive explanation of these findings, neither explanation can be definitively ruled out given the correlational nature of these data and the absence of information about possible mediating mechanisms.

Differences Among Children in Susceptibility

While children can readily identify differential teacher treatment, it is likely that not all children are equally vulnerable to teacher expectancy effects, either as recipients of low teacher expectations or as confirmers of such teacher predictions.

Our research group has also been exploring differences in child susceptibility to teacher expectancy effects. In this work we have considered developmental or grade-level differences, and individual differences in race, gender, and self-concept as moderators of children's exposure to and response to expectancy effects. We are also exploring different achievement pathways in response to inaccurately (relative to prior achievement) high or low teacher expectations: that is, vulnerability or resilience in the face of low expectations and susceptibility or resistance in the face of high expectations.

Developmental Factors

In a number of our studies we have found developmental or grade-level effects in children's responses to teacher expectations. With regard to self-expectations, for example, younger children were more likely to show greater congruence or match between teacher and self-expectations in child-identified high versus low differential treatment classrooms whereas by fifth grade, older children were less likely to be influenced by the perceived classroom climate (Weinstein et al., 1987). Older students for whom teachers held low expectations had more negative ability perceptions in both high and low differential treatment classrooms. These differences are perhaps indicative of a more stable and enduring self-view (in general, less responsive to information from the teacher) with increasing age.

However, when we examine child susceptability as measured by achievement outcomes, we see a more complex relationship between developmental as well as individual differences in child response to teacher expectations. With regard to differential achievement patterns, in a sample of first-, third-, and fifth-grade urban elementary school children, Soulé (1992) classified children who were the target of inaccurate expectations by teachers and charted their achievement across a school year, toward or away from teacher expectations. First, he found that although the classroom level of differential teacher treatment was a significant predictor, individual child characteristics predicted response to inaccurate teacher expectations above and beyond classroom environment. Second, Soulé found developmental differences in patterns of susceptability to teacher expectations. The classroom level of differential teacher treatment was most salient as a predictor in first grade and individual differences among children became a more important predictor in third and fifth grades.

Third, Soulé found developmental differences in the child characteristics that distinguished confirmers from disconfirmers of teacher expectations. First graders who confirmed low teacher expectations (that is, demonstrated decreased achievement) generally evaluated themselves highly, misperceived teacher expectations, had discrepant self- and teacher-expectations, and viewed teacher expectations as positive. Consistent with Botkin's study (1990), this suggests that young children who confirm teacher expectations may defend against the anxiety of low expectations by endorsing extremely positive self-views and by unrealistically and defensively misappraising low teacher expectations, thereby preserving self-esteem. For third and fifth graders the most salient predictors of susceptibility to low teacher expectations included extreme teacher expectations, vigilence to differential teacher treatment, and congruence between self-expectations and perceived teacher expectations. Thus, in contrast to the younger susceptible children, older susceptible children were more accurate in their appraisals and less protected by positively skewed views of teacher behavior and expectations for them.

These differential achievement patterns in response to teacher expectations underscore complex relationships between environment, developmental differences, and individual characteristics of children. While older and younger children do not differ in their capacity to identify differential teacher treatment (that is, in perceptions of others), developmental differences play a larger role in the pattern of factors that predict individual child susceptibility to teacher expectations (that is, in perceptions and actions of self). Among the subgroup of children who respond to teacher expectations, younger children appear more responsive to the treatment climate of the classroom and more defensive in their self-perceptions whereas older children appear more reponsive to specific qualities of teacher expectations for them and more realistic in their self-appraisals. Most importantly, Soulé's study underscores that both younger and older children can show susceptibility to teacher expectations.

Race and Self-concept

Characteristics such as race and qualities of the self-conceptions of children have also been investigated in our laboratory as factors at various stages in expectancy processes. With regard to the formation of teachers' expectations, in a study by Jones (1989), African-American students were found to be the recipients of lower academic expectations than Caucasian students. After controlling for achievement differences among students, teachers' expectations were found to be lower for African-American students than for Caucasian students in third and fifth grade (but not in first grade). Furthermore, this tendency to underestimate the ability of African-American students (relative to measured achievement) was more common for high rather than low differential treatment teachers (as identified by students).

We have also explored relationships between the ways in which children think about themselves and the potential for teachers' expectations to influence children's achievement outcomes. Focusing on a sample of third and fifth graders, Brattesani (1984) found that children with moderate (neither high nor low) self-evaluations were most susceptible to teacher expectations. It appears that children who have more extreme self-evaluations (regardless of the valence of those views) are less likely to shift in response to contrasting teacher expectations, perhaps because they are more sure of their own self-evaluations and thus rely less on information from teachers in forming their self-image. In contrast, children who view themselves as average in cognitive ability may be less sure of themselves and may look to authority figures, such as teachers, for validation.

Also documented with two different self-perception measures (cognitive competence and self-expectations) is the finding that children whose self-views were moderately discrepant from perceived teacher expectations were more likely to confirm teacher expectations (Brattesani, 1984; Madison, 1991). When teacher expectations are not too far removed from children's self-perceptions, children may find the message more plausible than when expectations depart more radically from self-evaluations. In cases where the discrepancy between children's self-views and perceived teacher views was large, the less susceptible children were more likely than the more susceptible children to rely on their own self-evaluations (for example, to make 'I' statements rather than referring to the teacher's judgment) to support their appraisals.

PUTTING WHAT WE KNOW TO WORK: PREVENTIVE INTERVENTIONS

While it is clear from our findings that children's responsiveness to teacher expectations may vary with age and other characteristics, it is also apparent that expectancy effects are heightened in classrooms where teachers' beliefs about ability and teaching practices highlight ability differences among students. Thus, in thinking preventively, working toward the creation of classroom environments with equitable treatment (as described by children) is an important place to begin. Empirically derived and evaluated preventive interventions to enhance positive expectancy practices in schooling have been relatively scarce, despite a long history of research on expectancy effects and despite the call for higher expectations. As Babad (1993) has underscored, we need to increase research on the application of expectancy findings to the improvement of practice but in ways that reflect the theoretical and empirical state of knowledge in the field. Babad (1993) has characterized the few empirical efforts as reflecting three types of interventions: the provision of research-derived recommendations (Brophy, 1983; Cooper & Tom, 1984; Smith & Luginbuhl, 1976), the use of controlled studies where empirically based feedback is given to teachers (Babad, 1990; Good & Brophy, 1974), and school-wide in-service programs based on research (Gottfredson, Weinstein, & Marshall, 1995; Kerman, 1979; Penman, 1982; Proctor, 1984; Weinstein et al., 1991). Where evaluative data are available, there is some but limited support for positive change in teacher practices and student achievement.

In explaining the often equivocal results, we find it useful to distinguish between these efforts in terms of the underlying theoretical models of expectancy communication and expectancy change applied. With the exception of our own intervention, other efforts have adopted a largely behavioral model focused on increasing teacher awareness of and change in patterns of teacher-student dyadic interaction in the classroom through prescriptive methods. While the level targeted (whole school versus individual teachers) or the precision of the method (research recommendations versus individually targeted feedback) may vary, the focus for change remains a prescribed set of behaviors.

In contrast, in applying what we learned from the student perspective, we address the problem of expectancy change culturally and systemically. Our research findings identify the structural and interactional features of schooling practice and policy targeted for change. Importantly, the findings also alert us to the cultural embeddedness of these practices in the classroom and in the beliefs about ability that undergird them. Thus, any change process must actively involve teachers in new learning and engage teachers and administrators in collaborative and systemic efforts to examine interrelationships between levels of the school culture that might support or undermine expectancy processes in schooling.

Here, we provide a brief overview of the findings from two studies: the first in a high school where we applied our research-derived expectancy communication-change model toward raising expectations for ninth graders at risk for school failure (Weinstein et al., 1991; Weinstein, Madison, & Kuklinski, 1995; Weinstein, 1998); the second in a small but diverse private elementary school where we analyzed the nature of principal leadership in creating and guiding a school-wide environment where "supported" high expectations were already in place for students, teachers, and parents (Butterworth & Weinstein, 1996). There is much to be learned about expectancy processes from our efforts to change them as well as from analyses of good practices already in place.

The Problem of Expectancy Change

The Model

Our research on children's perspectives provided the empirical direction for intervention: to target the eight features (outlined earlier) of the instructional environment for assessment and change toward more positive expectancy policies and practices. Research on the disconfirmation of stereotypes (psychological change) and on organizational reform (social system change) guided us toward the method of bringing about change: a collaborative school-based model. Social psychological studies point to conditions under which negative perceptions are successfully challenged: where disconfirming evidence is systematically available, analyzed, and generalized (Bar-Tal, 1989; Olsen & Zanna, 1993; Rothbart & John, 1985), where motivational goals stress accuracy and accountability (Neuberg, 1989), and where interactions between individuals are cooperative, co-equal, successful, and lacking conflict (Desforges, Lord, Ramsey, Mason, & Van Leeuwen, 1991). Research on school culture and reform underscores the importance of involved participants and collaborative working conditions (Beer & Walton, 1987; Berman & McLaughlin, 1978; McLaughlin, 1990) as well as of systematically targeting interrelationships between levels of school culture that might undermine the change effort (Epstein, 1985; Maehr & Midgley, 1991). Sarason (1996) has long warned about the need for change efforts to address the school culture.

These findings suggest a model of intervention that moves beyond prescriptive and behavioral approaches involving teachers alone toward collaborative and systemic approaches that include administrators as well as teachers. An ongoing context within schools needs to be created in which negative beliefs about students can be disconfirmed, positive beliefs developed, and changes made in practices and policies. Thus, we identified the critical components of the expectancy change process to include: (1) collaborative methods, (2) a diversity of membership across teachers, administrators, and university researchers, (3) a regularized, weekly school-based meeting, with additional planning time, (4) a long-term perspective, (5) shared responsibility for students, (6) the translation of research findings into practice and policy, and (7) the monitoring of multilevel outcomes. The ways in which this model asked teachers and administrators to work together mirrored the ways that research findings suggest teachers work with students.

Features of the Study

This model was implemented and evaluated over a two-year period in a mid-sized urban and ethnically diverse high school. Eight teachers and four administrators (principal, vice-principal, counselor, and dean) joined the collaborative team. As negotiated, all incoming ninth graders assigned to the lowest track of English classes were targeted for the intervention. This yielded a sample of 158 students over the two years of the project as contrasted with an archival sample of 154 comparable students from the two previous years of classes.

Students were programmed into one or more of the classes of participating teachers, who shared a common group of students as well as a common preparation period each day, scheduled around the lunch hour. This created in effect, a school within a school. Teachers, administrators, and researchers met for two hours weekly where they read the research literature targeting the eight features of expectancy communication, observed practices and policies currently in place, and systematically implemented and evaluated changes. The evaluation design (also collaboratively negotiated) captured both the course of the intervention over time and evidence for change at the multiple levels of student, teacher, and school policy. As sources of data, we utilized narrative records of meetings, pre-post ratings of practices by teachers, and pre-post data collected from student records in an archival cohort design.

Selected Findings

The coded narrative records of meetings documented that, over time, teachers shared more complex, differentiated, and positive views of student abilities. Early talk was focused on the deficits of these students. As the collaboration unfolded, as rated by the coders, talk shifted more to the capabilities of students and to what teachers could do to solve the problems they experienced with students in their classrooms. Teachers became more active agents for change in their own school, as mentors to other teachers in the district, and at a national level in writing about their work (e.g., Cone, 1993). The teacher self-ratings also documented a trend for the predicted increase in use of positive expectancy practices across the eight domains. Further, new school policies were implemented to untrack these low-achieving students and to broaden the curricular challenge offered. Thus, what began as noncollege-bound classes became integrated into college preparatory classes, with an open enrollment policy (based on interest and a work contract rather than test scores) introduced for advanced placement classes in English.

Changes in students were also evident. After controlling for entering achievement differences, project students (in contrast to comparison students from previous years) earned higher GPAs and received fewer disciplinary referrals after one year in the program. Qualitative records underscored the more positive presence of these students within the school. For example, for the first time in school history, two of the elected officers on the student council came from project classes. At the one-year followup (by the end of tenth grade), only half as many of the project students as compared to the contrast students (19 to 38%) had transferred out of the school. This greater holding power of the school (a behavioral confirmation) was not accompanied by significantly higher grades at followup although the difference was in the predicted direction. Significant achievement differences were not demonstrated perhaps because of the more challenging curricula provided and because the lower attrition rate had produced a wider band of achievers among the project students than among the comparison students.

Because of the largely consistent changes across levels as predicted by expectancy theory, we suggest that this approach holds promise, particularly given that the project addressed the hardest core of low-achieving students and at a relatively late stage in their school careers (ninth grade). However, critical to the promotion of a positive expectancy climate was the provision of consistent, stimulating, and supportive conditions for school staff where they could grow together in reframing their work with students, teachers, and the administration. With regularized collaboration, perceived obstacles to change were translated into opportunities. With a sense of efficacy (Ashton & Webb, 1986), teachers took increased responsibility for the success and failure of their students. The most unyielding barrier to the spread of changes across the school lay in the difficulty of engaging the administration and district in consistent and systematic realignment of policy and practice.

Creating and Sustaining a Positive Expectancy School Climate

Case Study

There is much to be learned from an analysis of best practices already in place and from the perspective of the school leader, the principal. We identified an innovative K-six private school with a diverse student body (there because of admission policies and a generous scholarship fund) and a reputation for helping children who did not succeed in other settings. This school was noted for the diversity of motivational and learning opportunities available not only for students but also for teachers and parents.

The data we utilized consisted primarily of retrospective vignettes written by the principal which captured her observations and actions across representative times and days of the week. Corroborating evidence was found in the school's newspaper and the principal's weekly letter home to families. Utilizing this material, we analyzed the vision, policies, structures, and interactions, that is, the acts of principal leadership that created a climate of high expectations for all (Butterworth & Weinstein, 1996).

Ecological Principles of Leadership

We identified four ecological principles that were key in creating and sustaining a motivated community at all levels. The principal led the school community in (1) the creation of a diversity of niches, that is, activities that demanded student involvement and supported adaptation, (2) resource development, that is, the expansion of resources to include the students themselves, their parents, and the community, (3) monitoring interdependence, that is, reducing conflict and promoting mutual reinforcement between levels of the system, and (4) conserving and balancing resources and activities.

What did this look like in practice? Niche development meant that integrated with and beyond the lessons of the classroom, students participated in a diversity of activities such as the school economy (with student jobs, a student-run bank, and student-run stores), student government, community service, environmental education, and the arts (school plays, newspaper, literary magazine, and musical performances). The varied nature of these activities drew upon the diverse talents of the community. By design, the demand for involvement grew from the creation of "underpeopled" settings—there were more roles than individuals to enact those roles. For example, parts in school plays were double-cast and elections for the student council were held twice a year to enable greater student participation. Thus, in order to staff the number and variety of activities, in addition to tapping teachers' interests, the energy of students, parents, and community members was seriously cultivated.

By hiring part-time specialist teachers, by freeing teachers each day for planning time through the specialist offerings, by requiring school staff to meet weekly, and by setting aside Fridays as school-wide activity days, room was made for participation and development of a broad and exciting array of educational opportunities. The principal—by teaching classes, by inviting different students each Friday to share a brown-bag lunch, by holding an open hour to speak with parents at the end of the day, by requiring participation in family workdays at the school, by working closely with teachers—could first-hand view the potential for conflicts, assess the need for intervention and reinforcement, and monitor the outcomes of the school's efforts.

Confirming the Features of Expectancy Communication

We suggest that the qualities of this school's motivational environment *naturally* encouraged heterogeneous grouping of students (across classes and grades) and provided challenge, choice, and support to all students. The meaningful nature of the school activities, with real audiences to utilize and appreciate the fruits of the labor, underscored the importance of the task itself (not comparative accomplishments) and recognized individuality in the students while building a strong sense of community. These features, akin to those identified by children as critical in the communication of expectations, exist not as disparate factors but as integrally linked aspects of the classroom and school teaching environment. The vision here is one of "development in diverse contexts" and one that takes seriously the development of all the players in schooling.

What can be said about links between the motivational climate described and the motivational outcomes achieved? Here we can only speculate, drawing evidence from school publications. Staff turnover was virtually nil, despite lower teacher salaries than the public schools. A long waiting list existed for a place at the school. Parent participation was extremely high: in one school year reflected in 39 percent of the families serving as coordinators of activities, a 73 percent contribution rate to school fund-raising, and often 100 percent participation at school events (since each child and/or family had a role). Finally, 68 percent of the students participated in the school government during the course of a single school year. While such outcomes were achieved at a private rather than public school, it is important to remember that this school's population was more diverse than that of the surrounding community as a result of admission policies of first-come, first-serve, sibling priority, and ethnic-minority representation. Rather than a result of selection, in our view, the key to involvement lay in the by-design *constructed demand characteristics* of this setting.

IMPLICATIONS FOR TEACHER EDUCATION AND FUTURE RESEARCH

In this chapter we have described a program of research on basic processes and preventive interventions that explores the contexts in which expectancy effects (particularly negative effects) are heightened, diminished, or changed in schooling. The work speaks to children's awareness of teachers' expectations, their capacity to distinguish between classroom environments (those with more or less differential teacher treatment toward high and low achievers), and accompanying processes and outcomes that underscore the predictive validity of these classroom distinctions. That is, greater perceived differential teacher treatment is associated with an increasing gap between high and low teacher-expectancy students across a variety of outcomes—true not only within the academic domain and within a single school year but involving social and emotional consequences and carryover to subsequent school years. The work also suggests that beyond classroom environmental effects, individual differences in children (such as in development and self-view) play a complex role in moderating children's differential response patterns to teacher expectations. Given developmental changes in the stability of self-perceptions, we may have more success in the early years of elementary schooling with regard to shaping children's self-views, motivation, and achievement in positive ways.

Listening to the voices of students underscores a broader and more integrative view of a *culture* of academic expectations in the classroom and beyond to the school—an understanding that has steered our intervention efforts with teachers and administrators toward systemic as well psychological change achieved through school-based collaborative methods. Listening to the voices of teachers as they grapple with changing their expectations for students underscores that the road to the disconfirmation of negative expectations and the building of a positive expectancy environment involves not only a change in beliefs but also a realignment of policy and practice at both the school and classroom level. Teachers and administrators, alike, need opportunities to examine instructional efforts with students regularly, systematically, and with support and to increase their efficacy. Our findings suggest that such collaborative working conditions, which target for change empirically validated expectancy beliefs, practices, and policies, are associated with positive changes in teachers, students, and school policies, even for a hard-core population of older low-achieving students entering high school.

While the correlational and quasi-experimental design of these studies does not permit inferences about causal links in expectancy processes, there are strengths here in the ecological validity of the data, the controls for prior achievement differences between students, the temporal ordering of constructs, and the theoretically consistent findings. There is coherence in this set of findings that depicts a student role in expectancy processes in the classroom (both in awareness of and response to differential treatment by teachers) and a clear pattern of classroom and school differences in the degree to which teachers and schools make salient ability comparisons among their students, with associated consequences for students.

Educating Teachers about Expectancy Effects

Teacher education about the importance of academic expectations for students has focused largely on issues of psychological awareness, that is, sensitizing teachers to their sometimes unconscious beliefs about the differing abilities of students and ensuing differential treatment. This approach, operationalized in the widely adopted TESA program, grows from the earliest and largely behavioral paradigm of Brophy and Good. Relatively little empirical evaluation has been done, and what exists is largely not positive with regard to effectiveness (Gottfredson et al., 1995).

Although not explicitly linked to the teacher expectancy research literature, there are also systemic efforts at the policy level to promote positive and higher expectations and to change the behaviors of teachers. These policies include legislation regarding desegregation and mainstreaming, the development of state and district academic standards for subject matters by grade level, and recent calls for de-tracking the high school curriculum. Yet, as one example underscores, evaluations of desegregation efforts point to underlying avoidant behavior in carrying out the policies and and to evidence of resegregation of students in other ways, within tracks and special programs (e.g., Epstein, 1985).

Given our findings, we would argue that neither prescribed awareness training nor mandated policy change goes far enough in defining both the *targets for change* (the depth and breadth of what in teaching is addressed) and the *methods for change* (stimulating and supporting the active participation of teachers and administrators in reframing expectations and instructional approaches) (Weinstein, 1996). Interventions to promote positive expectancies must recognize and address the complexity of the institution of schooling within which expectancies are formed, played out, and changed.

As the work of Oakes and colleagues (1992, 1997) suggests, poorly indicated pedagogical practices are supported by institutional forces—forces that must be addressed before substantial change in teacher expectations will come about. Our own research also points to the interrelated structural as well as interactional features of instruction at both a classroom and school level that provide students differential or equitable opportunities to learn and the necessary supports (varied, flexible, and non-stigmatizing) to enable that learning to occur. Moving educators away from attributions to student deficits toward more positive beliefs about student potential and more effective classroom interventions and school policies requires ongoing institutional support. This includes support from school policies, collegial scaffolding, and the willingness to be held accountable for high expectations and for effective instructional strategies in response to diversity in student body (Weinstein, 1996).

Most critically, perhaps, this research on children's perspectives underscores the importance of recognizing students as sensitive and vital participants in a classroom culture where their self-esteem, motivation, and ultimate achievement are at stake. That students derive meaning from the differential treatment they observe suggests that their perspectives should be appraised and heard in our efforts to improve the education of teachers.

Directions for Future Research

Given our findings, we would suggest a number of implications for future research in the field. First, research efforts must distinguish between teachers (and classroom and school environments) and between students (by population, ages, and individual characteristics) in studies of expectancy effects in schooling. As our studies have shown, expectancy effects are stronger in some contexts and for some students. Thus, efforts to demonstrate all-or-none findings averaged across samples (e.g., Jussim, 1989; Jussim & Eccles, 1995) are likely to mask important evidence critical to evaluating the power of expectancy effects in schooling. We need to increase our knowledge about the conditions under which such effects are maximized, minimized, and changed.

Second, there is far more at stake here—beyond messages about ability, beyond student achievement outcomes, and beyond a single year. As our findings suggest, negative self-fulfilling prophecies at work reflect policy and practice, curricular and instructional choices (with both direct and indirect effects on student achievement), academic as well as social-emotional consequences, and effects that are cumulative within the child or the school culture. There is a conspicuous lack of research examining the dynamic interplay among these concurrent, interactive, and longitudinal conditions and consequences.

Third, there is much to be learned about how to prevent negative self-fulfilling prophecies in schooling. Careful and systematic studies are necessary to examine both *what* must be changed and *how* it can be changed. Efforts to raise academic standards and/or eliminate ability-based tracking will fail, as did racial integration efforts in schooling, unless the underlying and interrelated causes of negative self-fulfilling prophecies are identified and effectively addressed.

Finally, we are hampered in our efforts to improve understanding of the magnitude of expectancy effects by conceptual and methodological limitations inherent in the constructs themselves. Our characterization of what reflects an "expectancy effect" (that is, distinguishing between the accurate and inaccurate expectations of teachers) fails to confront the limits of measurement or to grapple with underlying philosophical questions about the purposes of education and the role of the teacher.

The theory of self-fulfilling prophecies proposed by Merton (1948) was predicated upon the *false* nature of a belief that became confirmed in reality. As translated in experimental studies of expectancy effects, false beliefs about students are generated and then induced in teachers. But in its translation to naturalistic studies in classrooms, the determination of a false belief rests upon finding an appropriate standard against which to compare teacher expectations. However limited in design, naturalistic studies are essential to providing an ecologically sound perspective on naturally unfolding expectancy effects in schooling, particularly negative prophecies, which for ethical reasons are not appropriate for manipulation.

Often forgotten is that the standards against which teachers' beliefs are judged accurate or inaccurate (that is, IQ or achievement scores) are themselves flawed in myriad ways (e.g., Neisser et al., 1996). Such measures are highly sensitive to educational exposure, language usage, preparedness, anxiety, and error. Such measures reflect relative rather than absolute standards of knowledge and are often more predictive of scores on other tests than of real-world functioning. Such measures may unwittingly incorporate a prior history of teacher expectation effects that in turn arose out of earlier testing. Thus, it is not easy to determine the accuracy of teachers' beliefs about students concerning their academic abilities or likely achievements, let alone to characterize specific beliefs simply as true or false.

We often forget, as well, that the performance measured is past performance, not future potential, and that the distinctions made between students are largely relative ones. If an important goal of education is to develop the talent of each student, not simply to select students for certain pathways (that is, maintaining differences among students), then our failure to focus on potential and our reliance on relative rather than absolute criteria prove problematic. While recognizing that the determination of such absolute standards carries its own set of complexities (that is, reflecting yet another set of expectations), the research suggests that when children are socialized to seek mastery as opposed to performance goals, they are ultimately more engaged and productive learners (Ames, 1992). Therefore, measuring level of mastery against an absolute standard (if progress toward that standard is emphasized), may in the long run be healthier for children than measuring performance comparatively.

Given these methodological and philosophical concerns, the "accuracy" of teacher expectations appears a problematic construct. So too is what we call "appropriate" teacher treatment in response to student differences on so-called ability or achievement tests. If our capacity to measure student potential via test scores is flawed and if our teaching goals are to help all children reach agreed-upon standards, then the kind of differential treatment documented in our studies is *never* "appropriate." Furthermore, the effects we now call sustaining effects (keeping children where they are, relative to their measured achievement) may really be best understood as another form of self-fulfilling prophecies—that is, schooling that does not move children forward beyond the capacities they bring to the classroom.

Thus, in raising questions about the language and measurement of expectancy effects, we suggest that current conceptualizations, even when taking contextual factors into account, may lead to conservative and probable underestimates of magnitude. Current conceptualizations may lead us to assume that there is little to study in the arena of "accurate" expectations and "appropriate" treatment. Indeed, within this definitional frame, we would argue that a degree of positive inaccuracy is necessary to accomplish the main aim of education—to move all children forward.

To Conclude

Can teachers' more positive expectations for all students play a role in the provision of more equitable treatment in the classroom and in the lessening of the discrepancy between student potential and performance? We think so. A key to how it can be accomplished lies in the creation of stimulating school conditions that enable teacher learning, as Sarason forewarned, and given this support, in the capacity for teachers to take responsibilty for student learning and *for learning from stu-*

dents, as was underscored by Rabinovitch. Such shifts in accountability and support, targeted toward the empirically derived features of expectancy communication in schooling, would facilitate the development, rather than the selection of, talent in all students.

REFERENCES

Alvidrez, J. (1994). *Early teacher expectations and later student achievement.* Unpublished masters' thesis. University of California, Berkeley.
Ames, C. (1992). Classrooms: Goals, structures, and student motivation. *Journal of Educational Psychology, 84,* 261-271.
Aronson, E. (1978). *The jigsaw classroom.* Beverly Hills, CA: Sage.
Ashton, P. T., & Webb, R. B. (1986). *Making a difference: Teachers' sense of efficacy and student achievement.* White Plains, NY: Longman.
Babad, E. (1990). Measuring and changing teachers' differential behavior as perceived by students and teachers. *Journal of Educational Psychology, 82,* 683-690.
Babad, E. (1993). Pygmalion—25 years after interpersonal expectations in the classroom. In P.D. Blanck (Ed.) *Interpersonal expectations: Theory, research, and application* (pp. 125-153). London: Cambridge University Press.
Babad, E., Bernieri, F., & Rosenthal, R. (1989). Nonverbal communication and leakage in the behavior of biased and unbiased teachers. *Journal of Personality and Social Psychology, 56,* 89-84.
Babad, E. Y., Inbar, J., & Rosenthal, R. (1982). Pygmalion, Galatea, and the Golem: Investigations of biased and unbiased teachers. *Journal of Educational Psychology, 74,* 459-474.
Babad, E., & Taylor, P. J. (1992). Transparency of teacher expectancies across language, cultural boundaries. *Journal of Educational Research, 86,* 120-125.
Barker, R. G. & Gump, P. V. (1964). *Big school, small school: High school size and student behavior.* Stanford, Ca: Stanford University Press.
Bar-Tal, Y. (1989). Can leaders change followers' stereotypes? In D. Bar-Tal, C.F. Graumann, A.W. Kruglanski, & W. Stroebe (Eds.), *Stereotyping and prejudice: Changing conceptions* (pp. 225-242). New York: Springer-Verlag.
Beer, M., & Walton, A. E. (1987). Organizational change and development. *Annual Review of Psychology, 38,* 339-367.
Berman, P., & McLaughlin, M. (1978). *Rethinking the federal role in education.* Santa Monica: Rand Corporation.
Botkin, M. (1985). *Perceived competence and friendship choice as a function of differential teacher treatment.* Unpublished Masters thesis. University of California, Berkeley.
Botkin, M. (1990). *Differential teacher treatment and ego functioning: The relationship between perceived competence and defense.* Unpublished doctoral dissertation. University of California, Berkeley.
Brattesani, K. A. (1984). *The role of initial self-evaluation in student susceptibility to teacher expectations.* Unpublished doctoral dissertation, University of California, Berkeley.
Brattesani, K. A., Weinstein, R. S., & Marshall, H. H. (1984). Student perceptions of differential teacher treatment as moderators of teacher expectation effects. *Journal of Educational Psychology, 76,* 236-247.
Braun, C. (1973). "Johnny reads the cues: Teacher expectations." *Reading Teacher, 26,* 704-712.
Braun, C. (1976). Teacher expectations: Sociopsychological dynamics. *Review of Educational Research, 46,* 185-213.
Brophy, J. E. (1983). Research on the self-fulfilling prophecy and teacher expectations. *Journal of Educational Psychology, 75,* 631-661.

Brophy, J. E., & Good, T. L. (1970). Teachers' communication of differential expectations for children's classroom performance: Some behavioral data. *Journal of Educational Psychology, 61*, 365-374.
Brophy, J. E., & Good, T. L. (1974). *Teacher-student relationships: Causes and consequences.* New York: Holt, Rinehart & Winston.
Butterworth, B., & Weinstein, R. S. (1996). Enhancing motivational opportunity in elementary schooling: A case study of the ecology of principal leadership. *The Elementary School Journal, 97*, 57-80.
Cone, J. (1993). Learning to teach an untracked class. *The College Board Review, 169*, 20-31.
Cooper, H. M. (1979). Pygmalion grows up: A model for teacher expectation communication and performance influence. *Review of Educational Research, 49*, 389-410.
Cooper, H. M., & Good, T. L. (1983). *Pygmalion grows up: Studies in the expectation communication process.* White Plains, NY: Longman.
Cooper, H. M. & Tom, D. Y. H. (1984). Teacher expectation research: A review with implications for classroom instruction. *The Elementary School Journal, 85*, 77-89.
Corno, L. (1993). The best-laid plans: Modern conceptions of volition and educational research. *Educational Researcher, 22*, 14-22.
Covington, M. V. (1992). *Making the grade: A self-worth perspective on motivation and school reform.* Cambridge: Cambridge University Press.
Darley, J. M., & Fazio, R. H. (1980). Expectancy confirmation processes arising in the social interaction sequence. *American Psychologist, 35*, 867-881.
Desforges, D. M., Lord, C. G., Ramsey, S. L., Mason, J. A., & Van Leeuwen, M. D. (1991). Effects of structured cooperative contact on changing negative attitudes toward stigmatized social groups. *Journal of Personality and Social Psychology, 60*, 531-544.
Dweck, C., & Leggett, E. (1988). A social-cognitive approach to motivation and personality. *Psychological Review, 95*, 256-273.
Epstein, J. L. (1985). After the bus arrives: Resegregation in desegregated schools. *Journal of Social Issues, 41*, 23-43.
Foo, L. (1991). *Cultural differences in children's mediation of classroom expectancy effects.* Unpublished honor's thesis, University of California, Berkeley.
Gamoran, A. (1987). The stratification of high school learning opportunitiies. *Sociology of Education, 60*, 135-155.
Gardner, H. (1983). *Frames of mind: The theory of multiple intelligences.* New York: Basic.
Good, T. L., & Brophy, J. E. (1974). Changing teacher and student behavior: An empirical examination. *Journal of Educational Psychology, 66*, 390-405.
Good, T. L., & Weinstein, R. S. (1986). Schools make a difference: Evidence, criticisms, and new directions. *American Psychologist, 41*, 1090-1097.
Goodlad, J. I. (1990). *Teachers for our nation's schools.* San Francisco: Jossey-Bass.
Gottfredson, D. C., Marciniak, E. M., Birdseye, A. T., & Gottfredson, G. D. (1995). Increasing teacher expectations for student achievement. *The Journal of Educational Research, 88*, 155-163.
Jones, L. (1989). *Teacher expectations for black and white students in contrasting classroom environments.* Unpublished master's thesis, University of California, Berkeley.
Jussim, L. (1989). Self-fulfilling prophecies: A theoretical and integrative review. *Psychological Review, 93*, 429-445.
Jussim, L., & Eccles, J. (1995). Naturally occurring interpersonal expectancies. *Review of Personality and Social Psychology, 15*, 74-108.
Kerman, S. (1979). Teacher expectations and student achievement. *Phi Delta Kappan, 60*, 716-718.
Kuklinski, M. (1992). *The longitudinal influence of perceived differential teacher treatment.* Unpublished master's thesis, University of California, Berkeley.
Madison, S. (1991). *Pathways to the disconfirmation of teacher expectations.* Unpublished master's thesis, University of California, Berkeley.
Maehr, M. L., & Midgley, C. (1991). Enhancing student motivation: A school-wide approach. *Educational Psychologist, 26*, 399-427.

Marshall, H. H., & Weinstein, R. S. (1984). Classroom factors affecting students'self-evaluations: An interactional model. *Review of Educational Research, 54*, 301-325.

Marshall, H. H., & Weinstein, R. S. (1986). The classroom context of student-perceived differential teacher treatment. *Journal of Educational Psychology, 78*, 441-453.

Marshall, H. H., & Weinstein, R. S. (1988). Beyond quantitative analysis: Recontexualization of classroom factors contributing to the communication of teacher expectations. In J. Green, J. Harker, & C. Wallat (Eds.), *Multiple analysis of classroom discourse processes*. Norwood, NJ: Ablex.

McCaslin, M. M., & Good, T. L. (1996). *Listening in classrooms*. New York: HarperCollins.

McLaughlin, M. W. (1990). The Rand change agent study revisited: Macro perspectives and micro perspectives. *Educational Researcher, 19*, 11-16.

Merton, R. K. (1948). The self-fulfilling prophecy. *Antioch Review, 8*, 193-210.

Mitman, A. L., & Lash, A. A. (1988). Students' perceptions of their academic standing and classroom behavior. *The Elementary School Journal, 89*, 55-68.

Moos, R. H. (1979). *Evaluating educational environments*. San Francisco: Jossey-Bass.

Neisser, U., Boodoo, G., Bouchard, T. J., Boykin, A. W., Brody, N., Ceci, S. J., Halpern, D. F., Loehlin, J. C., Perloff, R., Sternberg, R. J., & Urbina, S. (1996). Intelligence: Knowns and unknowns. *American Psychologist, 51*, 77-101.

Neuberg, S. L. (1989). The goal of forming accurate impressions during social interactions: Attenuating the impact of negative expectancies. *Journal of Personality and Social Psychology, 56*, 374-386.

Oakes, J. (1992). Can tracking research inform practice?: Technical, normative, and political considerations. *Educational Researcher, 21*, 12-21.

Oakes, J., Wells, A. S., Jones, M., & Datnow, A. (1997). Detracking: The social construction of ability, cultural politics, and resistence to reform. *Teachers College Record, 98*, 482-510.

Olsen, J. M., & Zanna, M. P. (1993). Attitudes and attitude change. *Annual Review of Psychology, 44*, 117-154.

Penman, P. R. (1982). *The efficacy of TESA training in changing teacher behaviors and attitudes toward low achievers*. Unpublished doctoral dissertation, Arizona State University, Phoenix.

Proctor, C. P. (1984). Teacher expectations: A model for school improvement. *The Elementary School Journal, 84*, 469-481.

Raudenbush, S. W. (1984). Magnitude of teacher expectancy effects on pupil IQ as a function of the credibility of expectancy induction: A synthesis of findings from 18 experiments. *Journal of Educational Psychology, 76*, 85-97.

Rosenthal, R. (1973). *On the social psychology of the self-fulfilling prophecy: Further evidence for Pygmalion effects and their mediating mechanisms*. New York: MSS Modular Publications.

Rosenthal, R., & Jacobson, L. (1968). *Pygmalion in the classroom*. New York: Holt, Rinehart and Winston.

Rothbart, M., & John, O. P. (1985). Social categorization and behavioral episodes: A cognitive analysis of the effects of intergroup contact. *Journal of Social Issues, 41*, 81-104.

Sarason, S. B. (1988). *The making of an American psychologist: An autobiography*. San Francisco: Jossey-Bass.

Sarason, S. B. (1996). *Revisiting the culture of the school and the problem of change*. New York: Teachers College Press (Originally published in 1971, 2nd edition, 1982).

Slavin, R. E. (1983). *Cooperative learning*. New York: Longman.

Smith, F. J., & Luginbuhl, J. E. R. (1976). Inspecting expectancy: Some laboratory results of relevance for teacher training. *Journal of Educational Psychology, 68*, 265-272.

Soule, C. (1993). *Predictors of children's susceptibility to teacher expectations*. Unpublished doctoral dissertation, University of California, Berkeley.

Sroufe, A., & Rutter, M. (1984). The domain of developmental psychopathology. *Child Development, 55*, 17-29.

Walberg, H. J. (1976). Psychology of learning environments: Behavioral, structural, or perceptual? In L. S. Shulman (Ed.), *Review of Research in Education* (Vol. 4, pp. 142-178). Itasca, IL: Peacock.

Weinstein, R. S. (1976). Reading group membership in first grade: Teacher behaviors and pupil experience over time. *Journal of Educational Psychology, 68,* 103-116.

Weinstein, R. S. (Ed.) (1983). Students in classrooms. Special Issue of *Elementary School Journal, 82,* 397-540.

Weinstein, R. S. (1989). Perceptions of classroom processes and student motivation: Children's views of self-fulfilling prophecies. In R. Ames & C. Ames (Eds.), *Research on motivation in education* (Vol. 3, pp. 187-221). Academic Press.

Weinstein, R. S. (1990). The universe of alternatives in schooling: The contributions of Seymour B. Sarason to education. *American Journal of Community Psychology, 18,* 359-369.

Weinstein, R. S. (1993). Children's knowledge of differential treatment in school: Implications for motivation. In T. Tomlinson (Ed.), *Motivating student to learn* (pp. 197-224). Berkeley: McCutchan.

Weinstein, R. S. (1996). High standards in a tracked system of schooling: For which students and with what educational supports? *Educational Researcher, 25,* 16-19.

Weinstein, R. S. (1998). Promoting positive expectations in schooling. In N. Lambert & B. L. McCombs (Eds.), *How students learn: Reforming schools through learner-centered education.* Washington, DC: American Psychological Association.

Weinstein, R. S., Madison, S., & Kuklinski, M. (1995) Raising expectations in schooling: Obstacles and opportunities for change. *American Educational Research Journal, 32,* 121-159.

Weinstein, R. S., Marshall, H. H., Brattesani, K., & Middlestadt, S. E. (1982). Student perceptions of differential teacher treatment in open and traditional classrooms. *Journal of Educational Psychology, 74,* 678-692.

Weinstein, R. S., Marshall, H. H., Botkin, M., & Sharp, L. (1987). Pygmalion and the student: Age and classroom differences in children's awareness of teacher expectations. *Child Development, 58,* 1079-1093.

Weinstein, R. S., & Middlestadt, S. E. (1979). Student perceptions of teacher interactions with high and low achievers. *Journal of Educational Psychology, 71,* 421-431.

Weinstein, R. S., Soule, C. C., Collins, F., Cone, J., Mehlhorn, M., & Simontacchi, K. (1991). Expectations and high school change: Teacher-researcher collaboration to prevent school failure. *American Journal of Community Psychology, 19,* 333-363.

WHEN STIGMA BECOMES SELF-FULFILLING PROPHECY: EXPECTANCY EFFECTS AND THE CAUSES, CONSEQUENCES, AND TREATMENT OF PEER REJECTION

Monica J. Harris, Richard Milich, and Cecile B. McAninch

THE STORY OF DANNY THOMPSON[1]

Danny Thompson's problems began on the first day of school in third grade. Danny's teacher, Mrs. Wells, called Danny's parents, and said "I need to talk to you about Danny." Apparently Danny had poked Mrs. Wells in the rear end with his pencil. The Thompsons were shocked and said that Danny had not acted up this way before. (When they asked him about the incident, he admitted poking Mrs. Wells with the pencil but said he had done so only because she had ignored repeated attempts to gain her attention.)

Advances in Research on Teaching,
Volume 7, pages 243-272.
Copyright © 1998 by JAI Press Inc.
All rights of reproduction in any form reserved.
ISBN: 0-7623-0261-5

In the first grade Danny had been on the active side and sometimes had trouble sitting still and finishing his assignments. His teacher had suggested the possibility that he might be hyperactive, but he did well in school that year and had no problems in second grade. He was clearly a very bright young boy, and in fact the third-grade class he had started was a special program for mentally gifted children.

Over the next six weeks, Danny's teacher e-mailed Mr. Thompson at least three times and called home another two occasions. Danny was causing problems in the class, she said. He was too restless, too talkative, and would not focus on classwork. Mrs. Thompson acknowledged that he had problems with his homework as well, dawdling and taking over an hour to do an assignment that he should have been able to complete in 15 minutes. Mrs. Wells told the Thompsons that she was convinced Danny had ADHD (attention deficit-hyperactivity disorder) and urged them on several occasions to see a pediatrician about prescribing medication.

Mrs. Wells also tried various techniques she had been led to believe were effective in dealing with ADHD children. She thought Danny wasn't finishing his classwork because he was too easily distracted by the other children, so she set up a cubicle in the back of the room. Danny had to do his work in the cubicle while wearing headphones, so he could not see or hear the other children in the class. She instructed the Thompsons to supervise Danny's homework closely every night, sitting nearby and making sure that Danny stayed on task.

Around this same time, Danny began experiencing social difficulties. Because this was a special program for gifted children, Danny came into class knowing very few students and leaving his friends behind in the regular classes. Danny had always been on the shy, quiet side, preferring to read books rather than going out and playing ball with the other kids. Danny's mother was concerned about Danny's relative lack of a social life and encouraged him to call his neighborhood friends. On one such occasion Danny replied, "I just don't feel like doing that; I don't know why," and "I guess I'm just one of those people who are cut out to be a loner." The children in Danny's class mostly either ignored Danny or teased him for getting in trouble and having to sit in the cubicle.

At this point, with Mrs. Wells still encouraging the Thompsons to try Danny on a course of Ritalin, the Thompsons had Danny undergo a thorough psychological and neurological evaluation. Several weeks later the verdict from the intensive tests came in: Danny was indeed exceptionally bright, although he had a rather large gap between his verbal and performance IQ, suggesting that he might have a learning disability. This could account for some of the frustration he had encountered when doing math homework. More to the point, both the psychologist and neurologist agreed that Danny did not have ADHD and that the use of medication was not warranted. They recommended merely that Danny be given extra time on math assignments and to pursue counseling if his anxiety worsened.

So basically the Thompsons were told that nothing was really wrong after several months of worry. Mrs. Wells stopped asking for Danny to be put on Ritalin, dismantled the cubicle, and started treating Danny just like all the other students. After a couple of months the other children had stopped teasing Danny, he had made a few friends, and he was happy and doing well in school.

TEACHER AND PEER EXPECTANCIES AS A CAUSAL FACTOR IN CHILD BEHAVIOR PROBLEMS

Danny Thompson's story points to the powerful impact of teacher and peer perceptions on the behavior—appropriate or inappropriate—of a student. When Danny's teacher *thought* that Danny had a problem and was hyperactive, Danny did in fact act up. The actions of Danny's teacher led the other children in the classroom to believe that Danny was a troublemaker, and they ignored or teased him, with the end result of Danny being rejected and having no friends in the class. But when Danny was "cleared" of the ADHD charge, his behavior and experience in the classroom changed drastically. Because obviously nothing had changed about Danny himself (he never really was ADHD), the difference was most likely caused by changes in his teacher's and peers' perceptions of and behavior toward him. The goal of the present chapter is to describe theory and research on the causal role of interpersonal expectancies in children's peer interactions, focusing especially on expectancies about children with behavior problems.

Stories like Danny's occur often in schools across the country. A large number of children are diagnosed with or suspected to have emotional and behavior problems, such as ADHD; in fact, incidence estimates of ADHD range as high as 10 percent in some studies (Ross & Ross, 1982). Because federal regulations require offering the "least restrictive" learning environment for students with problems, and because most school systems have difficulty funding special education classes for students with emotional and behavior disorders, children with behavior problems today are more often mainstreamed than not.

Children with emotional and behavior problems or who are disruptive and aggressive can wreak havoc in a classroom. They encounter significant amounts of peer rejection (Asher & Coie, 1990; Whalen & Henker, 1985), and their teachers—who are often already overextended with larger than optimal class sizes—do not have the time, resources, or training to deal with them effectively. Consequently, a considerable amount of attention has been devoted to studying the classroom and interpersonal problems of children at risk for behavior disorders and to developing interventions to minimize or eliminate those problems (Barkley, 1995; Mash & Barkley, 1989).

In addition to children with diagnosed, severe emotional or behavioral problems, there are even more children like Danny whose problems do not reach clinical levels yet who are still disruptive in the classroom and/or are rejected by their peers. Every class has its share of rejected or unpopular children, and these children are at pronounced risk for adjustment problems later on in life (Burks, Dodge, & Price, 1995; Kupersmidt, Coie, & Dodge, 1990; Parker & Asher, 1987). Understanding and treating the social skills problems of unpopular and rejected children is therefore also a pressing concern in the field (Schneider, Rubin, & Ledingham, 1985).

The traditional approach in the clinical literature has been to focus on the child as the locus of all his or her difficulties, an approach termed the "deficit hypothesis" (Asher & Renshaw, 1981; Hymel, Wagner, & Butler, 1990). However, the classroom is a *social* setting, and the problems that these children encounter are *interpersonal* in nature. Thus we believe it is both desirable and necessary to look at the contributions of the teacher's and peers' thoughts and behavior when trying to understand the behavior problems and rejection of a particular child. Indeed, the focus of this chapter is to present the argument that teachers' and peers' expectations and beliefs about a target child can be a direct *cause* of subsequent interpersonal problems. In other words, teacher and peer expectancies about the disruptive or unlikable behavior of a student can act as self-fulfilling prophecies.

We begin by offering a brief definition of the self-fulfilling prophecy construct and a history of expectancy research in the classroom. We follow with a review of social developmental research that offers indirect support for the importance of peer expectancies in determining target behavior. We then review at length our research program demonstrating causal effects through the experimental manipulation of peer expectancies and the observation of their effects on target self-concept and behavior. We conclude by discussing how research on expectancy effects can be used to help explain the difficulties involved with developing effective interventions for rejected children.

INTERPERSONAL EXPECTANCY EFFECTS IN A CLASSROOM CONTEXT: A BRIEF HISTORY AND REVIEW

Readers of this volume are presumably familiar with the history of research on teacher expectations, so our background here will be brief (see Jussim, this volume, and Harris, 1991 for reviews). A self-fulfilling prophecy, or interpersonal expectancy effect, occurs when one person's (the perceiver) originally false beliefs about another person (the target) lead the perceiver to act in such a way as to elicit the expected behavior from the target (Darley & Fazio, 1980; Merton, 1948). Expectancy effects have been documented in a wide range of settings and have been shown to be of practical importance (Rosenthal, 1994; Rosenthal & Rubin, 1978).

The domain that has attracted the most research attention and controversy is that of the classroom. This area had its start with Rosenthal and Jacobson's (1968) classic *Pygmalion in the Classroom* study where teachers' expectations were experimentally manipulated and shown to affect pupils' later academic performance. Although the results of that study were initially greeted with great skepticism and criticism, subsequent research demonstrated convincingly that teachers' expectations about their students' academic performance can act in a self-fulfilling manner across a broad range of grades, academic subjects, and dependent variables (Dusek,

1985; Rosenthal, 1994; Rosenthal & Rubin, 1978). Teacher expectancy effects have been found to be stronger at earlier grade levels (Rosenthal & Jacobson, 1968) and when teachers and students have been acquainted for shorter durations prior to the manipulation or measurement of the expectancy (Raudenbush, 1984).

Studies of the mediation of teacher expectancy effects have helped to illuminate the processes by which teacher expectancies operate as self-fulfilling prophecies. Meta-analyses of the expectancy mediation literature have identified a large number of behaviors differentially expressed by teachers holding positive or negative expectancies (Harris & Rosenthal, 1985, 1986). These behaviors can all be categorized along the lines of Rosenthal's affect-effort theory (Rosenthal, 1989): Teachers holding positive expectancies about a student will display a more positive socio-emotional *affect*, for example, smiling more and displaying greater nonverbal warmth in their voice and facial expressions. Similarly, teachers holding positive expectancies will invest more *effort* in their teaching of the student, for example, teaching more material, spending more time with the student, and teaching more difficult material (Harris, 1993).

EFFECTS OF LABELS ON PEER PERCEPTIONS

In short, a great deal is known about how teachers' expectations can affect their students' academic performance. What we don't know much about is whether and how teachers' expectations about other aspects of pupil behavior (e.g., their personality or social skills) affect students and their peers, or how peers' expectations about target children affect them.

With respect to the first question, we do know that children are sensitive to a teacher's verbal and nonverbal behavior toward other students (Weiner, Graham, Stern, & Lawson, 1983). In a study that investigated the effects of teacher statements on students' impressions of a peer, White and Kistner (1992) created a videotape showing three children behaving appropriately and two children behaving inappropriately. The teacher's feedback to one of the disruptive students was experimentally manipulated, and analyses showed that the feedback affected children's judgments about the disruptive peer above and beyond that of the child's actual behavior. A follow-up study replicated these findings, especially with respect to harsh or derogatory teacher criticism (White, Sherman, & Jones, 1996). In sum, it appears that one of the avenues by which children develop negative expectancies about particular peers is by witnessing teachers criticizing them and calling attention to their disruptive behavior (Coie, 1990).

A large literature exists in the social developmental area on labeling effects, showing that diagnostic labels such as "mentally retarded" can yield substantial adverse effects on children's perceptions of the labeled peer (Hobbs, 1975; MacMillan, Jones, & Aloia, 1974). For example, Bromfield, Weisz, and Messer (1986) showed children a videotape of a target child working on a puzzle. Half of the participants were told that the target child was retarded, whereas the other participants were given no label. Analyses indicated that the label affected the children's perception of the target child's ability such that they were less likely to encourage the target to persist with the puzzle if they believed the target to be retarded.

The labeling studies cited above provide clear and convincing evidence that children with behavior and learning disabilities are stigmatized by their disorder. Other studies have shown that this derogation can occur literally within minutes (Bickett & Milich, 1990; Pelham & Bender, 1982). However, although these studies are compelling in showing the adverse interpersonal effects of possessing a stigmatizing disorder, they were not designed to and cannot demonstrate the causal effects of the stigmatizing label itself. In other words, when we read research documenting that hyperactive children are rejected by their peers, we cannot determine to what extent the rejection is caused by the hyperactive child's disruptive behavior or by the peers' awareness that something is "different" or "wrong" with the hyperactive child.

This causal ambiguity is inevitable in real-world studies that investigate peers' reactions to a labeled other. In order to investigate how much the stigmatizing label itself contributes to the interpersonal problems encountered by these children, experimental studies must be conducted where target children are randomly assigned to the label or control conditions—that is, studies using the classic expectancy effects paradigm.

However, this kind of research is strangely lacking. Prior to the experiments conducted by our laboratory (described at length below), we were able to locate only two experimental studies of expectancy effects using children as participants (Musser & Graziano, 1983 [cited in Miller & Turnbull, 1986]; Rabiner & Coie, 1989). In the Musser and Graziano (1983) study, children's expectations about the age of another child were manipulated such that the target was described as being either two years older or younger than the perceiver. The children then interacted via intercom and worked on puzzles. Analyses revealed no effect of the expectancy on the perceivers' verbal behavior; however, perceivers believing they were interacting with a younger child chose less difficult puzzles to build.

In the Rabiner and Coie (1989) study, children identified as rejected on the basis of sociometric ratings were asked to play with two other children. One week later the rejected targets were told that one of the other two children had enjoyed playing with them and was looking forward to playing with them again. The children then interacted again. Results showed that the interaction partners of rejected targets given a positive expectancy liked the target more than they did children in the control condition. Thus, when rejected children *believe* that they are liked, they will subsequently behave in ways that increase the likelihood that they will in fact be liked by their interaction partners.

Coding of the videotapes of the interaction in the Rabiner and Coie (1989) study revealed no significant differences between positive expectancy boys and control boys; for the girls, however, targets who received the expectancy induction were rated as behaving more competently and being more cooperatively engaged with the other children than were the control targets. This is consistent with a study that was conducted with adults using a similar experimental design (Curtis & Miller, 1986). This study found that people who believed they were liked engaged in more eye contact with the other person, spoke more warmly, and were more self-disclosing than those who expected to be disliked.

EXPERIMENTAL STUDIES OF CHILDREN'S INTERPERSONAL EXPECTANCY EFFECTS

The two studies described above therefore suggest that self-fulfilling prophecies can occur with children. Limitations of those studies, however, should be recognized. In the Musser and Graziano (1983) study the children did not interact face-to-face, and no effect of expectancy was found for the majority of the dependent variables. And in the Rabiner and Coie (1989) study, the expectancies involved were *self-expectancies*, not interpersonal expectancies. Thus, we still did not know whether one child's expectations about another child could lead the targeted peer to behave as expected.

In an effort to address this gap in the literature, we designed and conducted a series of studies wherein children's expectancies about a peer were experimentally manipulated and their subsequent interactions videotaped. Thus, the experiments described below possess two important strengths: (a) Because we experimentally manipulate expectancies and randomly assign perceivers to condition, we are able to show that peer expectancies have causal effects; and (b) because we have the perceiver and target children interact with one another, we are able to show effects of the expectancy on actual target self-ratings and behavior, rather than merely looking at children's ratings of hypothetical targets.

Our first study was designed to assess the impact of telling one child that the peer he would be playing with was hyperactive (Harris, Milich, Johnston, & Hoover, 1990). Participants were 80 boys between the ages of eight and 11. (We used boys only because the incidence of ADHD in girls is substantially lower.) Participants were run in dyads; because we wanted to ensure that the boys' attitudes and behavior would not be contaminated by prior knowledge of each other, we made sure that the boys were unacquainted with each other prior to the beginning of the experimental session, and they were placed in separate rooms so that they would have no contact until after the expectancy manipulation had been delivered. Participants were randomly assigned to perceiver/target roles, and perceivers were randomly assigned to the ADHD versus normal expectancy condition.

The expectancy manipulation was delivered orally by one of the experimenters. In the normal control condition, perceivers were told only their partner's name and grade. In the ADHD expectancy condition, perceivers were told also that their partner was in a special class for his behavior and that "he gets in trouble a lot for disrupting the class, talking when he shouldn't, not sitting in his chair, and acting silly." Lastly, the experimenter noted that because the target gets in trouble like this, the perceiver might have a hard time playing with him. We intentionally designed the manipulation to mention specific behaviors because we were concerned that some participants might not know or might have a wildly inaccurate conception of what "ADHD" or "hyperactive" meant.

Experimental studies of expectancy effects are only as good as the expectancies they manipulate, so we took considerable effort to ensure that the perceivers heard, understood, believed, and remembered the expectancy manipulation. Following the delivery of the expectancy manipulation, a second experimenter entered the room and gave instructions for the experimental task. The first experimenter then came back in to administer a manipulation check. To disguise the purpose of the check, perceivers were asked first what they remembered about the task instructions. They were then asked their partner's name and, in the ADHD condition, what else they had learned about their partner. The experimenter then reminded the perceiver of any of three salient symptoms (special class, disrupts class, and talks when shouldn't) that he may have failed to report. Lastly, a manipulation reminder was delivered to all ADHD perceivers approximately halfway through the experimental session. After the boys had finished the first task and had been separated, the experimenter popped into the perceiver's room and asked him who his teacher was, and then the experimenter said, "That's right; you were in (Mrs. Jones's) class, and your partner was in that special class for his behavior."

Studies of experimentally manipulated expectancies, especially when a negative expectancy is being manipulated, require special attention to ethical issues. Our research was carefully reviewed and approved by our university's Institutional Review Board. In this and the subsequent studies that will be described, we carried out several procedures designed to safeguard the rights of our participants and to minimize any adverse effects of the deception involved. First, all children were individually and fully debriefed at the conclusion of the experiment. Perceivers were told that the information we had given them about the target was false, and we explained the reason for the deception in detail. We did not conclude debriefing until we were convinced the perceiver understood fully that the expectancy manipulation was not true. Targets were told that their partners had been given erroneous information about them (although we did not explain the exact nature of the manipulation unless they asked), and they were reassured that we had told the perceiver the information was false. Following the individual debriefings, a "friendly reconciliation" of the two participants was held so that the two boys could interact in a deception-free setting. During this meeting the boys were given their payment and allowed to pick out a small toy as thanks for their participation. This invariably was met with much positive affect on the part of the boys; thus, we were able to ensure that the subjects left the experiment and each other in a good mood. In sum, we are confident that if any negative effects occurred in the targets, they were transitory, mild, and confined to the experimental situation.

Following the expectancy manipulation, the two boys were brought together and videotaped while working on an unstructured, cooperative task. The task required the boys to build any design of their choice with Lego blocks. The instructions stressed the necessity for cooperation, saying in part, "The design can look like anything you want; it's up to the two of you. But the two of you will have to decide what kind of design you want to build, and then work together to build it." Each dyad was allowed 10 minutes to complete the task.

Following the interaction, the boys were separated and asked to complete a questionnaire (using a Likert format) asking for their impressions of their partner and the interaction; resulting variables included liking for their partner, how much their partner acted out, how well they worked together, and how hard the task was. The boys were also asked a series of forced-choice attribution questions where they were asked to choose between effort, task, ability, and partner attributions for their own and their partner's performance; for example, "You said you did a good job playing this game. Was that because the game is easy for kids your age, or because you tried hard?" Finally, the videotapes were coded by raters blind to the condition of the subjects for a variety of verbal and nonverbal behaviors, such as talking, gaze, commands, global friendliness, and physical closeness.

Analyses focused on two central questions: (a) did perceivers' expectancies create a self-fulfilling prophecy in the targets?, and (b) if so, what was it about the perceiver's behavior or the nature of the interaction that influenced the target to behave as expected? To answer the first question, we compared targets of the ADHD expectancy to targets of the normal expectancy on their self-reports and behavior. We found that targets of the ADHD expectancy felt that the task was harder than targets of the normal expectancy. ADHD targets were also significantly less likely to say that they were good at the task when making attributions for their own performance. A significant expectancy by age interaction was also obtained for the global ratings of friendliness of the targets. For the younger targets (grades three and four), targets of the ADHD expectancy appeared less friendly than the control targets; this pattern was reversed for the older targets (grades five and six) such that the ADHD targets appeared to be more friendly than the control targets. This latter finding is consistent with literature suggesting that stronger expectancy effects are obtained with younger children (Rosenthal & Jacobson, 1968).

Analyses of how these expectancy effects were mediated helped to shed light on the processes by which the perceivers' expectancies affected the targets. Perceivers who believed that their partners had a behavior problem reported that the target was less good at the task, and they said they had helped the target more, than did perceivers holding the normal expectancy. When asked to name three things they didn't like about their partners, perceivers in the ADHD expectancy condition nominated traits or behaviors that were significantly more global in nature. In short, perceivers holding a negative expectancy derogated their partners' performance, took more personal credit for their partners' performance, and held more globally negative views of their partner.

With respect to the perceiver's behavior, an expectancy by age interaction for nonverbal friendliness was obtained similar to that for the targets: in the younger dyads, perceivers holding the ADHD expectancy were significantly less friendly toward the target than the control perceivers, whereas the pattern was reversed for the older dyads. Thus, it appears that the older ADHD-expectancy perceivers in our study may have actually engaged in compensatory behavior in the face of the anticipated problems posed by their partners.

Correlational analyses suggested that the negative expectancy seriously disrupted the interactions. Correlations were computed between partners for each of the behaviors coded from the videotapes; these correlations were computed separately within the normal and ADHD expectancy conditions. What we found was that behaviors were more strongly correlated between partners in the normal expectancy condition, indicating greater reciprocity of behavior for those dyads. In general, the behaviors of talking, gaze, friendliness, physical closeness, and commands tended to be highly positively related to each other in the normal expectancy condition, but they were negatively or not related in the ADHD expectancy condition. To give a specific example, perceiver's use of commands was related positively to ratings of friendliness in the control dyads, whereas perceiver commands were *negatively* related to friendliness in the ADHD dyads.

Thus, the ADHD expectancy produced interactions that were less reciprocal and, especially for the younger dyads, less positive. This withdrawal on the part of the ADHD perceivers led the targets to work harder, resulting in two unanticipated effects: The ADHD targets reported that they and their partners worked better together, and the ADHD perceivers reported finding the task easier, presumably because their partners were working harder.

In short, the Harris and colleagues (1990) study provided an encouraging first step in the experimental study of children's expectancy effects. We showed that manipulating one child's expectations that a target child had a behavior problem produced behavioral and self-concept changes in the target. However, these findings raise several important questions. First, how do the expectancy effects we obtained compare to actual effects of having a behavior disorder? In other words, we know that our ADHD expectancy manipulation disrupted the interactions among our participants, but we do not know if they were disrupted in ways similar to what would occur in interactions with a boy who truly had ADHD. Second, how would the ADHD expectancy affect interactions with a child who did in fact have ADHD? Would peer difficulties remain the same or be exacerbated?

UNTANGLING THE SEPARATE EFFECTS OF EXPECTANCY AND DIAGNOSTIC STATUS USING THE BALANCED PLACEBO DESIGN

Our next study was designed to answer the aforementioned questions by including a sample of ADHD boys in a balanced placebo design. The balanced placebo design, or expectancy control group design as it is also known (Rosenthal, 1976), represents an important methodological advance in research on expectancy effects and thus warrants a detailed description of its logic and structure. As discussed earlier, the problem with most studies of naturally occurring expectancies is that the expectancy is confounded with the actual status of the target: One has high expectations about the academic abilities of a student who possesses an IQ of 130, for example, or one expects disruptive behavior from a student diagnosed with ADHD. In such cases, it is impossible to determine to what extent it is the child's traits or the teacher's expectations about the student's behavior that is causing the relevant outcomes.

The balanced placebo design cleverly allows a researcher to untangle the separate effects of expectancy versus the target's actual status by crossing these two variables in a factorial design. For example, a teacher's expectation that certain of her students were academically gifted would be manipulated orthogonally to whether or not the students actually were gifted. Three effects are then analyzed: the main effect of the expectancy, the main effect of the target's actual status on the variable, and the interaction of expectancy and actual status.

Studies using the balanced placebo design have shown that the effects of expectancy can be just as great as the effects of actual status. For example, Rosenthal and Rosnow (1991) report the results of an unpublished study by Burnham where experimenters' beliefs that a rat had been brain-lesioned were manipulated orthogonal to whether or not the rat was actually lesioned. Analyses showed that the magnitude of expectancy effects was essentially identical to the magnitude of the main effect of lesioning. In other words, and rather astonishingly, rats who were inaccurately labeled as missing part of their brains performed just as poorly on a maze-running task as did rats who were actually missing parts of their brains! The balanced placebo design in alcohol research has yielded similar results, with people who believed they had consumed alcohol (but had not) showing similar behavioral effects as people who had actually consumed alcohol.

By using a balanced placebo design, we are thus able to assess the extent to which a normal child's expectations about a stigmatized peer exacerbate or even cause the social difficulties experienced by the target. Moreover, as discussed above, only this design permits an assessment of whether stigmatizing expectations disrupt interactions in a manner qualitatively and quantitatively similar to that of the actual stigma, in this case, hyperactivity.

Figure 1 shows the balanced placebo design as it applies to the Harris, Milich, Corbitt, and colleagues (1992) study we describe below. There were 68 perceivers, half of whom had been led to believe that their partner had a behavior problem, and 68 targets, half of whom were in fact diagnosed with ADHD. Perceiver expectancy was crossed factorially with actual target status. In other words, there were four groups: (a) ADHD targets whose perceivers had been led to believe were ADHD; (b) ADHD targets whose perceivers thought they were normal; (c) normal targets whose perceivers believed they had ADHD; and (d) normal targets whose perceivers thought were normal.

Actual Diagnostic Status

		Normal	ADHD
Peer Expectancy	Normal	N = 18	N = 20
	ADHD	N = 15	N = 15

Figure 1. An Illustration of the Balanced Placebo Design Crossing Perceiver Expectancies about ADHD with Actual ADHD Status of the Target

In other respects, the procedure of the Harris and colleagues (1992) study was similar to that of the Harris and colleagues (1990) study described above. The primary difference is that half of the targets were boys who had been recruited from a hyperactivity clinic associated with the university medical center. The ADHD boys were originally referred for behavior problems and met conventional criteria for a diagnosis of ADHD, defined as the child displaying at least eight out of 14 relevant symptoms (e.g., inattention, overactivity, impulsivity).

The expectancy manipulation was identical to the one used in the previous study, and manipulation checks and reminders were conducted in the same way. In the second study we used two experimental tasks. One was the cooperative Lego task used before, but the other task was competitive in nature. Boys were given identical color-by-number pictures of a dinosaur and were instructed to finish their pictures as quickly and neatly as possible, with the best picture winning a prize at the end of the study. To enhance the competitiveness of the task, only one box of crayons was provided for use. The boys were videotaped while engaging in the two tasks (in counterbalanced order); after each task they answered a series of questions asking about their impressions of their own and their partner's performance. The videotapes were subsequently coded for a number of verbal and nonverbal behaviors.

Analyses indicated several strong effects of both peer expectancy and actual diagnostic status. In the Lego task targets of an ADHD expectancy rated the interaction less positively than did normal expectancy targets. This was qualified by a significant expectancy by status interaction: Among normal boys, those who had been labeled ADHD found their interactions with their partners to be quite unpleasant. However, how an actual ADHD boy was labeled appeared to have less impact on his later feelings about the interaction.

As in the first study, targets of the ADHD expectancy were less likely to offer internal ability attributions for their performance; that is, they were less likely to say that they had done a good job because they were good at the task. Targets of the ADHD expectancy also were more likely to say that their dyad had not done well at the task, although this result reached only the $p < .10$ level of significance.

Analysis of the perceiver data revealed several findings that replicated the first study and showed that perceivers' attributions for their partner's performance varied as a function of their expectations. Perceivers given the ADHD expectancy were less likely to attribute their target's performance to ability than perceivers in the control condition. A significant expectancy by status interaction was found for helping attributions; perceivers interacting with an actual ADHD boy who was also given the ADHD expectancy were more likely to say that the target did well because the perceiver helped him.

Effects of the expectancy manipulation were also found for other perceiver variables. When asked to name three things they didn't like about their partner, perceivers chose symptoms that were subsequently rated more consistent with ADHD symptomatology (e.g., "he's bossy"), and their nominations were more global in nature, which again replicates the first study. Perceivers in the ADHD expectancy condition were observed to be significantly less friendly toward their partners than were perceivers in the normal expectancy condition, and there was a trend for them to spend less time talking with their partners. Lastly, perceivers who believed their partner was hyperactive were more likely to say that the task was easier than did control perceivers. Although seemingly counterintuitive, this finding replicates our first study and presumably reflects the fact that ADHD-labeled targets had to work harder at the task, making it easier for the perceivers. Alternatively, the perceivers could have reported that the task was easier because it had in fact turned out to be easier than they had anticipated given the negative expectancy manipulation, which had explicitly warned the perceivers that they might have a hard time playing with the target.

With respect to effects of actual diagnostic status, several findings were obtained showing that the ADHD boys had generally positive reactions compared to the control participants. They reported liking the crayon task and their partners more than did normal boys. ADHD boys also talked more often and were rated friendlier than the normal targets and were somewhat more likely to issue more commands. In the crayon task ADHD targets drew sloppier pictures, talked more often, and checked their partner's progress more often than normal targets.

Although the ADHD targets' self-reported experience was, if anything, more positive than the control targets' experience, main effects of target diagnostic status on perceiver data revealed that partners of ADHD boys generally had more negative interactions. Partners of ADHD boys did not like the task and their partners as much as did partners of normal boys, a finding that has been reliably shown in other studies of ADHD children's peer interactions (Barkley, 1995; Diener & Milich, 1997). Moreover, partners of ADHD boys were more likely to say that the ADHD boys were mean and had performed less well, and they were less likely to credit their partner's ability for their performance. In summary, the analysis of actual status revealed that although hyperactive boys liked their interactions better and were judged to be acting friendlier, this did not translate into positive affect on the part of their partners; in fact, the ADHD boys were rejected in comparison with the normal boys.

An important theoretical implication from the Harris and colleagues (1992) study, then, is that stigmatizing effects of a label such as ADHD may be qualitatively different than the actual effects of the label itself. Our ADHD expectancy manipulation did not make the targets more active or impulsive, although it did disrupt the interaction and led to negative consequences for the target. In short, a stigmatizing label may lead to a self-fulfilling prophecy with respect to peer rejection but not necessarily with respect to symptomatic behaviors of a given disorder.

The effects of the expectancy manipulation in our study were also weaker in magnitude than the actual behavioral differences between ADHD and normal targets. Thus we should stress that we are not arguing that the adverse consequences of a stigma are due solely to expectancy effects. Rather, our findings should be interpreted as reflecting that the expectancy process can be a causal *contributor* to the stigmatization-peer rejection cycle.

OTHER TYPES OF STIGMA: EFFECTS OF AN ACADEMIC ABILITY EXPECTANCY ON PEERS' INTERACTIONS

The studies reviewed above make a convincing case that children's expectancies about a behavior-disordered peer can seriously disrupt their interactions with and result in negative outcomes for the peer. One possible limitation of this set of findings is generalizability. We know that the ADHD expectancy manipulation wreaks fairly substantial interactional consequences, but would other, nonbehavior-oriented manipulations similarly influence peer relations? Our next study addressed this question by manipulating children's beliefs that the child they would be interacting with either was or was not particularly smart.

We chose academic ability as the basis for our manipulation for several reasons. First, the extremely large extant literature on teachers' expectancies about children's academic performance provides an interesting comparison for the results we would obtain with children. Second, children of elementary school age perform the bulk of their socializing in academic contexts. A child's academic standing therefore could very well influence his or her social standing. Third, we know that children engage in social comparison and are highly sensitive to where they stand academically in their class. Students pick up on differential teacher behavior toward high and low achievers (Weinstein, 1985, this volume), and they are acutely aware of whom the teacher considers to be the brightest and dullest students in the class. Lastly, the academic expectancy made it feasible to enroll both boys and girls in the study, allowing us to explore possible sex differences in reaction to negative expectancies.

In the McAninch, Milich, and Harris (1996) study, 36 boys and 34 girls between the ages of seven and 12 participated in same-sex, same-age dyads. Participants were randomly assigned to either perceiver or target role, and perceivers were then randomly assigned to either positive or negative expectancy conditions. The children were given instructions for the first experimental task, and then perceivers were delivered the expectancy manipulation. In the positive condition, perceivers were told their partner's name and that "his/her teacher says he/she is one of the smartest kids in the class and gets the highest grade, so he/she should do well on this game." In the negative expectancy condition, the perceiver was told that the target's teacher had said the target doesn't do very well in class, gets the lowest grades, and so may not do well on the game.

The children then engaged in two experimental tasks, each performed first individually and then together as a dyad. The first task was the NASA Moon Landing Survival Task, a task that has been commonly used in studies of team decision making and which requires the participants to rank a list of 15 items in order of their usefulness in surviving a crash landing on the moon. The second task was to create stories about pictures from the Children's Apperception Test (Bellak & Bellak, 1975); these pictures depicted children and adults engaged in various individual and group activities. Participants were encouraged to use their imagination and to make up a story about what was happening in the picture, what led up to the scene, and how the story ends. The two tasks were administered in counterbalanced order, and participants completed questionnaires following each team interaction. Videotapes of the interactions were coded for a variety of global and specific verbal and nonverbal behaviors.

Analyses indicated that telling the perceiver that his or her partner was especially smart or not smart had a large impact on their perceptions of the target's behavior and their own behavior toward the target. Perceivers in the "smart" expectancy condition reported that their partners were indeed smarter, that they worked better with their partners, and that the tasks were easier. Perceivers expecting a "not smart" partner controlled the interaction more; that is, they spoke more, made more disagreements, and behaved in a more globally dominant manner nonverbally than did perceivers expecting a "smart" partner. Thus, it is clear that the expectancies resulted in perceptual confirmation (Miller & Turnbull, 1986) on the part of the perceivers and that the expectancies affected how the perceivers treated the target. It is noteworthy that these effects were obtained despite ample opportunity during the two 10-minute interactions for the target to dispel the incorrect expectancy.

Of course, from an expectancy effects perspective, the question of greater interest is whether the targets themselves exhibited any change as a function of their perceiver's expectancies. Analyses showed evidence of an expectancy effect. Targets who were labeled "not smart" deferred more to their partners, as indexed by making more agreements, asking more questions, and looking more at their partners. Targets of the negative expectancy also were rated as more nonverbally friendly and reported that their partners did more of the talking. Interestingly, targets in the negative expectancy condition also attempted to control the interaction more, through talking and disagreeing more often and being rated as more dominant. Remember that the perceivers in this condition were more controlling themselves; the increase in controlling actions on the part of the target could therefore reflect the target's attempts to assert themselves in the interaction.

Thus, the McAninch, Milich, and Harris (1996) study provides further supporting evidence for the existence of children's expectancy effects and in a markedly different context than in the ADHD expectancy studies. The academic expectancies given the perceivers in the current study affected not only the perceivers but also the targets' subsequent impressions and behaviors.

It is interesting to note how the perceivers in the various studies were differentially affected by the academic versus behavior problem expectancies. In the ADHD expectancy studies perceivers largely withdrew from the interaction and disliked their partners to a substantially greater degree. In the academic expectancy study perceivers actually became more involved in the interactions through increased efforts at controlling what was going on. There were also divergent effects on liking toward the peer; perceivers actively disliked the ADHD expectancy target, yet there were no differences in liking between the "smart" and "not smart" targets.

The differential impact of the two types of expectancies on perceiver behavior can probably best be explained through a consideration of task demands and requirements for a smooth social interaction. The ADHD expectancy may create more negative affect toward the target because it is a behaviorally salient manipulation where the target will presumably be disruptive. Children might be more likely to dread potential interactions with such a peer because they are afraid the behavior-disordered peer would cause trouble. However, one could still have fun with and like a peer who was not very smart. Children with behavior problems are often socially unpleasant, but unintelligent people are not necessarily so.

The negative academic expectancy did lead the perceivers to assume a more controlling role in the task, but as noted above it did not translate into more negative affect, perhaps because the tasks were described as a game and there was nothing at stake. Had the academic expectancy been coupled with a context that attached more serious consequences to the team performance (e.g., if the participants were being graded as in a cooperative learning system or competing for a reward of some sort), the not-smart expectancy could very well have led to greater disliking and more interpersonally unpleasant behavior toward the target.

USING RESEARCH ON EXPECTANCY EFFECTS TO EXPLAIN WHY PEER INTERVENTIONS DON'T WORK

The history of research on social skills interventions for rejected and unpopular children has been marked by puzzling inconsistencies and dashed hopes. A wide variety of social skills training programs has been designed and implemented (Asher & Coie, 1990; L'Abate & Milan, 1985). Validation research on these programs indicated that they were actually highly successful in training rejected children to display more socially appropriate behaviors. In short, it seemed as if the problem was solved: We could take children who were ostracized and rejected by their peers, teach them to behave like popular children, and—voila—they would be rejected no more.

Much to the consternation of researchers in this area, it wasn't that easy. Although rejected children could be reliably taught to improve bothersome aspects of their behavior with peers, they remained just as rejected as before (Hymel, Wagner, & Butler, 1990; Krehbiel & Milich, 1986). In short, as Bierman, Miller, and Stabb reflected, "Improving the social behavior of rejected boys is easier than improving their reputations or increasing peer acceptance" (1987, p. 199). Thus the question becomes, why is it that improvements in interpersonal behavior do not lead to improvements in social standing?

Answers to this question can be found in the program of research by Shelley Hymel and her colleagues (Hymel, 1986; Hymel, Wagner, & Butler, 1990). Essentially, Hymel argues that the social skills deficit perspective attacks only one part of the peer rejection process—the maladaptive behaviors of the target. However, inadequate social skills may not be the only or primary cause of peer rejection; instead, the contribution of the peer group needs to be considered. Hymel argues that children's reputations create strong impression formation biases, as shown in biased evaluation of behavior, selective recall of reputation-congruent behavior, and biased attributional interpretations (Hymel, Wagner, & Butler, 1990). For example, in one study Hymel (1986) found that children display a marked positivity bias with respect to liked children. That is, popular children are given the benefit of the doubt: They are assigned internal attributions for positive behaviors and less blame for negative behavior. Positive behavior on the part of a disliked target, however, may be perceived as unstably caused and therefore does not change the peer's liking for him or her.

Research on interpersonal expectancy effects can shed light on the processes through which negative peer impressions persist despite improvements in the rejected child's behavior. After all, the situation is directly analogous to when a perceiver's negative expectancy persists despite contradictory behavior on the part of the target. Our laboratory has conducted a series of studies that focus more closely on what happens to a perceiver's expectancy or impression as it is confronted with expectancy-consistent or inconsistent material.

In our first experiment (McAninch, Manolis, Milich, & Harris, 1993) we manipulated target child gender and expectancy about the target's personality (shy versus outgoing). Participants were 64 boys and 50 girls between the ages of eight and 12. The expectancy manipulation was delivered both orally and in writing; the children were told that they were going to be watching a videotape of another child about their age and were given general information about the target's name, grade, and school. Then, in the shy expectancy condition participants were told that the target "is not very outgoing and does not have many friends." In the outgoing expectancy condition the target was described instead as "very outgoing and has lots of friends."

The children then were asked to rate their impressions of the target based on what they had just heard, and then they were shown a videotape presumably of the target child. The tape, which we had prepared using actors recruited from a children's theater group, showed either a male or female target describing him- or herself on videotape. The scripts addressed topics such as hobbies and family background and were written to include six outgoing phrases (e.g., "I made a lot of good friends right away"), six shy phrases (e.g., "My favorite subject in school is reading because I like to work by myself"), and 14 neutral phrases (e.g., "My mom's a teacher and my dad sells insurance").

Following the videotape participants completed two measures, administered in counterbalanced order. One was an impression questionnaire similar to the one they completed prior to the videotape, and the second measure was a free recall task. The children were asked to write down as much as they could remember from the videotape. The recall protocols were later coded by research assistants for the proportions of correctly recalled shy, outgoing, and neutral statements. We also coded the number of errors (information incorrectly recalled) and intrusions (recall of information that was not in the original script).

Analyses revealed strong expectancy by time interactions for the ratings of how friendly and shy the target appeared. For both ratings there were highly significant and strong main effects of expectancy on the post-manipulation, pre-videotape rating, indicating that the outgoing target was indeed considered more friendly and less shy than the shy target. Following the videotape, however, these differences disappeared, and the target was rated equally friendly and shy regardless of preexisting expectancy, which is how the target should have been rated considering that all participants were viewing the exact same male or female videotape. Thus it appears that children are able to disregard their initial expectancies and rely more heavily on the individuating information available in the videotape when reporting their final impressions of the target.

Although at first these findings would seem to do little to explain the mystery of why peers' impressions of rejected children do not improve after social skills interventions, the findings of the next analysis helps bring clarity to this problem. Participants were also asked post-manipulation and post-videotape how much they *liked* the target. The results for this variable were extremely interesting. For the ratings taken directly after the expectancy manipulation was delivered, a strong main effect of expectancy was obtained, such that targets of the shy expectancy were liked to a significantly lesser degree than targets of the outgoing expectancy. More interesting, there was also a significant difference between expectancy conditions for the liking ratings obtained after the videotape. The effect of expectancy was diminished at the second rating though still moderate in magnitude, $r = .27$ versus $r = .48$ for the first rating.

Thus, even though children adjusted their descriptions of the target's behavior to accommodate the expectancy-incongruent material they witnessed on the videotape, their subjective *affect* toward the target remained significantly affected by the expectancy. And that may explain why it is that social skills interventions do not succeed in making rejected children better liked by peers. The peers may very well notice and attend to the changes in the rejected child's *behavior*, even to the extent of affecting their impressions when asked specific questions about the target's behavior, but their subjective *preference* for the target doesn't change.

The analyses of the free recall data indicate possible memory distortions that could be contributing to the persistence of the expectancy. Participants in the shy expectancy condition made significantly more errors in recall than did participants in the outgoing expectancy condition. They also made significantly more intrusions of a shy nature, meaning they were more likely to remember that the target had said and done shy things that were not actually in the videotape. In addition, children given the shy expectancy were less confident in their impressions of the target than children given the outgoing expectancy.

McAninch (1997) replicated and extended these findings, adding two refinements: (a) A no-expectancy control group was included in addition to the shy and outgoing expectancy conditions, thus allowing a determination of whether giving any sort of label at all affects impression formation; and (b) participants' self-schemas for introversion/extraversion were assessed, to allow a determination of how one's own standing on a particular domain affects how one reacts to an expectancy about a peer on the same domain.

In other respects the procedure was similar to McAninch and colleagues (1993). Participants were 110 boys and 77 girls between the ages of seven and 12. As in the previous study, participants received an expectancy manipulation that led them to believe that the target was either shy or not shy at all, or it stated nothing at all about the target's personality but only mentioned name, age, and grade in school. Participants gave their impressions of the target and then watched the same videotape as used in the previous study that included a mixture of outgoing, shy, and neutral phrases. Following the videotape the children once again rated their impressions of the target and reported everything they could remember from the videotape. Participants also completed the Shyness Self-Report (Lazarus, 1980), a measure of shyness designed specifically for children, along with other measures of shyness and schematism for shyness (Fong & Markus, 1982).

McAninch (1997) included several new dependent measures that were not in the previous study. Following the post-videotape measures, participants completed an on-line recognition task for phrases from the videotape script. Shy, outgoing, and neutral phrases were selected from the script as well as an equivalent number of distractor phrases *not* from the script but were matched for shy, outgoing, and neutral content. The phrases were shown in random order on a computer, with two irrelevant phrases included at the beginning and end to control for practice and fatigue effects. Participants were instructed to read each phrase as it was presented and to press a key indicating whether or not the child on the tape had said the phrase.

Lastly, participants were asked to write down why they did or did not like the target. These responses were coded by judges as to whether they referred to the target's personality, physical appearance, or perceived similarity to the target. Participants were also asked to write down why the target was either shy or outgoing; these responses were rated on a nine-point scale for overall positivity by judges.

Analyses replicated the McAninch and colleagues (1993) study, with significant expectancy by time interactions obtained for nearly all the impression ratings. Following the expectancy manipulation but prior to the videotape, participants rated the outgoing target significantly less shy than the shy target (with the neutral target rated in the middle), and they were more certain of their impressions. After the videotape, however, differences on those two dependent measures disappeared, indicating that the children were able to adjust their impressions to take into account the information they received on the videotape.

However, analyses of other dependent variables revealed that although differences among expectancy conditions diminished following the videotape, they remained significant. Specifically, participants rated targets in the outgoing expectancy condition as being more popular and more similar to themselves relative to participants given a neutral or shy expectancy. The popularity result is particularly interesting, as it suggests that the children recognized that relative social standing is slow to change in response to behavioral changes in a targeted peer.

Moreover, there was only a significant main effect of expectancy and no expectancy by time interaction for children's ratings of how much they liked the target. The target was liked to a significantly greater degree in the outgoing condition than in the shy condition (with the neutral condition falling in the middle) both immediately after the expectancy manipulation as well as following the videotape. A significant main effect of expectancy was found also for the ratings of the children's responses to the open-ended question of why the target was shy or not; children given the outgoing expectancy were more positive in their explanations than children given the shy expectancy. Thus, consistent with the McAninch and colleagues (1993) findings, children did not change their subjective liking for a target child in spite of adjustments in their more objective ratings of the target's behavior.

With respect to the memory data, only one significant effect of expectancy was obtained, but it is a direct replication of the previous study: Participants given the shy expectancy made significantly more intrusions of a shy nature than did participants in the other two expectancy conditions. None of the other memory measures was significant, however, indicating that the expectancy exerted strong effects on children's impressions of a target peer but only minimal effects on their memory.

There are a couple of possible explanations for this disparity in effects. The first is that the salience and the brevity of the videotape (it lasted about two minutes) led to ceiling effects for the recall measures, with most participants recalling the script with high rates of accuracy, and this is especially true for the on-line recognition task. This explanation becomes less likely in light of the fact that significant effects on memory were obtained for the self-schema variables (described below). A more theoretically oriented (and more plausible) explanation is that expectancies may have their greatest influence on the most subjective or affect-related judgments.

Across both studies the strongest effects of expectancy were found for liking judgments, then for trait assessments, and the weakest effects were found for the more cognitively based variables of memory for specific statements. In other words, perhaps people are more likely and able to strive for objectivity and disregard preconceptions when making relatively more cognitive judgments of what a target has said or how he or she is behaving, but they cannot or are not motivated to disregard their prior feelings when arriving at a subjective judgment such as degree of liking for the target individual.

One of the enhancements in the McAninch (1997) study was to determine if children's schematism for the manipulated trait interacted with expectancy to affect their judgments of the target. People are considered *schematic* for a trait if they rate themselves highly on the trait *and if* they rate that trait as being highly important to their self-concept; people who do not rate themselves high on a trait or for whom the trait is not particularly important are considered *aschematic* (Fong & Markus, 1982). Several interesting effects of self-schemas were obtained on the rating and memory data. For example, participants who were schematic for extraversion liked the target of the shy expectancy less than did the aschematics, and they remembered more outgoing items of information about the target of the outgoing expectancy. Children who received an expectancy manipulation that was consistent with their self-schemas had higher total recall of videotape script items and their recognition latencies were shorter.

The self-schema results point to the importance of considering how important a domain is to a child in predicting his or her judgments of peers with respect to that domain. If a target expresses behavior on a dimension that a child considers important, it is likely that the perceiver would pay greater attention to and expend more effort in processing information about that person. This conclusion is consistent with what has been predicted and found by other social psychological theories, particularly Fiske and Neuberg's (1990) model of categorical versus individuating processing and the Elaboration Likelihood Model (Petty & Cacioppo, 1986).

To summarize the main implications of the McAninch (1997) study, further support was found for the notion that perceivers who are given a prior expectancy about a target and then view target behavior that contains both expectancy-congruent and expectancy-incongruent elements will make appropriate adjustments in their more objective judgments of the target's behavior and personality, but their liking for the target is still significantly affected by the expectancy. You might find yourself more convinced of the intuitive appeal of this finding if you think about your reaction when an individual whom you intensely dislike does something uncharacteristically generous; you are probably quite surprised, and you may grudgingly give him or her credit for the action (simultaneously wondering what ulterior motives might account for it), but you will probably still dislike the individual. It takes a very long time to erase a negative impression of someone, regardless of how pleasantly he or she might be behaving.

In the two studies just described the expectancy manipulation was always followed by a videotape segment showing the target child making expectancy-congruent, incongruent, and neutral statements. In the most recent study to come from this line of research (McAninch, Joffrion, Downs, & Randolph, 1997), the procedure was extended so that participants viewed a second videotape showing the target behaving either cooperatively or disruptively. In other words, the prior studies focused on how ambiguous evidence was interpreted by children; this particular study allowed a determination of how children reacted when provided evidence that unambiguously either confirmed or disconfirmed the expectancy.

To sum up the major findings of this study, strong effects of expectancy were found following the manipulation (consistent with the prior studies) but disappeared following the first and second videotapes. This is not consistent with the past research and may reflect the fact that new videotapes had been prepared that showed the target behaving very much like a child with ADHD, fidgeting and squirming constantly. The behavior was evidently so salient that it made children derogate the target, regardless of expectancy condition. This interpretation is supported by the fact that ratings of the ADHD-expectancy target remained unchanged following the two videotapes, but ratings of the neutral-expectancy target became more negative, leading to no differences between the groups. These findings therefore point to the stigmatizing power of disruptive behavior, even in as mild a form as fidgeting and squirming. Analyses of the recall data revealed that recall was highest for those children who saw a second videotape disconfirming the initial expectancy, further evidence that children do in fact attend to expectancy-inconsistent behavior even if that does not change their subjective liking for the target.

PRACTICAL SUGGESTIONS FOR IMPROVING INTERVENTIONS WITH REJECTED CHILDREN

To recap the major points covered so far: Our research has shown that children's expectations about a stigmatized peer can act as self-fulfilling prophecies, and we have found that the stigma persists even in the face of objective behavioral improvement on the part of the target. The bottom-line question thus becomes how can these findings be applied to improve treatment of and interventions with rejected children?

Coie (1990) makes the important distinction between the *emergent status* phase and the *maintenance* phase of peer rejection, and he notes that different processes operate at each of these two phases. That is, children's social status may fluctuate quite markedly between rejection and acceptance for a while before stabilizing, and it is instructive to consider the factors that determine when a child first encounters peer rejection as distinct from the factors that maintain the rejected status once established. This distinction is a useful one to make when considering interventions as well.

With respect to the emergent status phase, the key would be to identify children at risk for rejection and to intervene before their rejected status stabilized. We have argued that some degree of peer rejection may occur because children pick up on a stigmatizing label of a peer and react negatively. The label could be generated by overt characteristics or behavior of the target (e.g., obesity or aggressive behavior) or through observation of more subtle features such as teacher statements or behavior toward the target.

One approach to intervention, therefore, would be to minimize children's awareness of the existence of a potential stigma in a targeted peer. While this might prove impossible for stigmas caused by highly visible, salient physical attributes such as obesity or facial disfigurements,[2] in such cases interventions could focus on helping the stigmatized child to prepare for the negative reactions he or she is likely to incur and to generate socially adaptive responses to them. Our laboratory has recently begun a program of research looking at children's responses to teasing, and we have found that arming a target child with a humorous response to the kind of taunts they are likely to encounter is considered by children to be more effective than traditional responses to teasing such as ignoring (Scambler, Harris, & Milich, in press). For example, a child who is teased for having been held back in school a year can be trained to reply, "I'm not dumb; I just got the chance to learn the stuff twice as well as you did."

In many cases, however, the stigma that provides the basis for eventual peer rejection is not immediately obvious to a child's peers; learning disabilities would be a good example. In these cases, a little increased sensitivity on the part of educators and parents could go a long way in minimizing the negative effects of the stigma. We have described earlier how children are acutely aware of the things teachers say and do to their students. Consequently, teachers should make efforts not to call attention to a child's disorder or disability. For example, in Danny Thompson's story, the teacher had him sit in a cubicle in the back of the room. This is frequently done, but no research exists to document its effectiveness in getting children to stay more on task. It does, however, send a telling message to the other children in the class that this particular student has a problem.

As another example, the common practice of telling a child with ADHD to go take his or her medication (or, worse yet, having an announcement come over the school loudspeaker) should be discontinued. Instead, children on medication should be instructed to excuse themselves quietly at the appropriate time, with teachers providing unobtrusive reminders if needed. Even more desirable would be for children with ADHD taking Ritalin to switch to the now-available time-release version of the medication, so they would not need to leave the class at all (Barkley, 1995). Although these recommendations may sound trivial, interviews with ADHD children indicate that they intensely dislike being singled out to take their medication; Barkley (1995), for example, provides the poignant story of a 15-year-old ADHD boy who tells of the classroom reaction when his physician increased his medication: "Everyone in the room would say, 'Dummy has to go take his pill' " (p. 208).

The changes we recommend here are easy to introduce yet can make a big difference.

The point is that any contact a child with behavior or emotional problems has with mental health personnel in the school has the potential to evoke the self-fulfilling cycle of stigmatization (Landau, Milich, & Diener, 1998; Milich, McAninch, & Harris, 1992). Obviously we are not recommending that children with problems therefore should not receive treatment for them; rather, we are stressing the need for teachers and school psychologists to demonstrate heightened sensitivity to these issues and to avoid unnecessarily calling attention to the child's treatment.

This recommendation has implications for the training of teachers, school psychologists, and counselors. There is considerable variability across curricula with respect to what extent the basic social psychological processes of expectancy effects and stigma are addressed. It goes without saying that teachers should be highly familiar with the teacher expectancy literature and educated in ways to avoid establishing and communicating inaccurate negative expectancies. More generally, there is a need for better training for both teachers and school psychologists in how to deal with children who are rejected and/or have behavior problems. Rather surprisingly, a recent master's thesis demonstrated that the average teacher has a better understanding of preventive classroom management strategies than do school psychologists (LaRoque, 1997). In accordance with Kurt Lewin's maxim that "there is nothing as practical as a good theory," we believe that a solid grounding in social psychological research and theory on interpersonal and small-group processes would help to improve the services provided by teachers and school psychologists.

With respect to the maintenance phase of peer rejection, the expectancy effects perspective would lead one to conclude that it is easier to intervene at the early stages of stigmatization and peer rejection than it is to change a child's reputation once it has become firmly entrenched. The problem, as we have seen, is that even if you can change the problematic behaviors that led to the target child becoming rejected, he or she will still continue to be rejected. In short, most social skills interventions to date have concentrated on training the rejected peer to behave in more socially appropriate ways; however, the theory and data described in this chapter would suggest that, to be successful, interventions need also to address the perceptions of the peer group.

Considerably less attention has been paid to this possibility, although there have been some efforts to involve peers actively in interventions (e.g., Bierman & Furman, 1984; Vitaro & Tremblay, 1994). In one study (Bierman & Furman, 1984), Gordon Allport's prescription for reducing prejudice—equal status contact in the pursuit of a common goal—was compared as one form of intervention against a more traditional social skills deficit type of intervention. In the peer involvement intervention, rejected children worked with classmates together on a group task of producing videotapes. Analyses showed that the social status of the rejected children improved only for those who participated in the peer involvement intervention; however, this effect was found only at the immediate posttesting and was not present at the follow-up testing. Subsequent studies confirmed that traditional social-skills training was more effective when implemented in the context of a peer involvement intervention (Bierman, 1986; Bierman, Miller, & Stabb, 1987).

Interventions that include a peer involvement component therefore represent a promising avenue to pursue when intervening with children whose rejected status is already well-established. However, it should be acknowledged that these types of interventions are not without risk. In order for cooperative learning programs to improve peers' impressions of a rejected child, the learning endeavor needs to succeed. But that may not always happen. Rejected children often have different agendas, and a child with ADHD, for example, may prefer to goof off or clown around in order to gain attention rather than quietly cooperate. This child may indeed receive more attention, but it will be of a negative sort and will not improve his or her social standing. McCaslin and Good (1996) note that the "literature is replete" with examples of cases where students are cruel to other group members in small group situations (p. 129). They further acknowledge that small groups may therefore help children to "recognize and cope with social frustration and conflict" (McCaslin & Good, 1996, p. 111). The problem, of course, is that stigmatized children are more vulnerable to victimization by others and less equipped to do such coping.

Another possibility is that the rejected child may not have the requisite skills and abilities to perform well in the cooperative task. Significant disparity in abilities across students in small groups can set up a "caste system" where the lower-ability student is relegated to a passive role (McCaslin & Good, 1996). When the rejected child is of lower ability, then, the very setting that was designed to improve the target's social status becomes instead a lens that magnifies the target's deficits and can even increase the amount of teasing and rejection experienced. Roger Brown (1986) raises this possibility in explaining the relative lack of success of the "jigsaw classroom" technique (Aronson et al., 1978) in reducing prejudice, stating "the very interdependence that is the point of the jigsaw plan would become a force for group hostility, so the jigsaw classroom contains, I think, a 'self-destruct' feature" (Brown, 1986, p. 620).

In sum, if the rejected child does not want to cooperate or if he or she tries to cooperate but is not capable at the task, the forced interaction involved in small group settings could backfire and result in even more negative attitudes by the peers. Thus, teachers or therapists wishing to use cooperative peer interactions as an intervention face the delicate challenge of designing a group task where all involved are guaranteed to succeed yet is not so simplistic as to alienate or bore the children. Although it may seem we are belaboring this point, we feel it is important to emphasize the disadvantages of group learning interventions because they are currently so heavily favored by educators. Teachers and administrators would do well to weigh carefully the risks and disadvantages of small-group interventions against the purported benefits (see Good, Mulryan, & McCaslin, 1992; McCaslin & Good, 1996 for reviews) before deciding to venture forth with them.

DIRECTIONS FOR FUTURE RESEARCH

The research described in this chapter could be extended in several important ways. First, although we continue to believe in the importance of studying experimentally manipulated expectancies, such experiments could and should be conducted in more naturalistic settings than the laboratory. A more realistic picture of how children react to stigma in everyday situations might be found if the children are not acutely aware that they are in a psychological experiment and facing a videocamera 10 feet away.

A second important direction would be to investigate the effects of stigmatizing expectancies within the peer group context. Our research has focused on interpersonal expectancies within dyads. However, much peer rejection and teasing takes place in group settings such as the classroom or play group. It would be useful and important to know whether the effects of expectancies are exacerbated or mitigated by group interaction. Two promising avenues of research can be proposed to explore these possibilities. First, experiments could be conducted in which a child perceiver's expectancies are manipulated about a target within a dyadic interaction in the usual method, but then the two children are placed in a subsequent larger group interaction. Such a study would permit the assessment of the degree of generalization of the stigmatizing expectation: Does it carry over and bias the perceptions of the other group participants? Or would interacting with other children who are unaware of the stigmatizing expectations provide sufficient opportunities for the target to negate the expectancies and improve the perceiver's impressions of him or her?

A related possible study would be to examine group interactions where two of the children are randomly assigned to the perceiver/target roles, and the perceiver is given the expectancy manipulation prior to the group interaction; that is, all interaction would take place in a group setting. As in the preceding study idea, the goal would be to assess the social transmission of stigmatizing information, with the data of primary interest being change in both the bystanders'(who are kept unaware of the perceiver's expectancy) perceptions and behaviors toward the target as well as changes in the target him- or herself.

A third direction for future research would be to take a more developmental and longitudinal perspective. We know that peer groups are more important to children at older ages, so there may be important developmental differences in how children process and react to social information that might affect the expectancy process. Along those lines, it would be interesting to examine the persistence of stigmatizing expectancy effects across important life transition points, such as moving to a new city or entering middle school. Students moving away to college, for example, often seize with some relief the opportunity to shed undesirable nicknames, friends, and aspects of personal appearance such as clothing and hairstyles. In sum, longitudinal research would be helpful to confirm the prediction that people can sometimes escape a stigma through life transitions.

If this turns out to be the case, it would suggest that intervention programs should take advantage of natural transitions such as the move from elementary to middle school or from middle to high school as offering the best chances of success. In other words, we know from the research reviewed earlier that interventions do not result in improved peer perceptions despite real improvements in the rejected child's behavior. One solution, therefore, may be to place the child in an entirely new social setting following successful social skills training.

CLOSING THOUGHTS

There may be no beliefs that everybody in this country can agree with, but if there is, the belief that childhood should be a time of carefree happiness would be one of them. It would tear at the heart of any parent to see his or her child stigmatized and rejected by peers. We have tried to show in this chapter how theory and research on interpersonal expectancy effects can be fruitfully applied to the problem of peer rejection so as to understand better the causes, consequences, and treatment of rejection. The long-term adverse consequences and social costs of sustained rejection are certainly ample reason to pursue this line of research, but the look on Danny Thompson's face as he came home from school and told his mother that the other kids laughed at him when the teacher scolded him would be reason enough. Our hope is that understanding better how stigma acts as a self-fulfilling prophecy can ultimately help us discover ways to keep stigma from winnowing its way permanently into the souls of our children.

ACKNOWLEDGMENT

The ideas and research discussed in this chapter would not have been possible without the inspirational work and mentoring of Robert Rosenthal. We therefore dedicate this chapter to him.

NOTES

1. This is a true story. Names and other identifying features have been changed.
2. Some might advocate directing intervention efforts directly at the cause of the stigma. For example, on a recent daytime television talk show, an overweight boy was presented with a treadmill as a gift and was told to lose weight so he would not be teased anymore.

REFERENCES

Aronson, E., Blaney, N. T., Stephan, C., Sikes, J., & Snapp, M. (1978). *The jigsaw classroom*. Beverly Hills, CA: Sage.

Asher, S. R., & Coie, J. D. (1990). *Peer rejection in childhood*. Cambridge: Cambridge University Press.

Asher, S. R., & Renshaw, P. D. (1981). Children without friends: Social knowledge and social skills. In S. R. Asher & J. M. Gottman (Eds.), *The development of children's friendships* (pp. 273-296). New York: Cambridge University Press.

Barkley, R. A. (1995). *Taking charge of ADHD: The complete, authoritative guide for parents*. New York: Guilford Press.

Bellak, L., & Bellak, S. S. (1975). *Children's apperception test*. Larchmont, NY: C.P.S. Inc.

Bickett, L., & Milich, R. (1990). First impressions formed of boys with learning disabilities and attention deficit disorder. *Journal of Learning Disabilities, 23*, 253-259.

Bierman, K. L. (1986). Process of change during social skills training with preadolescents and its relation to treatment outcome. *Child Development, 57*, 230-240.

Bierman, K. L., & Furman, W. (1984). The effects of social skills training and peer involvement on the social adjustment of preadolescents. *Child Development, 55*, 151-162.

Bierman, K. L., Miller, C. L., & Stabb, S. D. (1987). Improving the social behavior and peer acceptance of rejected boys: Effects of social skill training with instructions and prohibitions. *Journal of Consulting and Clinical Psychology, 55*, 194-200.

Bromfield, R., Weisz, J. R., & Messer, T. (1986). Children's judgments and attributions in response to the "mentally retarded" label: A developmental approach. *Journal of Abnormal Psychology, 95*, 81-87.

Brown, R. (1986). *Social psychology: The second edition*. New York: The Free Press.

Burks, V. S., Dodge, K. A., & Price, J. M. (1995). Models of internalizing outcomes of early rejection. *Development and Psychopathology, 7*, 683-695.

Coie, J. D. (1990). Toward a theory of peer rejection. In S. R. Asher & J. D. Coie (Eds.), *Peer rejection in childhood* (pp. 365-401). Cambridge: Cambridge University Press.

Curtis, R. S., & Miller, K. (1986). Believing another likes or dislikes you: Behavior making the beliefs come true. *Journal of Personality and Social Psychology, 50*, 284-290.

Darley, J., & Fazio, R. (1980). Expectancy confirmation processes arising in the social interaction sequence. *American Psychologist, 35*, 867-881.

Diener, M. B., & Milich, R. (1997). The effects of positive feedback on the social interactions of boys with ADHD: A test of the self-protective hypothesis. *Journal of Clinical Child Psychology, 26*, 256-265.

Dusek, J. B. (Ed.) (1985). *Teacher expectancies*. Hillsdale, NJ: Lawrence Erlbaum.

Fiske, S. T., & Neuberg, S. L. (1990). A continuum of impression formation, from category-based to individuating processes: Influences of information and motivation on attention and interpretation. In M. P. Zanna (Ed.), *Advances in experimental social psychology* (Vol. 23, pp. 1-74). New York: Academic Press.

Fong, G. T., & Markus, H. (1982). Self-schemas and judgments about others. *Social Cognition, 1*, 191-204.

Good, T., Mulryan, C., & McCaslin, M. (1992). Grouping for instruction in mathematics: A call for programmatic research on small-group processes. In D. Grouws (Ed.), *Handbook of research on mathematics teaching and learning* (pp. 165-196). New York: McMillan.

Harris, M. J. (1991). Controversy and cumulation: Meta-analysis and research on interpersonal expectancy effects. *Personality and Social Psychology Bulletin, 17*, 316-322.

Harris, M. J. (1993). Issues in studying the mediation of interpersonal expectancy effects: A taxonomy of expectancy situations. In P. D. Blanck (Ed.), *Interpersonal expectations: Theory, research, and applications* (pp. 350-378). London: Cambridge University Press.

Harris, M. J., Milich, R., Corbitt, E. M., Hoover, D. W., & Brady, M. (1992). Self-fulfilling effects of stigmatizing information on children's social interactions. *Journal of Personality and Social Psychology, 63,* 41-50.

Harris, M. J., Milich, R., Johnston, E. M., & Hoover, D. W. (1990). Effects of expectancies on children's social interactions. *Journal of Experimental Social Psychology, 26,* 1-12.

Harris, M. J., & Rosenthal, R. (1985). Mediation of interpersonal expectancy effects: 31 meta-analyses. *Psychological Bulletin, 97,* 363-386.

Harris, M. J., & Rosenthal, R. (1986). Four factors in the mediation of teacher expectancy effects. In R. Feldman (Ed.), *The social psychology of education* (pp. 91-114). London: Cambridge University Press.

Hobbs, N. (1975). *Issues in the classification of children.* San Francisco: Jossey-Bass.

Hymel, S. (1986). Interpretations of peer behavior: Affective bias in childhood and adolescence. *Child Development, 57,* 431-445.

Hymel, S., Wagner, E., & Butler, L. J. (1990). Reputational bias: View from the peer group. In S. R. Asher & J. D. Coie (Eds.), *Peer rejection in childhood* (pp. 156-186). Cambridge: Cambridge University Press.

Krehbiel, G., & Milich, R. (1986). Issues in the assessment and treatment of socially rejected children. In R. J. Prinz (Ed.), *Advances in behavioral assessment of children and families* (Vol. 2, pp. 249-270). Greenwich, CT: JAI.

Kupersmidt, J. B., Coie, J. D., & Dodge, K. A. (1990). The role of poor peer relationships in the development of disorder. In S. R. Asher & J. D. Coie (Eds.), *Peer rejection in childhood* (pp. 274-305). Cambridge: Cambridge University Press.

L'Abate, L., & Milan, M. A. (Eds.). (1985). *Handbook of social skills training and research.* New York: Wiley.

Landau, S., Milich, R., & Diener, M. B. (1998). Disturbed peer relations of children with attention deficit hyperactivity disorder. *Reading and Writing Quarterly, 14,* 83-105.

Lazarus, P. J. (1980). *The assessment of shyness in children: Preliminary instrumentation.* Paper presented at the Annual Convention of the American Personnel and Guidance Association, Atlanta, GA.

LaRoque, S. D. (1997). *Behavioral consultation and the training of school psychologists.* Unpublished master's thesis, The University of Arizona.

MacMillan, D. L., Jones, R. L., & Aloia, G. F. (1974). The mentally retarded label: A theoretical analysis and review of research. *American Journal of Mental Deficiency, 79,* 241-261.

Mash, E. J., & Barkley, R. A. (Eds.) (1989). *Treatment of childhood disorders.* New York: Guilford.

McAninch, C. B. (1997). *Effect of self-perception and prior information on children's peer impression formation.* Manuscript submitted for publication.

McAninch, C. B., Joffrion, K. R., Downs, A. M., & Randolph, J. J. (1997). *Effect of an ADHD-related expectancy and self-perception on children's peer impression formation: Does misery always love company?* Manuscript submitted for publication.

McAninch, C. B., Manolis, M. B., Milich, R., & Harris, M. J. (1993). Impression formation in children: Influence of gender and expectancy. *Child Development, 64,* 1492-1506.

McAninch, C. B., Milich, R., & Harris, M. J. (1996). Effects of an academic expectancy and gender on students' interactions. *Journal of Educational Research, 89,* 146-153.

McCaslin, M., & Good, T. (1996). *Listening to students.* New York: Harper Collins.

Merton, R. K. (1948). The self-fulfilling prophecy. *Antioch Review, 8,* 193-210.

Milich, R., McAninch, C. B., & Harris, M. J. (1992). Effects of stigmatizing information on children's peer relations: Believing is seeing. *School Psychology Review, 21,* 399-408.

Miller, D. T., & Turnbull, W. (1986). Expectancies and interpersonal processes. *Annual Review of Psychology, 37,* 233-256.

Parker, J. G., & Asher, S. R. (1987). Peer relations and later personal adjustment: Are low-accepted children at risk? *Psychological Bulletin, 102,* 357-389.

Pelham, W. E., & Bender, M. E. (1982). Socialization and peer relations in hyperactive children. In K. D. Gadow & I. Bialer (Eds.), *Advances in learning and behavioral disabilities* (Vol 1., pp 365-436). Greenwich, CT: JAI Press.

Petty, R. E., & Cacioppo, J. T. (1986). The elaboration likelihood model of persuasion. In L. Berkowitz (Ed.), *Advances in experimental social psychology* (Vol. 19, pp. 123-205). New York: Academic Press.

Raudenbush, S. W. (1984). Magnitude of teacher expectancy effects on pupil IQ as a function of the credibility of expectancy induction: A synthesis of findings from 18 experiments. *Journal of Educational Psychology, 76,* 85-97.

Rabiner, D., & Coie, J. (1989). Effect of expectancy inductions on children's acceptance by unfamiliar peers. *Developmental Psychology, 25,* 450-457.

Rosenthal, R. (1976). *Experimenter effects in behavioral research* (enlarged ed.). New York: Irvington.

Rosenthal, R. (1989). *Experimenter expectancy, covert communication, and meta-analytic methods.* Invited address at the annual meeting of the American Psychological Association, New Orleans.

Rosenthal, R. (1994). Interpersonal expectancy effects: A 30-year perspective. *Current Directions in Psychological Science, 3,* 176-179.

Rosenthal, R., & Jacobson, L. (1968). *Pygmalion in the classroom.* New York: Holt, Rinehart, & Winston.

Rosenthal, R., & Rosnow, R. L. (1991). *Essentials of behavioral research* (2nd ed.). New York: McGraw-Hill.

Rosenthal, R., & Rubin, D. B. (1978). Interpersonal expectancy effects: The first 345 studies. *Behavioral and Brain Sciences, 3,* 377-386.

Ross, D. M., & Ross, S. A. (1982). *Hyperactivity: Current issues, research, and theory* (2nd ed.). New York: Wiley.

Scambler, D. J., Harris, M. J., & Milich, R. (in press). Sticks and stones: Evaluations of responses to childhood teasing. *Social Development.*

Schneider, B. H., Rubin, K. H., & Ledingham, J. E. (Eds.). (1985). *Children's peer relations: Issues in assessment and intervention.* New York: Springer-Verlag.

Vitaro, F., & Tremblay, R. E. (1994). Impact of a prevention program on aggressive children's friendships and social adjustment. *Journal of Abnormal Child Psychology, 22,* 457-475.

Weiner, B., Graham, S., Stern, P., & Lawson, E. (1983). Using affective cues to infer causal thoughts. *Developmental Psychology, 18,* 278-286.

Weinstein, R. S. (1985). Student mediation of classroom expectancy effects. In J. B. Dusek (Ed.), *Teacher expectancies* (pp. 329-350). Hillsdale, NJ: Lawrence Erlbaum.

Whalen, C. K., & Henker, B. (1985). The social worlds of hyperactive (ADHD) children. *Clinical Psychology Review, 5,* 1-32.

White, K. J., & Kistner, J. (1992). The influence of teacher feedback on young children's peer preferences and perceptions. *Developmental Psychology, 28,* 933-940.

White, K. J., Sherman, M. D., & Jones, K. (1996). Children's perceptions of behavior problem peers: Effects of teacher feedback and peer-reputed status. *Journal of School Psychology, 34,* 53-72.

RESEARCH ON THE COMMUNICATION OF PERFORMANCE EXPECTATIONS: A REVIEW OF RECENT PERSPECTIVES

Thomas L. Good and Elisa K. Thompson

INTRODUCTION

What you do speaks so loud that I cannot hear what you say
—*Ralph Waldo Emerson*

This volume is an authoritative, important, and comprehensive review of research on the communication of performance expectations in classroom settings. It actively captures the past, present, and future of this exciting research area that is now moving into its fourth decade of extensive inquiry. The volume provides an accurate and rich history of the field, current theoretical conceptualizations, and recent empirical findings as well as useful suggestions for research in twenty-first-century schools. This collection of work has important implications for scholars and practitioners, although as in any academic field, there are conflicts, contradictions, and arguments about research interpretations. In this discussion we will illustrate the multiple strengths and limitations of these individual chapters and the field. We will also advance our own suggestions for enhancing research on the communication of classroom- and school-level expectations.

We first describe each individual chapter and comment upon its strengths and limitations. As individual chapters are presented, we also comment, when appropriate, on the relationship between the chapter and other chapters in the volume. After reviewing individual chapters, we will analyze the strengths and weaknesses of the volume as a whole. Finally, we discuss ways for responding to existing limitations in the field and advance new conceptualizations and inquiry approaches that hold potential for improving research on how expectations are communicated and interpreted in educational settings.

The review framework divides the set of chapters into three parts. The first part, *the study of classroom expectations*, analyzes three chapters (by Jussim, Smith, Madon, and Palumbo; Weinstein and McKown; and Babad). These chapters have, for the most part, dealt with the enduring issues that were most centrally stimulated by the 1968 text, *Pygmalion in the Classroom*. The second part of our review, *expectations in special contexts*, examines three chapters (Ennis; Harris, Milich, and McAninch; and Salonen, Lehtinen, and Olkinuora). These chapters extend the study of teachers' expectancy and students' mediation into new contexts—the urban school and specific types of students. The final section, *teacher efficacy*, includes two chapters (Soodak and Podell; and Ross). The focus in these chapters is upon teacher efficacy effects and our review describes potential bridges between teacher efficacy and teacher expectation conceptions. Finally, we will present several suggestions of our own for improving methodology and for expanding the scope of research on teachers' beliefs about student performance.

THE STUDY OF CLASSROOM EXPECTATIONS

Teacher Expectations: Jussim, Smith, Madon, and Palumbo

The Paradigm

The authors of this chapter weave contemporary research findings on teacher expectancies with what have become accepted "truths" about teacher expectancies in such a way as to present a moderated vision of teacher expectancy effects. Specifically, they address: the moot debate surrounding the Pygmalion study; the belief that teacher expectations are often founded on negative or biased perceptions; the special cases under which teacher expectancies might be expected to exert large effects on student outcomes; the tenability of positive teacher expectation effects; and the likelihood that negative teacher expectancy exerts cumulative, deleterious effects over time.

The Debates

Pygmalion. Jussim and colleagues commence the chapter by reviewing Rosenthal and Jacobsons's seminal Pygmalion study. They simplify the controversy surrounding this study to a debate between two diametrically opposed and inaccurate stances: teacher expectation effects exert dramatic effects on students versus teacher expectation effects as unimportant. Jussim and colleagues point out that the original Pygmalion study provided evidence against dramatic and enduring teacher expectation effects (expectations created by the experimenters), while subsequent research has amply demonstrated that teacher effects, while often not dramatic or long-standing, do unequivocally exist.

How do teachers predict performance? In their effort to dispel misunderstandings surrounding teacher expectation effects, Jussim and colleagues first address the mechanisms by which teacher expectations predict student outcomes. The query here is whether teacher expectancies predict student outcomes due to the fulfillment of negative self-fulfilling prophecy or biasing mechanism due to the veracity of these expectations. The authors suggest that the negative connotations of associations between teacher expectations and student outcomes may be unwarranted as their research indicates that teacher expectations are most frequently confirmed because they are accurate reflections of students' ability. Specifically, their research indicates that while teachers may be subject to a perceptual bias that causes them to inaccurately gauge how much effort high- and low-ability students put into classroom assignments, there tend to be high levels of correlation between teachers' initial impressions of student ability and students' final grades in the class. In other words, where teacher initial impressions are coupled with reliable and objective measures available in student files (e.g., prior grades and standardized test scores), teachers are able to accurately predict student achievement level in their initial meetings with students.

Moderators of Effect Size

Leaving behind the debate about whether the relationship between teacher expectancies and their relationship to student outcomes are always based on negative mechanisms, Jussim and colleagues address the issue of size of teacher expectancy effects. They argue that research suggests that most teacher expectation effects tend to be small. However, they acknowledge that certain student, teacher, or classroom characteristics might dramatically increase the magnitude of teacher expectation effects. Students in vulnerable groups, such as ethnic minority students, students from low SES backgrounds, and students negotiating transitions (i.e., start of kindergarten, middle school, high school) are more likely to be influenced by teacher expectations than are less vulnerable students. Being subject to multiple vulnerabilities compounds the risk. Classrooms with large numbers of students and a lack of corresponding resources may foster the type of environment in which teacher expectancy effects are likely to emerge. This may be due to the taxing cognitive atmosphere in large, undersupplied classrooms. Further, negative affect may be more likely to be expressed inadvertently when the complexity of the teaching situation increases (see Babad, this volume). Within-classroom ability grouping has also been associated with higher levels of teacher expectation effects compared to between-classroom ability grouping practices. This runs counter to the popular conception of tracked classrooms raising the salience of ability level and thus resulting in differential teacher treatment toward students in low-track classrooms. According to research conducted by the authors, between-class tracking actually reduces differential teacher treatment because it allows teachers to form more accurate impressions of student achievement potential.

Jussim and colleagues also discuss the relevance of student and teacher goals as moderators of the effect size of teacher expectation effects. Students who have high needs to be liked are more susceptible to teacher expectancies, while teachers who have high needs for control and order are more likely to form erroneous expectations. Finally, student perceptions of differential teacher treatment correlate with susceptibility to teacher expectancies. Specifically, in classrooms where the students perceive high levels of differential teacher treatment, susceptibility to teacher expectations tends to be higher. In classrooms where students perceive differential teacher treatment to be lower, the salience of teacher expectancies is lessened.

Are Expectation Effects always Negative and How Long do They Last?

Jussim and colleagues address two final issues of controversy surrounding the nature of teacher expectancies: the debates over whether teacher expectation effects are always negative and whether they accumulate or dissipate over the course of a student's career. The authors argue that their research suggests that positive expectancy effects tend to be more powerful than negative expectancy effects, but that this finding is mediated by ability levels. Specifically, both positive and negative expectancy effects are increased in magnitude for low-ability students. In addressing the accumulate/dissipate debate, they report that as a general rule, teacher expectation effects tend to dissipate over time, such that having been subject to negative teacher expectation effects at one point in time should not be predictive of student outcomes at a later point in time.

Implications

In sum, the authors present research that suggests that teacher expectations tend to be accurate, exert a small effect on student outcome, exert more influence when positively skewed than when negatively skewed, and dissipate over time. Jussim and colleagues suggest that future research address the processes behind teacher expectations. Processes that need to be examined include: the process by which high teacher expectations result in improved student performance; the role of teacher affect in terms of liking or disliking certain students as a catalyst for the formation of teacher expectations; and the role of student beliefs in mediating teacher expectations, with a special emphasis on how crucial student awareness of teacher expectations is in explaining student outcomes.

Critique of Jussim and Colleagues

Jussim and his colleagues provide an important review that presents research on new issues and establishes a clear platform for future research. The paper is weakened by the use of a false dichotomy between those who are characterized as holding a righteous indignation toward teachers and their role in creating racial inequalities and those who feel that the research is hopelessly flawed and makes little contribution. Although the authors try to dispel the dichotomy they present, the range of positions they consider is limited and misleading. Some behavioral scientists have expressed views that are strikingly different from these two positions (Good, 1969; Brophy & Good, 1970; Good & Brophy, 1997; Ennis, this volume; Weinstein & McKown, this volume). Further, some behavioral scientists have felt that teacher expectation effects were real—and important—but did not blame teachers exclusively. For example, the first author of this review has progressively become more sensitive to the difficulty of classroom teaching and places more "blame" on systemic conditions than on teachers (i.e., lack of support for teachers, the complexity of the classroom). The range of positions that social scientists hold regarding the existence or importance of teacher expectation effects is richly diverse. Although Jussim and colleagues are correct in arguing that some people hold these views, the conceptualizations derived by users of this literature are considerably more varied than Jussim and colleagues suggest.

We are puzzled by the "seamless" integration of literature from 1968-1998. Although it would have been unreasonable to expect Jussim and colleagues to undertake systemic analysis of differences in schooling from the late 1960s to the 1990s, it seems important at least to allude to potentially different contexts in classroom and societal conditions. For example, has the quality of teaching remained constant? Are teachers today more likely to have been given information about expectation effects in their teacher education programs? Is the variation in student characteristics and/or resources provided to schools greater now than it was in 1968? Has the media's perception of the quality of teachers and schooling remained constant over time? Has society's view of youth as depicted by the media remained constant over the past three decades? It would appear that emphasis on cooperative learning and teacher collaboration is greater now than in the 1960s. Do these changes impact teachers' views of student potential? Further, in the academic community, conceptions of intelligence and what counts for important learning have changed. Part of the difficulty is that time effects can be illusionary. What is true for some individuals in a given era may be notably different for others. To add further to the complexity, "counterforces" may be operative at the same time. For example, in the 1980s society's interest in enhancing students' self-esteem reached almost epidemic proportions—such that some teachers might have found it difficult to provide critical but appropriate feedback. In contrast, other teachers in the late 1980s may have been influenced more by societal messages of "grade inflation" and "get tough on students." Although we recognize the difficulty of providing a historical perspective in classrooms and teachers and societal expectations for student performance, we feel that the chapter authors could have wrestled more with possible contextual changes over time and the importance of context within a particular time period. We will return to this point in our general analysis of the chapters as a set.

We are also puzzled by the authors' assumption that simply because teachers *report* expectations for students to researchers, that they also communicate them to students. The authors' data sets are impressive in terms of the numbers of students sampled and the use of longitudinal designs that make it possible to estimate the influence of expectations over time. However, to us, this design advantage appears more illusionary than real because of the omission of essential data: teacher enactment of reported expectations (see also the chapter by Weinstein & McKown, this volume). Thus, such "black box" studies of expectations are likely to *underestimate* the effects of reported expectations. For example, consider a classroom in which a teacher reports low expectations for a group of eight low-achieving students. Teacher perception is enacted—consistently and detrimentally—with only two of them. Average performance data on the group of low-achieving students, six of whom did not receive differential teacher behavior, would seriously underestimate the effects of enacted low teacher expectation on the two students' performance. Teacher perception of low student ability in this type of research is equated with teacher low expectation for these students' learning and inappropriate teaching behavior. This research strategy seems more about researcher expectations than the phenomenon purportedly under study.

Naturalistic studies have repeatedly indicated that many teachers do not communicate differential behavior toward *groups* of students they describe as high or low expectation. Thus, many students included in Jussim and colleagues's research may not be the recipients of differential teacher attention. Although the issue that the authors raise—cumulative effects of expectations—is important, to understand longitudinal data we need to know what students experience in their classroom environments and how they perceive performance expectations (e.g., "Does the teacher think I'm smart?"). The authors systematically underestimate the value of observational measures in describing teacher expectancies as "independent variables," and this disregard for verifying the presence or absence of communicated expectations represents a regression in the field.

As social scientists interested in improving classroom learning conditions, we find the definition of expectations as "accurate" to be very limited. For instructional and philosophical reasons, many of us would want teachers' expectations to be "inaccurate" such that students who initially achieve relatively poorly, sometimes recover in later school years. At least, we would hope the percentage of "positive changers" would at least exceed the error of measurement associated with achievement tests in the early school grades—especially kindergarten and first grade. Unfortunately we know that once students are placed in a low track or group, most remain there.

We are skeptical of the authors' decision to make policy conclusions—in some areas—on the basis of a limited number of studies. Further, we are bothered by the authors' tendency to provide methodological critique of studies that they disagree with or that do not fit their overall pattern of interpretation while not quibbling with methodology used in research that supports their positions. Perhaps a degree of selective perception is involved in their sifting of evidence?

Expectancy Effects In "Context": Weinstein and McKown

A New Look

This chapter also moves away from the traditional discussions of whether or not teacher expectation effects exist, and instead addresses conditions under which teacher expectation effects are minimized or exacerbated. The perspective taken is that of the student, with an emphasis on the role students play in mediating the outcomes exerted by teacher expectation effects. Additionally, the research presented and recommendations made discourage the perspective that teachers and students resemble "black boxes" into which expectancies are inserted and outcomes produced with no intervention or negotiation on the parts of the individuals involved. In other words, this chapter abandons the search for the "existence" of teacher expectation effects in favor of delineating those variables that address the conditions under which teacher expectation effects render meaningful effects on student outcomes. The research presented in this chapter addresses how to design research-informed interventions and evaluate the effectiveness of these interventions.

Roots in the Potential-Performance Gap

Weinstein and McKown commence the chapter by discussing the theoretical underpinnings of their interactional approach to investigating teacher expectation effects. Their interest in the student's perspective and role in this phenomenon is rooted in the works of Sarason and Rabinovitch. Both of these researchers investigated the conditions under which children for whom high levels of performance were not expected (i.e., mentally retarded and learning disabled students), could be made to perform beyond their expected potential. Both emphasized the role of the teacher in facilitating the accelerated performance. Sarason emphasized the need for teachers to work in stimulating and supportive environments in order to have the resources to challenge students from whom not much was expected. Rabinovitch emphasized the need for teachers to listen to their students; to pay close attention to and capitalize on the strengths that students bring to the interaction. Weinstein and McKown's own research programs also have been concerned with promoting classroom environments that foster the development of talent. They state that while the research on teacher expectancy effects has been largely concerned with the development of student talent, the student's contribution and mediation of the phenomenon has been neglected. They believe that positive classroom expectancies can only be accomplished when students are considered co-players in the process.

The Student Perspective: What it Does to our Understanding of Classrooms

Weinstein developed the Teacher Treatment Inventory (TTI) in earlier work to test the notion that traditional teacher-observation systems of measuring teacher expectation effects fail to capture the students' experience and mediation of communicated teacher expectations. The TTI asks children to project onto a hypothetical high- or low-achieving student the types of teacher behaviors that would be directed toward this student if she or he were in the student's classroom. The difference in magnitude between the reported treatment for the hypothetical low- and high-achieving students suggests the level of differential teacher treatment occurring in the classroom. Results obtained with this measure support the implicit assumption of the chapter: Students are very sophisticated observers and interpreters of teacher behavior, often drawing on subtle nuances and inferences in their understanding of classrooms.

Eight Interactive Features of Instruction

The authors expand in important ways the traditional work in teacher expectations that focused mainly on teacher behaviors. In doing so they suggest a variety of dimensions that help to broaden our knowledge of how expectancies are communicated through structures, practices, and messages other than teacher behavior. Weinstein and McKown suggest eight ways in which performance information might be communicated to students: (1) the ways in which students are grouped for instruction; (2) the tasks and materials through which the curriculum is enacted; (3) the motivational strategies that teachers use to engage students; (4) the role that students play in directing their own learning; (5) how students are evaluated; (6) the quality of classroom relationships; (7) the quality of parent-classroom relationships; (8) the quality of classroom-school relationships.

Critique of Weinstein and McKown

Weinstein and McKown carefully attend to context and argue persuasively that school environments, teachers' educational philosophies, individual differences in students, and other variables are powerful mediators of expectancy communication. However, these authors are subject to criticisms at the other end of the continuum—that they ignore normative conditions. That is, by exploring a rich set of variables in only a single institution, it is difficult to know if any of the particular findings in this context would generate to other settings. Important next steps for these authors, and the field, would be to become more explicit in characterizing the variation in contexts at the broader ecological or school level. Are some high schools more amenable to change in expectations and relative practice? We suspect so. Is the high school the authors worked in more or less susceptible to change than other high schools with similar overt characteristics?

Further, in their strong, if not total, denial of behavioral measures, the authors may be guilty of throwing out the baby with the bath water. Ironically, it appears that these two psychologists who emphasize mediation and the need for dynamic reinterpretation appear to underemphasize the potential importance of particular behaviors (notably, primacy effects) in determining how affective patterns are formed. It has long been understood that a given teacher behavior can mean different things to different students and can have more or less effect on particular students. For example, Good in his 1969 dissertation wrote, "...students vary greatly in their thresholds for decoding information and realizing that a particular statement means 'I am a good student.' Equally apparent is the phenomena that students vary with regard to sensitivity and criticism. Some children are thick and impenetrable brick walls, seemingly unaffected by teacher comment; other children are fragile egg shells capable of being cracked by a slight frown from the teacher" (p. 24).

In essence, these and other authors in this volume interpret the earlier behavioral research findings in an extremely mechanistic fashion. Some reviewers tend to suggest that only one type or set of behaviors is involved in the communication of low expectations. Lists of ways in which research has shown teachers to communicate differentially to high- and low-achieving students are one way of thinking about expectations. There are other ways to conceptualize the process. Patterns of teacher behavior directed toward particular students are an important part of how students conclude that a teacher is "fair" or not, or infers that the teacher expects students in general or only a select set of students to perform academically. We know that some teachers do not differentiate negatively in their behavior toward students even though they may report that some students are performing at higher levels than are other students. Further, it is the case that those teachers who *do* differentiate pejoratively in their behavior toward less capable students may do so in *different* ways. For example, some teachers gratuitously praise low-achieving students in ways that indicate low performance expectations; whereas, other teachers communicate low expectations by criticizing low-achieving students disproportionately more often for incorrect answers than is the case for high-achieving students. In some classrooms, low-achieving students get frequent but easy questions; whereas in other classrooms, low-achieving students receive infrequent, easy questions and feedback that is too critical.

It seems important to begin to conceptualize and measure the cluster of teacher behaviors, classroom assignments, and environmental variables that operate in a particular context. As one starting point, it seems critical to recognize that the "independent variable," communication of teacher expectations, assumes many different forms. As a case in point, Weinstein and McKown are forceful in their critique of interventions based upon changing teacher behavior, particularly the Teacher Expectations and Student Achievement (TESA) intervention program. Part of the difficulty with the TESA (1993) intervention is that it teaches teachers only one way to respond to improve communication with low-achieving students (call on all students equally, praise equally). If indeed these issues are a problem in a teacher's classroom, then such advice might be helpful to this teacher. In contrast, if these behaviors are not the problem (i.e., the teacher's expression of low expectations takes a different form), the advice/treatment is not going to be very helpful. Further, this treatment may be counterproductive because those teachers who have reasonable patterns of interactions with high- and low-achieving students and who are given inappropriate advice (call on low-achieving students more frequently) may disrupt reasonable patterns of classroom behavior if they follow the advice. Thus, a problem with the TESA intervention, and other work based on the simple categorization of behaviors, is the assumption that there is a stable "set" of behaviors that needs to be altered. We suspect that the TESA program is valuable for many teachers in helping them to increase their awareness of potential communication problems with low-achieving students. However, if some teachers conclude after being exposed to the TESA program that there is only one poor communication model, then the program would be ineffective for these teachers.

Preferential Affect: The Crux of the Teacher Expectancy Issue: Elisha Babad

Babad suggests that the crux to understanding the processes mediating teacher expectancy effects lies within the domain of differential teacher affect toward high and low expectancy students. Babad divides the construct of teacher expectancies into two teacher behavioral patterns: teacher learning support and teacher emotional support. Babad points out that contemporary expectancy research has demonstrated that differences in teacher learning support are appropriate when they are expressed as increased learning support for low expectancy students. Additionally, Babad points out that most teachers are aware of the appropriateness of directing extra learning support toward low expectancy students and this is now a common teacher behavior.

Problems arise, though, when examining the effects of differential teacher emotional support. In current thinking, it is assumed to be appropriate that all students are offered similar degrees of emotional support (i.e., the emotional climate of the classroom is similar for all students). According to Babad, the most common outcome is that teachers provide a more positive classroom climate for high expectancy students than for low expectancy students. This results in a classroom environment in which even if low expectancy students receive needed and appropriate extra learning support, they do not enjoy the same positive emotional climate that high expectancy students do. This is a very difficult problem to address because while students are likely to perceive and report differential classroom climate, teachers are unlikely to believe that they treat high and low expectation students differentially. Part of the reason for this is that teachers often make a very concerted effort to respond to all students in a similar manner. However, emotional support tends to be expressed in subtle ways that are often difficult for the teacher to observe, but easily detected by students.

Another classroom phenomenon involving differential teacher learning and emotional support is the occurrence of teachers' pets. Again, this differential teacher behavior is more clearly perceived, or at least more openly reported, by students than by teachers. The teacher's pet phenomenon causes low expectancy students not only to receive less emotional support but also less learning support as well.

Babad discusses the classroom level results of differential teacher learning support/emotional support given to high and low expectancy students and teacher's pets. In classrooms where low expectancy students receive higher levels of learning support and emotional support, the class as a whole exhibits higher levels of morale, more positive reactions to their teacher, a stronger willingness to continue with this teacher in future grades, and less satisfaction with substitutes. In terms of future research in this area, Babad suggests that the emphasis switch from cognitive outcomes to social and affective outcomes and that research be directed toward identifying individual characteristics of teachers, students, and classrooms where teachers' pets emerge.

Critique of Babad

Babad has effectively argued the importance of teacher affect in communicating expectations with students. However, Babad, like Weinstein and McKown, expresses little interest in teacher behaviors—at least those that were studied in the 1970s. He argues that all differential teacher behaviors then were considered negative and that the monolithic norm called for equal treatment of students even when low-achieving students (as well as high-achieving students) might need unequal treatment. Although some individuals overinterpreted "differential behaviors," there were many others who did not fall into this trap and Babad's conclusions concerning past positions are sweeping and erroneous in some instances.

The historical comments made in the paper are often extremely instructive and interesting, although they are also, for the most part, speculation. For example, Babad notes that a decade ago he tended to believe the students with regard to differential emotional support, and to disbelieve teachers—viewing them as defensive and hypercritical. Today he believes that most teachers are generally warm and trying to be supportive of low expectancy students. These dichotomous descriptions are too general and both types of realities can be found in some classrooms. The existence and degree of differential teacher behavior depends on the particular context and classroom. Teachers communicate with students in different ways. Many are warm toward students but some are not—particularly when they are dealing with difficult and challenging students (see chapter by Ennis).

To extend this argument further, it would seem that the expression of teacher affect will differ markedly from classroom to classroom as a function of factors such as the ethnicity of teachers and students. For example, is the teacher Hispanic and the students African American and/or Anglo? Is the teacher teaching a subject in her or his area of speciality, or a subject for which the teacher is not well-prepared to teach? Given the author's description of notable variation in the number of students described as "teacher's pets" across classrooms, why do more or less exclusive classrooms exist? Further, what is the effect on students if you are in the majority (there are many other pets as well) versus when you are receiving more differential treatment? Teachers' teaching style has been shown to vary as a function of their perceived comfort in teaching a topic (Carlsen, 1991). Thus, teacher's affect in dealing with students' response may be influenced by characteristics other than the teacher's degree of liking a student. Indeed, it is possible that environmental factors may frame how teachers express affect.

EXPECTATIONS IN SPECIAL CONTEXTS

Shared Expectations: Sharing a Joint Vision: Catherine Ennis

Ennis argues for the creation of shared goals and expectations among teachers and students. The paradigm she explores is one of teacher-controlled versus teacher-student negotiated expectations. Ennis promotes the teacher-student negotiated perspective as a way to create an environment in which teachers can find fulfillment in teaching and students can find personal meaning in learning. Ennis commences the chapter by examining the consequences experienced by both teachers and students in classrooms where teachers control the expectations. This type of classroom environment is dissatisfying and unrewarding to all involved. The chapter concludes with suggestions for an approach to shared negotiation of classroom goals.

Ennis explains that historically teachers have had the authority and right to establish classroom expectations and that students more or less unquestionably accepted these expectations. This is no longer the case, and in fact many students today believe that they should have some say regarding classroom goals, management style, and material to be learned. This loss of moral authority, coupled with the increasing teaching demands associated with working with low SES and minority students in urban settings, results in many teachers blaming decreasing student performance on factors outside of the school environment. This disowning of responsibility for student outcomes lends itself to lowering of teacher expectations. The resulting classroom environment is demoralizing to teachers and students. Negative teacher assumptions are associated with low-expectation classrooms. These negative assumptions include statements such as: Students do not care about education, have little regard for themselves, view school as a place to cultivate their social life, and are no longer motivated to make high grades, but only care about passing (i.e., there is a lack of striving for excellence). Students respond to these negative expectations and their concomitant assumptions with despondency or defiance, and a negative teacher-student relationship cycle ensues.

When asked to respond to the typical urban classroom environment, students complain about the poor teaching efforts to which they are exposed. They report that the majority of their learning experiences involve straight lecture or repetitive drill exercises that do not provide opportunities for questions, debate, or the development of independent thought. Overall, the students report that their teachers tend to be "absent" from the classroom and uncaring in terms of students' educational and personal experiences, and the material they are required to learn is not made relevant to these experiences.

Ennis argues that creating an environment in which teachers and students create shared expectations can reduce teacher dissatisfaction and student disenfranchisement. In such an environment, students have a say in the material to be learned and classroom policy. Teachers, in turn, are rewarded with students who are compliant and engaged. Ennis argues that adopting a constructivist paradigm in the classroom can provide the scaffolding necessary for revising classroom expectations. Constructivism relies on a vision of learning that is mutually involving for student and teacher. The teacher must work to keep the curriculum within the learners' zones of proximal development and in doing so must be responsive to student needs, motivations, and ability levels. Other avenues to creating shared classroom expectations within a constructivist paradigm include: the development of trusting relationships, opportunities for second chances when students violate classroom expectations (e.g., another interpretation of this could be working with students one-on-one when they violate or cannot meet the new classroom expectations), encouraging student ownership of the negotiated expectations, and ensuring that expectations result in outcomes that are personally meaningful for students.

Critique of Ennis

The Ennis chapter presents an exciting vision—classroom expectations that are jointly articulated by teachers and students as opposed to teacher controlled or negotiated expectations. It argues persuasively for an enlightened view in which expectations are reciprocally created and maintained at reasonably high levels. The disappointing aspect of this chapter is that the research inquiry that yielded the vision is not fully described nor adequately documented in practice.

Often we found ourselves wanting more detailed information. For example, we are told that in one study, 10 teachers were participants in the inquiry and that in the following year in the same school district, 10 different teachers from 10 different urban high schools were sampled. Unfortunately we are told little about these teachers and their professional dispositions or why different teachers were sampled over consecutive years.

Later in the chapter we are told that data were collected from 18 teachers, representing eight different subject areas and six administrators. Were these an additional 18 teachers, or were these 18 of the 20 teachers that were referred to earlier? At times, the author uses national data when it would be at least as important as to include local data. For example, using a broader survey, the author reports that 82 percent of the 2,000 school districts surveyed indicated that student violence was increasing—but what were the perceptions of the 20 principals in these schools? Were these schools seeing violence as even more salient or perhaps less salient than schools in other settings?

In many places the author did not systematically argue a case. Often, only an isolated quote was used to support a viewpoint. For example, the author writes, "Administrators and teachers often respond by attempting to control expectations through a proliferation of rules and policies that are hard to enforce and that students perceive as unfair. Students may respond with a loss of interest in learning and an unwillingness to engage in the educational process, leading to detentions, suspensions, or expulsions." Although we suspect that students do see many rules and policies as unfair, it would be important for the author to present student statements to verify that point. This could be exciting data because at present many policy makers and citizens are urging a get-tough policy on students (e.g., zero tolerance). This appears to us to be extremely problematic and more likely to lower student perceptions of adult support than to facilitate more desirable patterns of socialization.

Ennis argues that teachers lower their academic standards to avoid feelings of professional failure, and that students in these classrooms then find the revised curriculum material irrelevant and hence become more resistant to conforming to teachers' expectations. However, it is not clear as to why changing expectations necessarily has to lead to less relevant activity. The argument for why teachers who want an easier or more engaging curriculum would pick tasks that are irrelevant seems inconsistent. If teachers were trying to make material easier, wouldn't they pick tasks that have more immediate transfer to students' interests and lives outside of the school setting? Why do teachers translate easier into boring?

Although many of the students' quotes are interesting, often they are not developed in a way that helps us to understand whether the teacher was acting reasonably or unreasonably. For example, one music teacher was quoted as saying, "this year I tried to include Japanese, Russian, and Austrian composers, but it was a battle from the start." The teacher then suggested that the students were not interested in learning the characteristics of music in cultures other than African American. One wonders how the teacher attempted to scaffold the move from a study of local music to international music and whether or not students inferred expectations from the teacher that she or he thought that international music was better. Further, one wonders if the African-American music unit was finished prematurely, or if more appropriately students began other units with, at least initially, a continuing comparison between the themes and the expressions in music from other cultures with those from African-American music.

The teacher seemed to feel that she or he was making special accommodations for students but most teachers design units that initially use experiences that are close to students. Hence, it would seem that students should have an extended part of the semester dealing with music they know best, learning how to analyze and think about it and to develop a language so that ultimately they can compare that music to music that differs in certain respects. Although the author alludes to this basic principle when discussing Vygotsky and the zone of proximal development later in the chapter, no connection is made to this theoretical perspective in discussion of the vignettes that are provided throughout the chapter.

More data from students would be useful(along the lines presented by Weinstein and McKown; and Babad). For example, Ennis's concept of second chances is insightful and potentially important. It would be useful to understand how students interpret behaviors that suggest that the teacher is offering them a new opportunity and continuing efforts to teach them as opposed to simply giving up on them. We would think that from certain standpoints a *second chance* could be interpreted as "it really doesn't make any difference if I try, the teacher will always give me another chance" or as "there are no standards in this classroom." Although we find the issue fascinating and suspect that the author is onto something important it would be useful to define and document more fully what is meant by a *constructive second chance*.

Similarly, there are some interesting quotes from teachers suggesting how they engage students in classrooms after students have experienced years of failure and embarrassment. It would have been helpful to supplement teachers' reports with students' accounts of how their thinking was changing over time, or at least interviews with students suggesting that they were becoming more organized, less embarrassed by failure, and so on.

Children's Stigmatizing Expectations: Harris, Milich, and McAninch

This chapter explores the role that interpersonal perceptions play in negative expectancy effects. Specifically, it addresses how interpersonal perceptions play into peer rejection. Harris and colleagues argue that teacher and peer perceptions can become a direct cause of the interpersonal problems associated with rejected peer status. For example, a teacher may contribute to peer rejection by calling attention to or criticizing a student for a behavioral or cognitive problem in front of the stigmatized child's peers. Harris and colleagues report experimental findings that improve on previous research in this area by implementing methodological procedures that allow for causal statements and examination of effects of the perceiver's expectancy on the target's self-ratings and behavior. Two studies examined negative expectancies associated with ADHD, and a third study examined negative expectancies associated with low academic ability.

Age played an important role in the outcomes of the first study. There was as expectancy X age interaction for nonverbal friendliness such that younger perceivers were less friendly to targets of negative expectancies, while in older perceiver/target dyads the perceivers tended to be more friendly toward the targets. Overall the results of the initial study indicated that manipulating the expectancies of one student in a dyad resulted not only in behavioral and self-concept changes for the target of the negative expectancies but also less reciprocal and less positive interactions for the dyad as a whole. The second study involved a balanced placebo design which allows for separation of the stigmatizing effects that occur as a result of the condition under study and stigmatizing effects that result from the perceiver's expectations of that condition. The results indicated that both the stigma associated with a negative expectancy for the ADHD target and the actual behavioral symptomology associated with ADHD created negative effects on peer interactions. Another interpretation of this finding is that the presence of a negative expectancy effect may create a situation of peer rejection independent of the actual symptomology. In the final study, Harris and colleagues extended their findings supporting the salient role of expectancies in peer rejection to the domain of academic achievement. Here, the stigmatizing expectancies related to low academic achievement rather than ADHD, but the results replicated the findings concerning the role of negative expectancies in peer rejection.

Having demonstrated that peer expectancies can result in peer rejection, Harris and colleagues shift the focus to reasons why interventions based solely on improving peer interactions are often not successful in reducing peer rejection. Harris and colleagues report that a primary reason that peer interventions do not work is that while they may result in changed behavior on the part of the target student, they usually do not increase liking on the part of peers. Even if perceiving students notice improved behavior on the part of the target student, their liking or tolerance of this student may not improve. It seems that liking is more determined by initial expectancy than proceeding behaviors. Bearing this in mind, Harris and colleagues suggest that the best solution to peer rejection caused by expectancies is to not allow the negative expectancy to form in the first place. The authors recommend simple and easy to implement tactics to help avoid negative expectancies. For instance, teachers should make a concerted effort to minimize attention called to the stigmatizing condition, including minimizing attention to medicine taking. In cases where the disorder cannot be concealed easily the student can be counseled in methods for positively deflecting negative reactions to his or her condition.

Critique of Harris and Colleagues

This research extends knowledge about self-fulfilling prophecies into a new and important area. These authors argue that the classroom is a social setting in which students often experience interpersonal issues. The chapter makes the argument that teachers' expectations and behavior influence peers' expectations and behavior toward a stigmatized student—teacher and peer expectancies about undesirable behavior of a student can enact as self-fulfilling prophecies. The data presented in this chapter are interesting and show some of the direct effect of expectations. Unfortunately, these data are collected in laboratory situations and the amount of time students spend interacting socially is limited.

The findings from Harris and colleagues are very discouraging. Their data suggest that a stigma is likely to persist even in the face of evidence showing objective behavioral change. Such findings again highlight the importance of primacy effects. That is, once a teacher or a peer firmly believes that a student is smart or hostile, it becomes very difficult to change that belief. Although the authors indicate the importance of understanding the classroom as a social setting and advise that it is inappropriate to conceptualize students' weaknesses as personal stigma, they do little with this important insight in designing social interventions. It seems likely that parents, teachers, community workers, school counselors, and coaches, working in tandem could be more successful in dealing with social stigma (e.g., bullies) than teachers working alone. More comment from the authors along these lines would be helpful.

Harris and colleagues cite the distinction that Coie (1990) makes between emergent status and maintenance status. Harris and colleagues write, "That is, children's social status may fluctuate quite markedly between rejection and acceptance for a while before stabilizing, and it is instructive to consider the factors that determine when a child first encounters peer rejection as distinct from the factors that maintain the rejected status once established." This is potentially a useful distinction but implications for intervention are not spelled out.

Several of the intervention ideas that the authors propose are interesting; however, most of these describe what teachers should not do. There also are many proactive behaviors that teachers might engage in to prepare students to deal more supportively with peers who have a disability or various potential stigmatizing characteristics (unattractive children, obese children, children who stutter, and so forth). It would seem that teachers could build classroom cultures that would be more robust in recognizing and appreciating individual differences. As Ennis suggests in her chapter, teachers can and do build environments in which students learn not to laugh or to be embarrassed at the initial difficulties that other students encounter. Teachers can help to socialize students to understand that physical characteristics are not predictive of character, persistence, smartness, or capacity for friendship. Given the insights of Harris and colleagues, it is unfortunate that they spend so little time analyzing strategies that teachers can use to attempt to reduce stigma issues, and help students cope. It is understood that they cannot offer "answers" based on research, but still teachers have to work with students who are in psychological pain and some advice about how to conceptualize and deal with these issues would be helpful.

Harris and colleagues suggest that when students receive special help from mental health personnel in the school, there arises the possibility that a self-fulfilling cycle of stigmatization will occur. They note, however, "obviously we are not recommending that children with problems therefore should not receive treatment for them; rather, we are stressing the need for teachers and school psychologists to demonstrate heightened sensitivity to these issues and to avoid unnecessarily calling attention to the child's treatment." Again, information about what one can do in these situations would be useful.

Expectations and Beyond: The Development of Motivation and Learning in a Classroom Context: Salonen and Colleagues

This chapter addresses in depth one mediating relationship between teacher expectations and student outcome variables. The mediating variable examined is the reciprocal relationship between student motivation orientation and teacher role expectations, which Salonen and colleagues call "frameworks." Salonen and colleagues describe at length three common student motivation orientations and the likely reactions that these motivations generate in teachers. These different student motivation orientations are postulated to result in differential student processing of teacher expectations.

Three Motivation Orientations

Task Orientation

The three orientations discussed by Salonen and colleagues can be viewed as involving complex and reciprocal relationships involving student, teacher, and task. In each orientation the emphasis is on a different pair-relationship, resulting in different teacher and student behaviors. In the first orientation, called task orientation, the emphasis is on the pair-relationship between the student and the task. The student's primary goal is explore and master challenging aspects of the learning environment. Students with a task orientation interpret difficult tasks as opportunities for self-evaluation, and they are not threatened by difficult tasks that may require them to call upon skills not yet mastered. Their mastery perspective is reinforced by high and stable academic self-concepts. These students tend to minimize the social aspects of the learning environment except as it pertains to obtaining resources necessary for successful task completion. When task-oriented students do implore teachers for assistance, the request tends to be instrumental in quality. These students are not likely to comply or defer to teacher opinions or suggestions when such feedback contrasts with their understanding of the task. Within this group of students these are two subtypes. There are "truth seekers" who pursue the meaning of a task and formulate a coherent understanding of it, if necessary, at the expense of social relationships. Salonen and colleagues describe this group as "intellectual dissidents." The second subgroup of task oriented students are more prosocial or conformist, not willing to risk social rejection by going against the social conventions that guide classroom behavior.

Teachers' reactions to task oriented students tend to depend on the teachers' own motivation orientations. Teachers who themselves are task oriented tend to form alliances with these students and are comfortable with deviating from the scheduled classroom activities in the quest for the truth. This type of teacher is stimulating to the task oriented student, but is likely to promote uneven learning in the classroom and not accomplish learning objectives laid out by the school or district administration. Other teachers who are less secure or rigid tend to view task-oriented students, especially those falling into the "intellectual dissident" category, as interruptive and disturbing. Often these teachers will start to avoid, pass off, or otherwise dismiss the constant inquiries of task oriented students in favor of emphasizing the traditional lesson plan. In between these two types of teachers are teachers who allow a minimal amount of lesson deviation while still making sure that curriculum goals are met in a timely manner by all students.

Social Dependence Orientation

In this orientation the primary relationship is between the student and the teacher, with the student's energies focused on obtaining social approval from the teacher. These students are not concerned with creating order and a coherent understanding of the material to be learned. Rather, they are concerned with adjusting to teacher expectations in order to receive help, approval, and reaffirmed affiliation with the teacher. Help seeking behavior tends to be executive in nature, indicating that the primary purpose of the behavior is to reinforce the student's lack of ability to take responsibility for his or her own learning behavior. There are two subtypes of social dependence orientation students: the first subgroup relies on regressive or "babyish" behaviors to elicit teacher assistance and belief in the student's inability to master tasks; the second subgroup relies on more subtle and socially gracious behaviors when attempting to ingratiate the teacher and reaffirm their inability to accept responsibility for personal academic outcomes.

Teachers react differently to socially dependent student behavior. Teachers who are socially dependent in motivation orientation or enablers themselves, and are especially comfortable with the care giving aspects of teaching, may completely reinforce the student's socially dependent orientation. In fact, this type of student is likely to form codependent relationships with teachers, parents, and siblings, such that their enabler perpetuates the student's perception of incompetence and the student in return graciously reinforces the enabler. These students seem quite apt at soliciting and accepting excessive help in such a way as to encourage the helper to sustain the helping behavior. Teachers who value high levels of control and order in the classroom are also likely to perpetuate student dependence behavior because socially dependent students are especially likely to bend their behavior to align with the behavior desired by the teacher. However, this type of teacher eventually is likely to come to ignore or reject the student's constant appeals for help, as this disturbs the teacher's planned schedule. This lack of teacher support could be devastating for a student who is not accustomed to working independently.

Ego Defensive Orientation

The dominating relationship in this orientation is self-protection on the part of the student. It results in a pattern of reciprocal avoidance and withdrawal between teacher and student. The student is motivated by the desire to avoid negative consequences of failure, particularly those consequences that bear social implications. These students are hypervigilant to signals of failure and, therefore, see novel and challenging tasks as something that must be avoided. These students rely on inhibited, passive, and avoidant behavior to meet these goals. When help is offered they tend to become highly recalcitrant to making even minimal gestures toward task completion. Teacher behavior toward this type of student is largely generated as a reaction to student behavior. Teachers, and parents too, often become defensive in the face of these students' repeated failure and inability to meet classroom goals. Teachers are likely to become defensive under these circumstances and to locate the problem external to themselves by looking for a stable and internal cause in the student. Such stable internal causes could involve low ability, cantankerous disposition, or poor home background. Whatever the cause generated, it relieves teachers of the self-doubt and guilt that can arise when working with a difficult student. In addition to looking for a cause outside the teacher (or parent), teachers are likely to start seriously avoiding these students and accepting low expectations and low levels of aspiration for them.

Summary of Implications of Motivation Orientations for Expectation Effects

In terms of how motivation orientations are likely to affect student mediation of teacher expectations, it seems safe to assert that students with high levels of task orientation are not likely to be strongly disturbed by teacher expectations. Not only are these students not extremely tuned into social convention, but their high and stable academic self-concept probably protects them even in situations where their teacher finds them irritating and attempts to extinguish their "truth seeking" behavior (see Weinstien & McKown, this volume). On the other hand, socially dependent students may be extremely vulnerable to expectation effects because they are highly malleable to teacher expectations and desires as a means of eliciting social approval from the teacher. Even when teachers express disapproval or ignore student implorations for help in an attempt to extinguish these behaviors and encourage more appropriate independent functioning, these students are likely to personalize teacher rejection. Also, since these students manipulate the environment to create a sense of low ability and legitimate need for help, they are likely to fall prey to the environmental contingencies they helped establish and eventually see themselves as low in ability and unable to function competently without help. Ego defensive students are probably more like task-oriented students in their imperviousness to teacher expectations. The problem here becomes that, even when teachers express positive expectations about student ability to master new tasks, students are unlikely to increase effortful behavior. In the short term these students may not be so influenced by teacher feedback (they seem quite convinced of their inability to deal with challenging tasks and possible failure generated by the unknown). Eventually, however, as teachers in a defensive reaction come to ignore or expect little of these students, it seems reasonable that the students would succumb to the low expectations for them. This presents a chicken and egg dilemma. Did the student create the expectation that no novel material could be mastered and therefore the ensuing teacher belief that not much should be expected of the student? Or did some early school or home experience teach the student that it would be best not to even attempt novel tasks?

Critique of Salonen, Lehtinen, and Olkinuora

Like other authors in this volume (Babad; Jussim et al.; and Weinstein & McKown) Salonen and colleagues argue that there is increasing evidence that academic performance cannot be explained in purely objective or cognitive terms because subjective beliefs, social pressures, and emotional processes need to be considered as well. These authors contend that the classroom situation is dynamic and that students and teachers attempt to fulfill their own needs while at the same time adapting to other participants' needs and expectations. Importantly, they argue that classrooms are not purely centered on cognitive learning or task focused, but are often construed in ways that meet students' personal and social needs.

Their argument is enriched by the delineation of specific types of students who may react differently to different environments. The three types of students that they analyze are interesting and their analyses are informative, but they present no formal argument for why these three types of students are used for analysis. There are a great many individual differences in students and ways to classify them (see for example, Cronbach & Snow, 1977; Good & Power, 1976; Snow, Corno, & Jackson, 1996).

The authors describe their intervention program but there is too little detail to allow us to understand or the evaluate the intervention. The conceptual orientation of the intervention strategy raises some questions. Essentially, the authors argue that multiple and simultaneous intervention strategies must be utilized to improve learning conditions—that is, systemic effects. However, the authors focus their intervention on children—why not involve teachers and parents or other school participants? Are there ways in which parent involvement in intervention efforts could be inappropriate or counter-productive (e.g., Casanova, 1996).

Teachers were involved in post-intervention extensions in the classroom; however, this description is insufficient for the reviewers to understand or comment upon. The chapter provides considerable analysis and insight into dilemmas of classroom communication but not enough information about the authors' own intervention strategies and the consideration of alternatives to those that they used. For example, given the systemic and social nature of the issues, would it be important to involve family-like arrangements where teachers share the same group of students or where an individual teacher even teaches the same students for two or three consecutive years?

Teacher Efficacy and the Difficult-to-Teach Student: Soodak and Podell

The first premise upon which this chapter is built is research indicating that teacher referral practices are affected by teacher efficacy. Specifically, low-efficacy teachers tend to make more referrals. The second premise is the observation that too many referrals have been made to remove difficult students from the regular classroom. These referrals often result in students ending up with stigmatizing diagnostic labels, and sometimes with permanent removal from mainstream education. Taken together, these premises suggest that low-efficacy teachers may make more referrals resulting in a poorer educational environment for the difficult-to-teach student. This chapter examines the outcomes that ensue when low-efficacy teachers are paired with difficult-to-teach students.

Soodak and Podell utilize a two-factor model of teacher efficacy comprised of teaching efficacy and personal efficacy. Teaching efficacy represents a teacher's generalized beliefs about the teaching profession and includes beliefs about the teacher role and the ability of school to overcome outside obstacles. The second factor, personal efficacy, relates to a teacher's beliefs about his or her personal ability to effectively manage students, educational goals, and problems that may arise in the classroom.

In a series of studies addressing teacher efficacy and difficult-to-teach students, Soodak and Podell found that teachers with high levels of personal efficacy are more likely to agree with regular classroom placement for students with learning problems than are teachers with low personal efficacy. Further, teachers with a low sense of personal efficacy are more anxious about including disabled students in their classrooms. Related to this, teachers low in teaching efficacy tend to be more hostile toward inclusion of students with learning problems, while teachers who are high both in teaching efficacy and in their differentiated teaching practices are most receptive to inclusion of students with learning problems. Teachers with low personal efficacy are more likely to consider a student's socioeconomic status in determining placement for low-achieving students. Additionally, teachers with low personal efficacy offer more non-teacher-based suggestions for remediating student learning problems, while teachers with high personal efficacy offer more teacher-based solutions. Finally, teachers who rate high on both teaching efficacy and personal efficacy are most likely to retain difficult students in their classrooms.

Based on these results, Soodak and Podell offer several suggestions for teacher education. First, teachers need more positive, preservice opportunities to work with difficult-to-teach students. These experiences would aid teachers in building perceptions of efficacy with regard to their ability to handle difficult students and would aid in preventing teacher disillusionment early in the teaching career. Second, in order to reduce the number of students in special education programs, it may be beneficial to address teacher beliefs about their abilities to effectively teach special education students. A more productive integration might focus on teacher efficacy beliefs with regard to inclusion and within this, attention should be given to the malleability of teacher efficacy beliefs.

Critique of Soodak and Podell

The chapter provides a succinct and effective review of the history of research on teacher efficacy. In contrast, it is less successful in developing new arguments and theory. One problem is the initially vague definition provided for difficult-to-teach children. Indeed, the definition of difficult-to-teach is not presented until relatively late in the chapter, and then at a very general level. It would seem that children who are conceived of as difficult to teach would vary from teacher to teacher and perhaps even from school to school. Some teachers are likely to find students who present management problems to be the most difficult to teach; others are likely to describe students who have difficulty in learning as most difficult; and still others would find the most difficult-to-teach children to be those who are smart but frequently challenge and show them up in the classroom. Yet, other teachers might be most concerned about trying to get withdrawn students to participate in the classroom, or dealing with students who are frequently subjected to ridicule or bullied by more dominant students. Further, there is no reason to believe that teachers' conceptions of difficult-to-teach students would remain stable over time. For example, beginning teachers might be more concerned about management difficulties presented by students; however with experience, those teachers might become concerned about other aspects of student behavior.

Late in the chapter the authors make it clear that they are concerned primarily about teacher efficacy and issues associated with inclusion. However, to reiterate, the authors do not directly make the point that even teachers in this sample find "inclusion students" to be among their most problematic or difficult students.

The set of four experiments that the chapter authors present are interesting, additive, and useful as examples of how programmatic research may inform this area in time. Still, we were disappointed that more details about the studies were not presented. The limited descriptions make it difficult to understand the context that was investigated. For example, it would be useful to know how much information and/or experience that the teachers they studied have had for dealing with inclusion students, in either teacher education programs or school-related in-service programs. In the first study, we only know that half (96) of the teachers were general educators and the other 96 were special educators and that all teachers had been teaching for a minimum of one year. It was found, as predicted, that regular teachers with a greater sense of efficacy were more likely to perceive regular education placement as being appropriate for students having difficulties. We wonder if this finding was due to their efficacy in general or because of earlier experience in working with special education students?

Soodak and Podell report that school climate was important. In particular they noted that only teachers who were unable to collaborate with other teachers *and* who reported low personal efficacy were against inclusion. These data remind us of the contention of Weinstein and McKown in this volume, that school climate is exceedingly important and interacts with individual classroom expectations. However, missing was any argument about how to change situations in which collaboration is "devalued."

This chapter, like most others in the volume, was a bit "thin" in its discussion of important specific implications for teacher education and its potential role in ameliorating efficacy and expectation issues. It would have been instructive to learn from the authors about how they would attempt to intervene in teacher education programs—are the vignettes that they use in their research potentially relevant for use in teacher education, or would they use vignettes that present different characteristics and decisions—why?

In closing, we want to note that Soodak and Podell as well as Salonen and colleagues (this volume) are helpful in establishing that teachers' expectations for, and communication with, students are based on aspects other than students' learning characteristics; teachers also react to students as social beings (see also McCaslin, 1996).

The Antecedents and Consequences of Teacher Efficacy: Ross

In this chapter Ross provides an overview of the teacher efficacy literature in terms of antecedents and consequences of teacher efficacy and how these antecedents and consequences inform interventions and future research. Before commencing a review of the literature, Ross defines teacher efficacy as an expectation exerting salient effects on important domains such as teacher attributes, workplace conditions, instructional practice, and student outcomes. Ross notes that teacher efficacy expectations become self-generating or self-fulfilling such that the beliefs encompassed under the domain of teacher efficacy come to determine behaviors and goals within the classroom in such a way as to reinforce the initial level of efficacy. As is accepted among many teacher efficacy researchers, Ross maintains a model of teacher efficacy comprised of general teaching efficacy and personal teaching efficacy.

Ross divides the antecedents of teacher efficacy into two categories: characteristics of the teacher and characteristics of the educational environment. Teacher characteristics, such as, teacher gender, level of experience and education, and locus of control are related to teacher efficacy. Specifically, female teachers tend to report higher levels of personal teaching efficacy than males, general teaching efficacy tends to decline with experience while personal teaching efficacy tends to increase with experience, more highly educated teachers tend to have higher levels of general teacher efficacy, and teachers with an internal locus of control tend to have higher levels of personal teaching efficacy.

In terms of the general workplace environment, the following are salient variables: grade level, student ability, social class, class size, teacher workload, and school culture. With regard to grade level, teachers at the elementary level report higher efficacy levels. With regard to student ability, social class, and class size, teachers tend to report higher efficacy levels when student characteristics lend themselves to more manageable students. For instance, working with high-ability students in a middle-class school tends to result in higher teacher efficacy than would be expected for a teacher working with low-ability, low-socioeconomic-status students. Class size works to increase teacher efficacy when it leads to more manageable students. For instance, at the elementary school level smaller classes are considered better. At the high school level, however, small classes may indicate students with special needs and therefore may not contribute to increased teacher efficacy. Appropriate workload levels and a school environment that encourages collaboration among teachers are associated with higher levels of teacher efficacy.

Teacher efficacy levels exert consequences in two domains: teacher outcomes and student outcomes. In terms of teacher outcomes, high levels of teacher efficacy are related to increased use of more complex and challenging teaching techniques, higher levels of willingness to attempt experimental teaching practices, higher expectations for successful teacher and student outcomes, and usage of less punitive classroom management techniques. In terms of student outcomes, higher levels of teacher efficacy are associated with higher achievement levels and more positive affective development. These student outcomes are mediated through high efficacy teachers' use of novel and demanding teaching techniques, more autonomous classroom management styles, greater attendance to the needs of low ability students, and greater persistence in pursuing student comprehension of material. Ross suggests that teacher efficacy can be increased through interventions that focus on increasing skills, changing teacher beliefs, and manipulating the work environment. When formulating teacher efficacy interventions it is particularly important to design multifaceted programs that enhance opportunities for collaboration, professional learning, decisional participation, and administrator support.

Ross recommends that future research address teacher efficacy over time with special attention to its cyclic nature over the course of a teaching career. Related to this, within-teacher efficacy variations should be examined so that fluctuations across a school day or across subject areas can be documented. Future research should utilize more specific measures that avoid defining teacher efficacy as a unidimensional construct. Finally, future research should address interventions that can enhance teacher efficacy.

Critique of Ross

Ross provides an admirable review of what is known about teacher efficacy effects. It is a nice analytical summary of different aspects of teacher efficacy—what factors influence the levels of teacher efficacy—and what are some of the consequences of teacher efficacy on classroom instruction. Previously, we raised the issue about possible historical effects on the conceptualization of teachers' expectations and their expression in the classroom. Similarly, we would have liked to have seen a historical perspective in terms of teacher efficacy. Are these scores increasing over time or decreasing? It would have been helpful if those authors addressing teacher efficacy would have incorporated the literature on students' beliefs about internality and externality. Is it the case that internality beliefs of students have decreased over time, especially for low-income students? Are contemporary teachers teaching students currently who have higher self-efficacy or self-esteem levels than students in the 1970s? In contrast to students in the late 1960s, are today's more likely to "suffer" from unrealistically high levels of self-esteem? Given the changing demographic characteristics, and perhaps the presence of students with lower self-efficacy beliefs, it would have seemed useful to ask "have teacher education programs stayed constant in their ability to select and to prepare teachers who have particular levels of efficacy?" Although Ross discusses the possibility of student ability as an influence on teacher efficacy, he does not discuss the possibility of student efficacy beliefs as determinants of teacher efficacy (perhaps in combination with student ability).

We found the section on consequences of teacher efficacy more problematic than the section on antecedent conditions that influence efficacy levels. Ross indicates that "higher teacher efficacy is consistently associated with the use of teaching techniques which are more challenging and difficult..." Although he does indicate that there are strategies that teachers are more likely to use if they have high efficacy, he does not present data or a rationale to illustrate that these techniques are more challenging and difficult. For example, some cooperative learning programs have a set curriculum (which makes teacher decision making less important) and one could argue that the instructional demand is no more difficult than leading a recitation or supervising a laboratory session.

We agree with Ross that higher levels of efficacy appear to be associated with teachers' willingness to experiment, but some of his conclusions are overargued. For example, Ross asserts, "high expectations of success enable teachers to set higher goals for themselves and others..." It is not clear if any data support this assertion. What data can be used to support the contention that high efficacy teachers set higher expectations for themselves or higher expectations for students?

We found the section on interventions to be a thoughtful and useful summary of attempts to increase efficacy. However, in our opinion, this section is hindered a bit by the author's willingness to assert positions without data. In particular, the author is willing to assume that variability in format is associated with quality of instruction. Reciprocal teaching strategies, for example, can be used effectively or poorly. Further, even a reasonably well-implemented reciprocal teaching plan may be more appropriate in some contexts and for certain goals than others.

We find Ross's concern about within-teacher differences and particularly his argument that we need to know more about teachers' confidence in particular abilities as they move from task to task to be an important line of research. Indeed, it would seem important to begin to move to more context-specific analysis within the teacher expectation literature. It may be that expectations that teachers communicate for students have more impact in some areas than others.

FUTURE DIRECTIONS

First, we reiterate our belief that this volume is a valuable summary of historical and contemporary research on teacher expectations and teacher efficacy. The volume also provides some useful suggestions for next steps for research in these related areas. Second, we want to acknowledge that some of the criticisms that we expressed go beyond the intent of chapter authors. Further, in some cases information that was omitted from these chapters can be found in the authors' previous publications. Still, our task is to raise issues about the field and to explore future research directions.

We present five topics for additional research. These are defining expectations, exploring the communication of expectations, conceptualizing expectations as clusters of behavior, and conceptualizing teachers and schools more broadly and more proactively as advocates for students. In addressing research issues and opportunities, we will initially stress micro case-level approaches and subsequently, the discussion of teacher expectations will be expanded to a macro-global perspective. We think that more work is warranted at various levels/units of measurement.

Defining Expectations

Teacher expectations have been defined in ways so varied that it is difficult to aggregate findings across studies. Although the measurement of teacher efficacy also is diverse and complex, at least the dimensions are codifiable because a few well-focused instruments have been used to define teachers' efficacy. In contrast, definitions of teacher expectations—whether high or low—are still at a very diffuse level of measurement. In some studies expectations are defined by false information that is given to teachers (as in the original Pygmalion study); whereas in other studies expectations for students are provided by teachers. In cases where teachers report their own expectations, how the researcher asks for the information—or subsequently decides to use it—has an enormous effect on the measurement of teacher expectations. For example, sometimes teachers are asked to rank all students in the classroom and then investigators make various decisions including: (a) they compare teacher behavior toward bottom-third students versus top-third students; (b) they confine the list to a smaller group of students—that is, comparing only the top four and the bottom four; (c) they exclude certain types of students—for example, educable mentally retarded students. In other cases it's difficult to tell what decision rules are used for deciding whether or not students are included. Studies vary widely in terms of the particular ranking instructions that teachers are given. In many studies it is difficult to tell whether or not teachers have low expectations for students. For example, among teachers who rank 30 students, some may believe that twenty-fifth ranked students are very teachable, whereas others may view students ranked that low as relatively impervious to instruction.

The TTI scale that Weinstein and McKown discuss is an example of instrumentation that has brought some degree of measurement coherence to the study of students' perceptions of differential classroom opportunity. The development of instrumentation along these lines might be helpful for achieving more coherent measurement of teacher expectations (toward the class, individual students, or groups of students—e.g., gender). In particular, some standard methodology would make it possible to explore more fully the intensity of teacher beliefs about the modifiability of student performance and to compare teacher expectations and their effects across studies.

These same issues are also present if we try to integrate the expectation and efficacy areas of research. For example, high and low-efficacy classifications suggest that teachers differ in their beliefs in their ability to bring about learning, but the range within a classification is likely to be very wide (e.g., I can teach four, 14, or 24 students?).

Given the social nature of classroom learning, it would seem advisable to explore the extent to which students feel supported or rejected as social beings. Most conceptions of teachers' expectations have focused on the extent to which teachers support students as learners. It is time to begin to explore how students see other aspects of teacher communication. For example, do students see the teacher as being fair to them individually or to the class as a whole (Nichols & Good, in press) and supporting them as social beings (McCaslin & Good, 1996a, 1996b)?

Another avenue worthy of exploration is the measurement of global personality variables that, via the creation of teacher liking or disliking of a student, contribute to teacher expectations and student reactions/outcomes to the expectations. For example, in trying to decipher teacher differential treatment within and between expectation groups, we may want to ask what student characteristics or what types of students cause a teacher to like or dislike a student (the work by Babad; Harris et al.; Salonen et al.; and Soodak & Podell, this volume are excellent starts on this issue). It is possible that some amount of teacher differential treatment within and between expectation groups is unrelated to teacher expectations for academic outcomes. In our personal relationships (which is what a teacher-student dyad may become, particularly in the early grades when teacher expectations are thought to exert more influence on student outcomes), we realize that our liking/disliking and expectations for an individual are based on more than our perception of the person's intellectual level. In other words, we could attempt to identify reasons other than teacher expectations for academic outcomes that cause teachers to treat students differentially. Babad's chapter represents a significant step in expanding the field in this direction.

Are Expectations Communicated to Students?

In addition to measurement issues concerning what expectations teachers hold for student performance, there are still many questions about how to determine if students are treated differently. Do teachers act on their expectations? It is possible to read research studies that explore differential teacher behavior toward students perceived to be high and low achieving and not be able to answer basic questions like: Given a mean difference on a particular dependent variable between highs and lows, were all six students treated differently, or only one out of six, or four out of six? Are some of the highs treated more like lows or vice versa? To argue that differential behavior or differential affect is a determinant of student performance, it would seem important to focus the analysis on students who are actually treated differently and to link students' individual patterns of communication opportunity with individual performance outcomes. Follow-up research along these lines could help clarify why some of the six low-ranked students appear to receive "self-fulfilling" communication but others do not. Are student variables informative in determining whether low expectations are communicated (e.g., race, gender, students' assertiveness).

To reiterate, in most extant research that includes observational measures, it is not possible to determine the extent to which teachers treated all identified low-achieving students differentially. This assessment, of course, is *impossible* in situations in which classroom behaviors and/or attitudes are not observed or sampled. In "black box" studies it has to be assumed that the general level of teacher expectations or the mean level of affect is received equally by all students in the group. However, we know from a long history of research (well-replicated) that individual differences in classroom communication are enormous—the mean number of higher level questions that a teacher asks does not predict the number of higher level questions that a particular individual student receives. Jussim and colleagues provide some important ideas for enhancing methodology in the field—longitudinal designs—looking for positive as well as negative effects, these design improvements will not be notably helpful unless they are also combined with observational work to verify and describe process communication.

Clusters of Behavior

In addition to looking at the communication environment that a particular student receives/negotiates in the classroom, it also would seem important to look for patterns of behavior that a teacher-student dyad share. Authors in this volume (e.g., Babad; Weinstein & McKown) have correctly criticized early researchers for placing too much reliance upon particular types of behavior. That is, frequent but gratuitous praise may communicate a low expectation for performance and if one is only coding the frequency of praise—with the assumption that teacher praise has positive effect on students—the interpretation is highly misleading. Furthermore, what constitutes gratuitous or supportive praise is subject to the mediation of individual students. However, as we have argued before, many authors in this volume have gone too far in dismissing the power of behaviors as communicators of performance expectancies; particularly clusters of behaviors.

If researchers collect multiple measures of classroom communication, it may be possible to develop and interpret communication patterns. For example, assume a teacher who (a) criticizes a given student proportionately more often than other students following a wrong answer, (b) praises the student proportionately less often than other students following a correct answer, (c) gives up (by providing the answer) when the student doesn't answer quickly, (d) asks the student to answer only simple recall questions, and/or (e) calls on the student only when he or she raises a hand, may be seen as communicating low performance expectations for the student. Further, this teacher would appear to *blame* the student for his or her inability to learn.

Low expectations can be expressed in highly divergent ways. For example, another pattern of communication might suggest that the teacher attempts to protect the student from humiliation and embarrassment by doing too much of the academic work for the student. One pattern here might be expressed by (a) frequently pairing the student for tutorial work with one of the better students in the class, (b) frequently praising the initiations of the student—even marginal ones, (c) frequently calling on the student to answer simple questions—especially about material that has already been discussed in the period, or (d) assigning the student easy books and reading assignments. Thus, by *protecting* the student from possible failure, the teacher doesn't allow the student to think or develop her or his own initiative and interest. Research in this area would benefit from the theoretical development of multiple communication patterns that likely convey low or high expectations, followed by assessment of the effect of those patterns or clusters on student performance. During the initial cluster identification process, it seems that an appropriate methodology could be the aggregation of case studies addressing antecedents and results of clusters of expectation-related behavior. In other words, perhaps intensive study of individual teachers and the students they interact with could illuminate clusters of behaviors more effectively than studies aimed at the examination of teacher expectation effects across many classrooms.

In attempting to define the composition of clusters of expectation behaviors, we could consult with teachers or collaborate with teachers to help define the framework. It seems reasonable that teachers may be quite knowledgeable and possess insight into the problem that others within the school setting may not possess. However, the research does not indicate that teachers' perceptions of this problem have been probed in detail. It may be that teacher's interpretation of teacher expectation effects gets dismissed as subjective or biased. Though there probably are some legitimate concerns here, by dismissing teachers' perceptions of the problem, we have designed a contradictory situation. Specifically, while expecting teachers will enact measures to prevent differential student treatment and outcomes, we do not incorporate their interpretation of the problem. It seems a bit like asking a doctor what she or he can do to remedy an illness without asking what the illness is.

Thus, it seems that the extant research is limited to some extent because the "independent variable" is often inadequately defined. Further, many studies fail to observe to see if teacher expectations are communicated to students—such "black box" studies link measured teacher expectations with measured student outcome. Such studies are the equivalent of testing a medicine by only measuring if a prescribed medication has effects beyond that of a placebo, without knowing whether patients actually *take* the prescribed drugs. Teachers administer many treatments in their classrooms, and looking only at individual behavior rather than clusters of behavior is likely to mask the effects of at least certain types of classroom communication on students.

Teachers' Beliefs about Students' Ability are Important

The authors of chapters in this volume provide a "normal distribution" of beliefs concerning the effects of teacher beliefs on student performance. None of the authors dismisses teacher beliefs as a trivial determinant of student performance, but some authors present the view that teacher beliefs are but minor determinants of students' performance. In contrast, other authors assert that teacher beliefs are of modest importance and yet other authors argue that teacher beliefs often have major consequences.

We confess to holding the view that teacher beliefs are often important determinants of student performance. Although these influences operate in numerous ways, we call attention to two major contexts. The first context is variation within normative practice of schooling (i.e., largely the viewpoint presented in this volume). The second context is what happens when normative practice is altered. The possibility of changing normative practice was an area largely ignored by authors in this volume (Weinstein & McKown are notable exceptions). Here we will illustrate that changes in the normative practice can have powerful impact on student performance.

Evidence exists to show that rather sizeable impacts on student performance can occur when students are provided with improved learning environments and/or opportunities. Mason and colleagues (1992) conducted a field experiment to examine the effect of providing students with access to new curriculum content. In particular they studied the performance of eighth-grade students who were assigned to pre-algebra classes rather than a traditional lower-level mathematics class. Even though the experimental students would not have had high enough test scores to be placed into the pre-algebra classes without intervention by the researcher, students placed in pre-algebra classes were found to benefit from their advanced placement in comparison to average-achieving eighth graders from previous years who were enrolled in general math. Further, the presence of average achievers in pre-algebra classes did not negatively impact other students. In addition, students who were exposed to new curriculum content went on to take more advanced mathematics classes than did comparable students in the past who were not given such opportunities. Ways to provide low-achieving students with exposure to new content—and content recognized as important by students—would appear to be an important area of research.

Kaufman and Rosenbaum (1992) explored the effects of community setting on student achievement. The U.S. Supreme Court ruled against the Chicago Housing Authority and the Department of Housing and Urban Development and concluded that these agencies had used racially discriminatory policies in managing low-rent public housing programs. One result of this ruling was the Gautreaux program which provided public housing residents with certificates that allowed them to move to other apartments either in mostly white suburbs or in other areas of Chicago. As Kaufman and Rosenbaum note, relatively few Americans have resided both in the inner city and in suburbia. Clearly, families who move to the suburbs had a distinctive perspective from which to compare relevant dimensions of these two environments.

The results of their study indicated that African-American youth who moved to white suburbs faired more favorably than those who moved to new urban locations. African-American youth in suburban schools were more likely to be in high school, in a college preparatory high school track, and employed with benefits. Thus, these students were *inside* of the educational and employment systems. Interviews conducted with them and their parents suggested that the new environment provided better educational opportunities and was more effective in stimulating students' motivation. In particular, students and their parents indicated that suburban schools had higher standards for achievement and provided better preparation for college. They also said that teachers in the two environments taught differently. In particular, suburban teachers not only held students accountable for high goals, but they also were available to provide extensive help for students as they needed it.

Other studies have also showed the importance of changing expectations, normative practice, and curriculum opportunities for students. Meahn (1997) describes the "achievement via individual determination" (AVID) program and its effects in detail. Briefly, the AVID program attempts to maintain a rigorous curriculum for all students but also provides increased support for low-achieving students. The AVID program allows previously low-achieving students (many of whom come from low-income and ethnic backgrounds) access to the same academic program that serves high-achieving students. Already, this program has achieved some notable effects. In terms of historical local averages, students from the underrepresented ethnic and linguistic backgrounds who participate in the AVID program enrolled in college at roughly 37 percent. Importantly, recent graduates of the AVID program have enrolled in four-year colleges at the rate of 48 percent. Clearly, changing the structural and classroom expectations for students can have important and enduring effects.

Advocates for Students: Expanding Normative Practice

One area not explored by chapter authors was the communication of expectations for students by school personnel other than teachers (Harris and colleagues touch briefly on the topic of mental health workers in the school and Weinstein & McKown describe possible roles for principals). It is likely that other people in the school serve (sometimes unintentionally) to label and validate the learning inadequacies of students. It would seem important to begin to challenge our assertion with data. For example, do guidance counselors provide equal or comparable time to discuss college or vocational options with low track students? And, if not discussions of higher education options, then do low-track students discuss with counselors strategies for becoming more successful and more independent learners? In essence, do guidance counselors provide a range of services equally to all students or do they exist more for certain students than others?

Do school psychologists help teachers to improve environments for students or do they simply verify labels of students' inadequacies—often basing their judgment on tests of intelligence that from certain perspectives have an antiquated or erroneous theoretical perspective (Sternberg, 1985; Gardner, 1983). LaRoque (1997) provided evidence to suggest that the average classroom teacher is apt to have more knowledge of proactive approaches to classroom management than does the average school psychologist. Hence, even if school psychologists are motivated to help design more supportive classroom environments, their lack of knowledge about proactive management skills would make them poor resource consultants for teachers who are trying to enhance learning environments. We suspect that the same issue would apply in the area of learning theory knowledge—although school psychologists would have more knowledge of behavioral theory, we suspect the average teacher to know as much if not more theory in other domains (social learning theory, information processing, other cognitive theories, etc.). Further, we suspect that teachers, in contrast to school psychologists, would be more sensitive to applying this knowledge to different developmental levels that students bring to the classroom. To begin even minimal planning for systemic reform, it would be important to upgrade the knowledge and dispositions of all school personnel—staff (secretaries, school psychologists, school counselors), teachers, administrators, parent organizations, and so on for wanting to transform the school from a labeling institution to a learning institution.

Surprisingly, some powerful forms of expectation have not been systematically explored even though they would seem to be critical venues for communicating enduring expectations to students. For example, when students are selected for the gifted program, what are the consequences for other smart students who are not selected? What happens when gifted students leave the classroom—do teachers provide meaningful activities or only time-filling activities? In elementary school, if the same students are selected for the lead in school plays and/or class honors of various types, what are the effects? Do these special opportunities for leadership in the early grades become enduring, enabling capacities that students take into the middle school and the high school? To what extent does the feeder elementary school that one comes from help to label and to provide opportunities for students at the middle school or in high school? For example, a middle school is often the receiving "home" for three or four elementary schools. Do student leaders inevitably come from the same elementary school? In what circumstances do exceptions arise?

It would seem important to begin to expand expectation work into other areas of possible communication and to systematically explore the ways in which expectations are communicated through school-level structures (e.g., the way teacher/parent meetings are conceptualized). Indeed, it is time to explore the effects of other societal issues and expectations for their possible effects on students' willingness to commit to educational excellences (e.g., What effects do dismal and unsafe school environments have on students' motivation?).

In designing systemic reform it may also be necessary to examine teachers', educators', and citizens' explicit and implicit philosophies and goals for educating low expectation students. For example, how or do we value the academic experiences of the less academically inclined? How do we define their purpose in the classroom? Are they somehow valuable, or there to just kill time? What is our responsibility to these students? Is it necessary to encourage their cognitive growth or can we be content to provide "babysitting?" In important ways the chapters in this volume by Ennis, Soodak and Podell, and Ross have addressed this issue at the class level. It is time to explore more broadly how communities and schools more generally communicate unintended messages to students. Further, following Jussim and colleagues, it will be important to more systematically examine the processes through which positive expectations are communicated in more macro levels (communities, newspapers, etc.).

In summary, we see that the field of teacher expectation research and studies of teacher efficacy have identified problematic aspects of teaching that occur in some settings and that serve to at least sustain student performance at levels below what students could achieve with more favorable climates. This volume and work elsewhere has presented strategies for helping teachers to become more aware of how they might construct more positive learning for all students (Good & Brophy, 1997). However, this volume presents a rich set of research opportunities and challenges that potentially can help teachers to better understand problems and to develop better learning environments for helping students to become successful learners.

ACKNOWLEDGMENT

The authors want to thank Sharon Nichols for typing the manuscript and for her helpful editorial and substantive comments and to acknowledge the helpful comments from Jere Brophy and Mary McCaslin.

REFERENCES

Brophy, J., & Good, T. (1970). Teachers' communication of differential expectations for childrens' classroom performance: Some behavioral data. *Journal of Educational Psychology, 61*, 365-374.

Carlsen, W. (1991). Subject-matter knowledge and science teaching: A pragmatic perspective. In J. Brophy (Ed.), *Advances in research on teaching* (Vol. 2, pp. 115-144). Greenwich, CT: JAI Press.

Casanova, U. (1996). Parent involvement: A call for prudence. *Educational Researcher, 25*, 30-32, 46.

Cronbach, L., & Snow, R. (1977). *Aptitudes and instructional methods.* New York: Irvington/Nailbury.

Gardner, H. (1983). *Frames of mind: The theory of multiple intelligences.* New York: Basic Books.

Good, T. (1969). *Differential opportunities for student response opportunity and achievement.* An unpublished dissertation study, Indiana University: Bloomington, IN.

Good, T., & Brophy, J. (1997). *Looking in classrooms* (7th ed.). New York: Longman.

Good, T., & Power, C. (1976). Designing successful environments for different types of students. *Journal of Curriculum Studies, 8*, 45-60.

Kaufman, J., & Rosenbaum, J. (1992). The education and employment of low-income Black youth in White suburbs. *Journal of Education and Policy Analysis, 14*, 229-240.

LaRoque, S. (1997). *Behavioral consultation and the training of school psychologists.* An unpublished master's thesis at the University of Arizona. Tucson: AZ.

Mason, D., Schroeter, D., Combs, R., & Washington, K. (1992). Assigning average-achieving eighth graders to advanced mathematics classes in an urban junior high. *Elementary School Journal, 92*, 587-599.

McCaslin, M. (1996). The problem of problem representation: The Summit's conception of student. *Educational Researcher, 25*, 13-15.

McCaslin, M., & Good, T. (1996a). *Listening in classrooms.* New York: Harper Collins.

McCaslin, M., & Good, T. (1996b). The informal curriculum. In D. Berliner & R. Calfee (Eds.), *Handbook of educational psychology* (pp. 622-673). New York: Macmillan.

Mehan, H. (1997). Tracking untracking: The consequences of placing low-track students in high-track classes. In P. Hall (Ed.), *Race, ethnicity, and multiculturalism: Policy and practice* (pp. 115-150). New York: Garland Publishing.

Nichols, S., & Good, T. (in press). Students' perceptions of fairness in school settings: A gender analysis. *Teachers College Record.*

Snow, R., Corno, L., & Jackson III, D. (1996). Individual differences in affective and conative functions. In D. Berliner & R. Calfee (Eds.), *Handbook of educational psychology* (pp. 243-310). New York: Macmillan.

Sternberg, R. (1985). *Beyond IQ: A triarchic theory of human intelligence.* New York: Cambridge University Press.

TESA (1993). *Teacher expectations and student achievement coordinator manual.* Los Angeles County, Office of Education, Downey, CA.

Advances in Cognition and Educational Practice

Edited by **Jerry Carlson**, *School of Education University of California, Riverside*

Volume 5, Conceptual Issues in Research in Intelligence
1998, 330 pp. $78.50/£49.95
ISBN 0-7623-0423-5

Edited by **Welko Tomic**, *Faculty of Social Sciences, The Open University, The Netherlands* and **Johannes Kingma**, *Department of Traumatology, University Hospital Groningen, The Netherlands*

CONTENTS: Preface, *Welko Tomic and Johannes Kingma*. Introduction: Issues in the Malleability of Intelligence, *Welko Tomic and Johannes Kingma*. The Schools: IQ Tests, Labels, and the Word "Intelligence," *James R. Flynn*. Intelligence as a Subsystem of Personality: From Spearman's G to Contemporary Models of Hot Processing, *John D. Mayer and Dennis C. Mitchell*. A Longitudinal Study of Factors Associated with Wechsler Verbal and Performance IQ Stores in Students from Low-Income, African American Families, *Frances A. Campbell and laura Nabors*. De Groot's Potentiality Theory of Intelligence: A Resume and a Validation Study, *Arie A.J. van Peet*. Relating Reading Achievement to Intelligence and Memory Capacity, *Ronald P. Carver*. Experimental Approaches to the Assessment and Development of Higher-Order Intellectual Processes, *Douglas H. Clements and Bonnie K. Nastasi*. The Detection of Interstimulus Relations: A Locus of Intelligence-related Differences, *Sal A. Soraci, Michael T. Carlin, and Richard A. Chechile*. Intelligence and Learning Potential: Theoretical and Research Issues, *Wilma C. M. Resing*. Inductive Reasoning and Fluid Intelligence: A Training Approach, *Karl Jozef Klauer*. Accelerating Intelligence Development through an Inductive Reasoning Training, *Welko Tomic and Johannes Kingma*. The Effects of Test Preparation, *Henk van der Molen, Jan te Nijenhuis, and Gert Keen*.

Also Available:
Volumes 2-4 (1994-1997) $78.50/£49.95 each
  Volume 1 (2 part set) $157.00/£99.90

JAI PRESS

JAI PRESS

Advances in Research on Teaching

Edited by **Jere Brophy**, *College of Education, Michigan State University*

Advances in Research on Teaching documents, and stimulates further work in emerging trends in research on classroom teaching. Topic selection for each volume will emphasize conceptualization and analysis of the processes of teaching (including not only the behaviors that can be observed in the classroom, but also the planning, thinking, and decision making that occur before, during and after interaction with students). Especially likely to be selected are topics that involve linking information about teaching processes with information about presage variables (especially teacher knowledge and beliefs), context variables, or student variables. Scholars who have made programmatic contributions to the emerging literature will be invited to prepare a chapter in which they not only describe their work but synthesize it, place it into the context of the larger body of research and scholarship on the topic, and give their current views of its meaning and implications.

Volume 6, Teaching and Learning History
1996, 336 pp. $78.50/£49.95
ISBN 0-7623-0104-X

CONTENTS: Introduction, *Jere Brophy*. Research and the Improvement of School History, *Peter Knight*. Narrative Simplifications in Elementary Students' Historical Thinking, *Keith C. Barton*. "I've Never Done This Before": Building a Community of Historical Inquiry in a Third-Grade Classroom, *Linda Levstik and Dehea B. Smith*. Making Learners and Concepts Central: A Conceptual Change Approach to Learner-Centered, Fifth-Grade American History Planning and Teaching, *Kathleen J. Roth*. Constructing Historical Interpretations in Elementary School: A Look at Process and Product, *Myra Zarnowski*. Designing Effective U.S. History Curricula for All Students, *Douglas Carnine, Jennifer Caros, Donald Crawford, Keith Hollenbeck, and Mark Harniss*. Closing the Gap Between School and Disciplinary History? Historian as High School History Teacher, *Bruce VanSledright*. Issues and Influences that Shape the Teaching of U.S. History, *Michael Romanowski*. Discussion, *Jere Brophy*.

Also Available:
Volumes 1-5 (1989-1995) $78.50/£49.95 each